Lecture Notes in Mathematics 2028

Editors:
J.-M. Morel, Cachan
B. Teissier, Paris

For further volumes:
http://www.springer.com/series/304

FONDAZIONE
CIME
ROBERTO CONTI
CENTRO INTERNAZIONALE MATEMATICO ESTIVO
INTERNATIONAL MATHEMATICAL SUMMER CENTER

Fondazione C.I.M.E., Firenze

C.I.M.E. stands for *Centro Internazionale Matematico Estivo*, that is, International Mathematical Summer Centre. Conceived in the early fifties, it was born in 1954 in Florence, Italy, and welcomed by the world mathematical community: it continues successfully, year for year, to this day.

Many mathematicians from all over the world have been involved in a way or another in C.I.M.E.'s activities over the years. The main purpose and mode of functioning of the Centre may be summarised as follows: every year, during the summer, sessions on different themes from pure and applied mathematics are offered by application to mathematicians from all countries. A Session is generally based on three or four main courses given by specialists of international renown, plus a certain number of seminars, and is held in an attractive rural location in Italy.

The aim of a C.I.M.E. session is to bring to the attention of younger researchers the origins, development, and perspectives of some very active branch of mathematical research. The topics of the courses are generally of international resonance. The full immersion atmosphere of the courses and the daily exchange among participants are thus an initiation to international collaboration in mathematical research.

C.I.M.E. Director
Pietro ZECCA
Dipartimento di Energetica "S. Stecco"
Università di Firenze
Via S. Marta, 3
50139 Florence
Italy
e-mail: zecca@unifi.it

C.I.M.E. Secretary
Elvira MASCOLO
Dipartimento di Matematica "U. Dini"
Università di Firenze
viale G.B. Morgagni 67/A
50134 Florence
Italy
e-mail: mascolo@math.unifi.it

For more information see CIME's homepage: http://www.cime.unifi.it

CIME activity is carried out with the collaboration and financial support of:
– INdAM (Istituto Nazionale di Alta Matematica)
– MIUR (Ministero dell'Universita' e della Ricerca)

Stefano Bianchini • Eric A. Carlen
Alexander Mielke • Cédric Villani

Nonlinear PDE's and Applications

C.I.M.E. Summer School,
Cetraro, Italy 2008

Editors:
Luigi Ambrosio
Giuseppe Savaré

Springer

FONDAZIONE CIME ROBERTO CONTI

Stefano Bianchini
SISSA-ISAS
Via Beirut 2-4
34014 Trieste
Italy
bianchin@sissa.it

Eric A. Carlen
Rutgers University
Department of Mathematics
Hill Center
Frelinghuysen Road 110
Piscataway, NJ 08854-8019
USA
carlen@math.rutgers.edu

Alexander Mielke
Weierstrass Institute for Applied Analysis
and Stochastics
Mohrenstr. 39
10117 Berlin
Germany
mielke@wias-berlin.de

Cédric Villani
University of Lyon
Institute Henri Poincaré
Rue Pierre et Marie Curie 11
75230 Paris, Cedex 05
France
villani@math.univ-lyon1.fr

ISBN 978-3-642-21718-0 e-ISBN 978-3-642-21861-3
DOI 10.1007/978-3-642-21861-3
Springer Heidelberg Dordrecht London New York

Lecture Notes in Mathematics ISSN print edition: 0075-8434
ISSN electronic edition: 1617-9692

Library of Congress Control Number: 2011934805

Mathematics Subject Classification (2010): 35-XX, 49-XX, 46-XX, 53-XX

Cover design: deblik, Berlin

Printed on acid-free paper

Springer is part of Springer Science+Business Media (www.springer.com)

Preface

This volume collects the notes of the CIME course *Nonlinear PDE's and applications* held in Cetraro (Italy) on June 23–28, 2008. The school consisted in 5 series of lectures, delivered by

Stefano Bianchini (SISSA, Trieste)
Eric A. Carlen (Rutgers University)
Alexander Mielke (WIAS, Berlin)
Felix Otto (Bonn University)
Cedric Villani (Ecole Normale Superieure de Lyon).

They presented a broad overview on some deep results and new exciting developments concerning in particular optimal transport theory, nonlinear evolution equations, functional inequalities, and differential geometry. A brief (and largely incomplete) account of the main topics considered here involves optimal transport, Hamilton–Jacobi equations, Riemannian geometry, and their links with sharp geometric/functional inequalities, variational methods for studying nonlinear evolution equations and their scaling properties, the metric/energetic theory of gradient flows and of rate-independent evolution problems.

The course aimed at showing the deep connections among all these topics and at opening new research directions, through the contribution of leading experts in these fields.

Stefano Bianchini gave a course on

Transport rays and applications to Hamilton–Jacobi equations

showing new recent results and applications of geometric measure/ disintegration theory to Hamilton–Jacobi equations and optimal transportation. The tools developed here lie at the core of new relevant achievements concerning the existence of transport maps for general Monge problems, leading eventually to a general solution of Monge's problem in Euclidean spaces, in the case cost=distance, when the distance is induced by an arbitrary norm.

The course of *Eric Carlen* on

Sharp functional inequalities and nonlinear evolution equations

discussed in a unified perspective the interplay between sharp functional inequalities (as the ones of Hardy-Littlewood-Sobolev, Onofri, Brascamb-Lieb) and the asymptotic behaviour of solutions to certain evolution equations. Many different techniques, involving rearrangement, symmetrization, entropy dissipation, gradient flows, and calculus of variations, enter as a crucial tool in the arguments and shed new light on this fascinating subject.

Alexander Mielke presented in his course

Differential, energetic and metric formulations for rate-independent processes

a general theory covering a wide spectrum of rate-independent problems arising in dry friction, elastoplasticity, magnetism, and phase transformation models. The notes cover many different approaches to this kind of evolutionary phenomena, starting from the "energetic" point of view, based on a recursive minimization of suitable quasi-static functionals driving the evolution. Different kind of solutions are then considered and the so called "viscous" approach leading to the recent notion of BV-solution has been investigated.

Felix Otto delivered a series of lectures on

Scaling laws by PDE methods

dealing with the problem of obtaining sharp asymptotic estimates on the evolution of suitable macroscopic quantities in some classes of nonlinear time-dependent problems. Even though the problems under consideration are quite different (Cahn–Hilliard equations, convection-diffusion Rayleigh-Bernard equations, Kuramoto-Sivashinsky evolution), in all cases lower bounds are obtained, which are sharp or almost sharp for generic solutions.

Cedric Villani gave a course on

Optimal transport and curvature

presenting the main results of optimal transport theory in Riemannian manifolds and their geometric counterparts. In particular, the link between lower bounds on Ricci curvature and the displacement convexity of integral/entropy functionals in the space of probability measures has been analyzed: it allows to prove stability of the lower Ricci curvature bounds with respect to measured Gromov-Hausdorff convergence and it leads to a synthetic formulation of lower Ricci bounds for metric measure spaces. The second part of the course has been devoted to the regularity theory of optimal transport maps, a theme now under active investigation, with deep links with the geometry of the ambient manifold: its cut-locus, its sectional curvature, and more.

This series of lectures attracted more than 80 participants, largely PhD students or post-docs, but also senior researchers; even though it is impossible to give a comprehensive account of the main themes of nonlinear PDE's, which cannot be exhausted in any kind of school or concentration period, we believe that this CIME course has been rich of useful suggestions and ideas for inspiring new researches and developments in the near future.

We wish to thank all the lecturers for their active participation and their valuable contribution, Alessio Figalli, Matteo Gloyer, and Riccarda Rossi for their worthwhile assistance in preparing the present volume, and the CIME foundation, in particular the director Prof. Pietro Zecca and the secretary Prof. Elvira Mascolo, for their helpful support and for the organization of such a remarkable event in Cetraro.

November 26, 2009

Luigi Ambrosio
Giuseppe Savaré

Contents

Contributors

Stefano Bianchini SISSA, via Beirut 2-4, 34014 Trieste, Italy, bianchin@sissa.it

Eric A. Carlen Department of Mathematics, Hill Center, Rutgers University, 110 Frelinghuysen Road, Piscataway, NJ 08854-8019, USA, carlen@math.rutgers.edu

Alessio Figalli Department of Mathematics, The University of Texas at Austin, 1 University Station, C1200, Austin, TX 78712-1082, USA, figalli@math.utexas.edu

Matteo Gloyer SISSA, via Beirut 2-4, 34014 Trieste, Italy, mgloyer@gmail.com

Alexander Mielke Weierstraß Institut für Angewandte Analysis und Stochastik, Mohrenstraße 39, 10117, Berlin, mielke@wias-berlin.de

Cédric Villani Institut Henri Poincaré, Ecole Normale Supérieure, 11 rue Pierre et Marie Curie, 75231 Paris Cedex 05, France, cvillani@umpa.ens-lyon.fr

Transport Rays and Applications to Hamilton–Jacobi Equations

Stefano Bianchini and Matteo Gloyer

1 Introduction

The aim of these notes is to introduce the readers to the use of the Disintegration Theorem for measures as an effective tool for reducing problems in transport equations to simpler ones. The basic idea is to partition $\mathbb{R}^d$ into one dimensional sets, on which the problem under consideration becomes one space dimensional (and thus much easier, hopefully).

Two main problems arise. On one hand, one has to verify that the partition under consideration satisfies the assumptions of the Disintegration Theorem: we say in this case that there is a Borel quotient map from $\mathbb{R}^d$ into [0,1].

On the other hand, the reduction on the one dimensional sets of the partition should preserve the absolute continuity of the Lebesgue measure: the weak derivative w.r.t. a singular measure has no meaning, in general. We thus have to check that the conditional probabilities are absolutely continuous, and this is an additional requirement on the quotient map.

These notes contain the main part of the course held in Cetraro, summer 2008.

2 Settings

We consider a σ-generated probability space (X, Σ, μ) and a partition $X = \cup_{\alpha \in A} X_\alpha$, where $A = X/\sim$ is the quotient space, and $h : X \to A$ the quotient map. We give to A the structure of a probability space by introducing the σ-algebra $\mathcal{A} = h_\sharp \Omega$, where Ω are the saturated sets in Σ (unions of fibers of h), and $m = h_\sharp \mu$ the image measure such that $m(S) = \mu(h^{-1}(S))$.

S. Bianchini (✉) · M. Gloyer
SISSA, via Beirut 2-4, I-34014 Trieste, Italy
e-mail: bianchin@sissa.it; mgloyer@gmail.com

S. Bianchini et al., *Nonlinear PDE's and Applications*, Lecture Notes in Mathematics 2028, DOI 10.1007/978-3-642-21861-3_1,

Remark 2.1 *$\mathcal{A}$ is the largest σ-algebra such that h is measurable.*

The following example shows that even though Σ is σ-generated, $\mathcal{A}$ in general is not.

Example 2.2 *Consider in $([0,1], \mathcal{B})$ (Borel) the equivalence relation*

$$x \sim y \quad \textit{iff} \quad x - y = 0 \mod \alpha,$$

for some $\alpha \in [0,1]$. If $\alpha = p/q$, with $p, q \in \mathbb{N}$ relatively prime, then we can take

$$(A, \mathcal{A}) = \left(\left[0, \frac{1}{q} \right], \mathcal{B} \right).$$

If $\alpha \in \mathbb{R} \setminus \mathbb{Q}$, then A is a Vitali set. If $\mu = \mathcal{L}^1|_{[0,1]}$, then $m = h_\sharp \mathcal{L}^1|_{[0,1]}$ has only sets of either full or negligible measure. Assume by contradiction that $\{a_n\}_{n\in\mathbb{N}}$ generates $\mathcal{A}$. Since $h^{-1}(x) = \{x + n\alpha \mod 1 : n \in \mathbb{N}\} \in \mathcal{B}$, it follows that each $a \in A$ belongs to a generating set of measure 0. But this leads to a contradiction:

$$1 = m(A) = m \left(\bigcup_{m(a_n)=0} a_n \right) \leq \sum_{m(a_n)=0} m(a_n) = 0.$$

We now define the measure algebra $(\hat{\mathcal{A}}, \hat{m})$ by the following equivalence relation on $\mathcal{A}$:

$$a_1 \sim a_2 \quad \text{iff} \quad m(a_1 \triangle a_2) = 0.$$

It is easy to check that $\hat{\mathcal{A}}$ is a σ-algebra and $\hat{m}$ is a measure on $\hat{\mathcal{A}}$.

Proposition 2.3 *$(\hat{\mathcal{A}}, \hat{m})$ is σ-generated.*

$\hat{\mathcal{A}}$ is isomorphic to a sub-σ-algebra of Σ.

Remark 2.4 *More generally, if (X, Σ, μ) is generated by a family of cardinality ω_α, then each sub-σ-algebra $\mathcal{A} \subset \Sigma$ is essentially generated by a family of sets of cardinality ω_α or less.*

This is a particular case of a deep result, Maharam's Theorem *([7], 332T(b)), which describes isomorphisms between probability spaces: if $(\hat{\Sigma}, \hat{\mu})$ is a probability algebra, then*

$$(\hat{\Sigma}, \hat{\mu}) \simeq \prod_i c_i \left[\bigotimes_{J_i} \{0,1\} \right], \quad \sum_i c_i = 1,$$

where $\bigotimes_{J_i} \{0,1\}$ is the measure space obtained by throwing the dice J_i times.

3 Disintegration

Definition 3.1 *We introduce the following relation on A:*

$$a_1 \sim a_2 \text{ iff the following holds:}$$

$$\text{for all } \hat{A} \in \hat{\mathcal{A}}, a_1 \in \hat{A} \text{ iff } a_2 \in \hat{A}.$$

The equivalence classes of this relation are the atoms *of the measure m. In particular, we can define the measure space*

$$(\Lambda = A/\sim, \hat{\mathcal{A}}, m).$$

The σ-algebra is isomorphic to the σ-generated $\hat{\mathcal{A}}$ constructed in the previous section.

Definition 3.2 *The* disintegration *of the measure μ with respect to the partition $X = \bigcup_\alpha X_\alpha$ is a map*

$$A \to P(X), \quad \alpha \mapsto \mu_\alpha,$$

where $P(X)$ is the class of probability measures on X, such that the following properties hold:

1. *for all $B \in \Sigma$, the map $\alpha \mapsto \mu_\alpha(B)$ is m-measurable;*
2. *for all $B \in \Sigma$, $A \in \mathcal{A}$,*

$$\mu(B \cap h^{-1}(A)) = \int_A \mu_\alpha(B)\,\mathrm{d}m(\alpha).$$

It is unique if μ_α is determined m-a. e.

Remark 3.3 1. *Since we are not requiring the elements of the partition X_α to be measurable, in general $\mu_\alpha(X_\alpha) \neq 1$ for those X_α which are measurable. In this case we say that the disintegration is not* strongly consistent *with h.*
2. *For general spaces which are not σ-generated, sometimes a disintegration nonetheless exists, but in general there is no uniqueness.*
3. *The disintegration formula can easily be extended to measurable functions:*

$$\int_X f\,\mathrm{d}\mu = \int_A \left(\int_X f\,\mathrm{d}\mu_\alpha \right) \mathrm{d}m(\alpha).$$

We now state the general disintegration theorem.

Theorem 3.4 *Assume that (X, Σ, μ) is a σ-generated probability space, $X = \bigcup_{\alpha \in A} X_\alpha$ a partition of X, $h : X \to A$ the quotient map, and $(A, \mathcal{A}, m)$ the quotient measure space. Then the following holds:*

1. *There is a unique disintegration $\alpha \mapsto \mu_\alpha$.*
2. *If $(\Lambda, \hat{\Lambda}, m)$ is the σ-generated algebra equivalent to $(A, \mathcal{A}, m)$, and $p : A \to \Lambda$ the quotient map, then the sets*

$$X_\lambda = (p \circ h)^{-1}(\lambda)$$

are μ-measurable, the disintegration

$$\mu = \int_\Lambda \mu_\lambda \, \mathrm{d}m(\lambda)$$

is strongly consistent $p \circ h$, and

$$\mu_\alpha = \mu_{p(\alpha)} \qquad \textit{for } m\textit{-a. e. } \alpha.$$

Definition 3.5 *$R \subset X$ is a* rooting set *for $X = \bigcup_{\alpha \in A} X_\alpha$, if for each $\alpha \in A$ there exists exactly one $x \in R \cap X_\alpha$.*

R is a μ-rooting set *if there exists a set $\Gamma \subset X$ of full μ-measure such that R is a rooting set for*

$$\Gamma = \bigcup_{\alpha \in A} \Gamma_\alpha = \bigcup_{\alpha \in A} \Gamma \cap X_\alpha.$$

Proposition 3.6 *If $\mu = \int_A \mu_\alpha \, \mathrm{d}m(\alpha)$ is strongly consistent with the quotient map, then there exists a Borel μ-rooting set.*

Example 3.7 *Consider again*

$$x \sim y \quad \textit{iff} \quad x - y = 0 \mod \alpha.$$

If $\alpha = p/q$ with p, q relatively prime, then the rooting set is $[0, 1/q)$, so that we know that the disintegration is strongly consistent with h. One can check that

$$\mu = \int_0^{\frac{1}{q}} \mathrm{d}\alpha \sum_{n=0}^{q-1} \delta \left(x - \alpha - \frac{n}{q} \right).$$

If $\alpha \in \mathbb{R} \setminus \mathbb{Q}$, then we know that

$$(\Lambda, \hat{\Lambda}, m) \simeq \Big(\{\lambda\}, \{\emptyset, \{\lambda\}\}, \delta_\lambda \Big),$$

so that

$$\mu = \int \mathrm{d}m\, \mu.$$

4 Hamilton–Jacobi Equation and Monotonicity

In the following we consider the Hamilton–Jacobi equation

$$\begin{cases} u_t + H(\nabla u) = 0, \\ u(0,x) = \bar{u}(x), \end{cases}$$

where $(t,x) \in \mathbb{R}^+ \times \mathbb{R}^d$ and $\bar{u} \in L^\infty(\mathbb{R}^d)$. We assume that $H : \mathbb{R}^d \to \mathbb{R}$ is C^1 and convex. We denote by $L = H^*$ the Legendre transform of H and assume that it has at least linear growth,

$$L(x) \geq \frac{1}{c}(|x| - c),$$

and is locally Lipschitz. By the properties of the Legendre transform, L is strictly convex.

For example, we can consider $H(x) = L(x) = \frac{1}{2}|x|^2$.

The viscosity solution is given explicitly by

$$u(t,x) = \inf \left\{ \bar{u}(y) + tL\left(\frac{x-y}{t}\right) : y \in \mathbb{R}^d \right\}.$$

Remark 4.1 *The following properties can easily be checked:*

1. *Finite speed of propagation: $u(t,x)$ depends only on the values of $\bar{u}(y)$ for $|x-y| \leq c$.*
2. *Uniform Lipschitz continuity: For fixed y, the function $\bar{u}(y) + tL((x-y)/t)$ is uniformly Lipschitz in x for $|x-y| \leq c$. Hence $u(t,x)$ is uniformly Lipschitz in x for all $t > 0$.*
3. *Semigroup property: For $t > s > 0$, we have that*

$$u(t,x) = \min \left\{ u(s,z) + (t-s)L\left(\frac{x-z}{t-s}\right) : z \in \mathbb{R}^d \right\}.$$

4. *If $D^2H \in [1/c, c]\mathbb{I}$, then $D^2L \in [1/c, c]\mathbb{I}$, and thus $u(t,x) - c|x|^2/2t$ is concave in x. Hence $u(t,\cdot)$ is quasi-concave for $t > 0$.*

We can solve the backward problem

$$\begin{cases} v_t + H(\nabla v) = 0, \\ v(1,x) = \bar{u}(1,x). \end{cases}$$

Then v has the same properties as above for $t < 1$, and the following duality holds:

$$u(1,x) = \min\Big\{v(0,y) + L(x-y)\Big\},$$
$$v(0,y) = \max\Big\{u(1,x) - L(x-y)\Big\}.$$

We say that $u(1,x)$ and $v(0,y)$ are *L-conjugate functions.*

Definition 4.2 *The couple* $[y,x] \in \mathbb{R}^d \times \mathbb{R}^d$ *such that*

$$u(1,x) = v(0,y) + L(x-y)$$

is called an optimal couple. *The corresponding segment*

$$\Big\{(1+t)y + tx : 0 \leq t \leq 1\Big\}$$

is called an optimal ray.

Remark 4.3 *Due to the strict convexity of L, two rays cannot intersect anywhere except at their end points.*

Remark 4.4 *In general, the duality of* $u(1,x)$ *and* $v(0,y)$ *does* not *imply that* $u(t,x) = v(t,x)$ *for* $0 < t < 1$.

Example 4.5 *In the case* $H(x) = L(x) = |x|^2/2$, *the optimal rays are the graph of the maximal monotone operator*

$$x \mapsto y(x) = \partial_x\left(\frac{|x|^2}{2} - u(1,x)\right).$$

It follows from Minty's Theorem ([2], p.142) that the map

$$x \mapsto tx + (1-t)y(x)$$

is surjective. Since the rays do not intersect, it follows that for all $0 < t < 1$ *and all* $z \in \mathbb{R}^d$, *there exists a unique optimal couple* $[y,x]$ *such that*

$$z = (1-t)y + tx.$$

Hence, using the explicit formula for the optimal ray, we obtain

$$u(t,x) = v(t,x) \quad \text{for all } t \in [0,1],\ x \in \mathbb{R}^d.$$

Hence the solution is both $(2t)^{-1}$*-concave and* $(2(1-t))^{-1}$*-convex, thus* $u \in C^{1,1}$.

The set $y(x)$ is convex, therefore the rays departing from a given point form a convex set.

The example can be generalized to the case $H(x) = 1/2\langle x, Ax\rangle$ by a linear change of variable.

Example 4.6 *A more difficult case is* $\mathrm{D}^2 H, \mathrm{D}^2 L \in [1/c, c]\mathbb{I}$. *One can use that* $v(0,y) + |y|^2/(2c)$ *is convex to compute the optimal rays for* $t \ll 1$*:*

$$u(t,z) = \min\left\{v(0,y) + tL\left(\frac{z-y}{t}\right)\right\}$$
$$= \min\left\{v(0,y) + \frac{|y|^2}{2c} + tL\left(\frac{z-y}{t}\right) - \frac{|y|^2}{2c}\right\}.$$

The last two terms together are convex for $t < c^{-2}$, *so that the minimizer is given by*

$$\nabla L\left(\frac{z-y}{t}\right) = \partial^- v(0,y),$$

where $\partial^- v(0,y)$ *denotes the subdifferential of* v *at* y*:*

$$\partial^- f(x) = \left\{p : \liminf \frac{f(x+h) + f(x) - ph}{|h|} \geq 0\right\}.$$

Similarly, we can introduce the superdifferential

$$\partial^+ f(x) = \left\{p : \limsup \frac{f(x+h) + f(x) - ph}{|h|} \leq 0\right\}.$$

These sets are convex, but in general they are empty. We thus obtain

$$z = y + t\nabla H(\partial^- v(0,y)) \quad \text{for } 0 \leq t \ll 1.$$

The strict convexity implies that the map $z \mapsto y$ *is single-valued and Lipschitz. For* $t \ll 1$, *the projections of the rays* $z(y)$ *are still convex sets. However, in general the rays do not extend to* $t = 1$.

Example 4.7 *Taking*

$$L(x) = 3|x| + x_1|x|,$$

$$u(1,x) = \min\left\{L\left(x - \begin{pmatrix}0\\1\end{pmatrix}\right), L\left(x - \begin{pmatrix}0\\1\end{pmatrix}\right)\right\},$$

and computing the corresponding $v(0,y)$, *one can prove that there is a gap between the rays, i. e. there exists a region inside of which one has*

$$u(t,x) > v(t,x).$$

Example 4.8 *In the general case, when L is only strictly convex, there is no notion of subdifferential or superdifferential. One then has no quasi-concavity or quasi-convexity, hence no BV regularity. The graph of the optimal rays $[y, x]$ is in general full of holes, and there is no interval where one could prove $C^{1,1}$ regularity.*

5 Regularity Properties of L-Conjugate Functions and Optimal Rays

We consider a pair of L-conjugate functions $u, v \in L^\infty \cap \mathrm{Lip}$,

$$u(x) = \min\Big\{v(y) + L(x-y) : y \in \mathbb{R}^d\Big\},$$
$$v(y) = \max\Big\{u(x) - L(x-y) : x \in \mathbb{R}^d\Big\},$$

where as before $L : \mathbb{R}^d \to \mathbb{R}$ is strictly convex and has at least linear growth.

We define the set

$$F = \Big\{[y, x] \in \mathbb{R}^d \times \mathbb{R}^d : [x, y] \text{ optimal couple }\Big\}$$

and its projection

$$F(t) = \Big\{z : z = (1-t)y + tx \text{ for some } [y, x] \in F\Big\}.$$

By the duality, we know that $F(0) = F(1) = \mathbb{R}^d$. On the set $F(t)$ we define the vector field

$$\begin{aligned}\mathbf{p}_t(z) &= (1, p_t(z))\\ &= (1, x-y), \qquad \text{where } [y, x] \in F \text{ such that } z = (1-t)y + tx.\end{aligned}$$

We also introduce the set-valued functions

$$y(x) = \Big\{y : [y, x] \in F\Big\},$$
$$x(y) = \Big\{x : [y, x] \in F\Big\}.$$

From the fact that u and v are L-conjugate, we obtain the following lemma.

Lemma 5.1 *The set F and its projections $F(t)$ are closed, and the set-valued functions $x(y)$, $y(x)$ have locally compact images.*

In particular $y(x)$, $x(y)$ are Borel measurable, because the inverse of compact sets is compact.

Example 5.2 *When ∇u, ∇v exist, then they are related to p by*

$$\nabla v(y) = \partial L(p_0(y)),$$
$$\nabla u(x) = \partial L(p_1(x)).$$

If we consider for example

$$L(x) = \frac{1}{2}|x|^2 + |x_1|,$$

$$u(x) = \min\left\{L\left(x - \begin{pmatrix}1\\0\end{pmatrix}\right), L\left(x - \begin{pmatrix}-1\\0\end{pmatrix}\right)\right\},$$

then it is easy to check that $u(0)$ has no subdifferential or superdifferential.

5.1 Rectifiability Property of Jumps

Define the sets

$$J_m = \left\{x \in \mathbb{R}^d : \text{ there exist } y_1, y_2 \in y(x) \text{ such that } |y_1 - y_2| \geq \frac{1}{m}\right\}$$

and

$$J = \bigcup_{m \in \mathbb{N}} J_m.$$

Lemma 5.3 *J_m is closed and countably $(d-1)$-rectifiable, i. e. it can be covered with a countable number of images of Lipschitz functions $\mathbb{R}^{d-1} \to \mathbb{R}^d$.*

The lemma can be proved by applying the rectifiability criterion [1], Theorem 2.61. The proof is analogous to Lemma 3.4 in [4].

In a similar way we obtain the following proposition:

Proposition 5.4 *The set*

$$J^k = \bigcup_{m \in \mathbb{N}} J^k_m$$
$$= \bigcup_{m \in \mathbb{N}} \left\{x \in \mathbb{R}^d : \text{ there exist} y_1, \ldots, y_{k+1} \in y(x) \text{ s. t. } B\left(0, \frac{1}{m}\right) \subset \text{co}\{y_1, \ldots, y_{k+1}\}\right\}$$

is countably $(d-k)$-rectifiable, i. e. it can be covered with a countable number of images of Lipschitz functions $\mathbb{R}^{d-k} \to \mathbb{R}^d$.

5.2 Some Approximations

To prove the estimates on the vector field $\mathbf{p}_t$ or p_t, we need an approximation technique. The following proposition will be an essential tool.

Proposition 5.5 *Assume that*

$$\bar{u}_n(y) \to \bar{u}(y), \qquad L_n(x) \to L(x)$$

locally uniformly, and that we have the uniform bound

$$L_n(x) \geq \frac{1}{c}(|x| - c) \quad \text{for all } n \in \mathbb{N},$$

where c does not depend on n.

Then the conjugate functions $u_n(1,x)$, $v_n(0,y)$ converge uniformly to $u(1,x)$, $v(0,y)$, and the graph F_n converges locally in Hausdorff distance for a closed subset of F.

5.3 Fundamental Example

Let $\{y_i : i \in \mathbb{N}\}$ be a dense sequence in $\mathbb{R}^d$, and define

$$u_N(x) = \min\Big\{u(y_i) + L(x - y_i) : i = 1, \dots, N\Big\}.$$

We can split $\mathbb{R}^d$ into at most N open regions Ω_i (Voronoi-like cells), inside which we have

$$u_N(x) = u(y_i) + L(x - y_i), \quad x \in \Omega_i,$$

together with the negligible set

$$\bigcup_{i \neq j} \Big(\bar{\Omega}_i \cap \bar{\Omega}_j\Big).$$

The boundary of each region is Lipschitz, and inside each region the corresponding directional field p_N is given by

$$p_N(x) = x - y_i, \quad x \in \Omega_i.$$

5.4 Divergence Estimate

In the points x where the field $p(x)$ is single valued, the approximate $p_n(x)$ converges to $p(x)$. This implies that

$$p_n(x) \to p(x) \quad \mathcal{L}^d\text{-a. e.}$$

Using this fact we can prove the following proposition:

Proposition 5.6 div p *is a locally bounded measure satisfying*

$$\operatorname{div} p - d\mathcal{L}^d \leq 0.$$

Proof. The approximating fields satisfy the bound, thus by the above convergence we get the bound for div p. It is a measure because positive distributions are measures.

6 Jacobian Estimates

As in the previous section, we take a dense sequence $\{y_i : i \in \mathbb{N}\}$ in $\mathbb{R}^d$. For a fixed time $t \in (0,1)$, we consider the approximation with finitely many points at $t = 0$,

$$u_N(t,x) = \min\left\{u(0,y_i) + tL\left(\frac{x-y_i}{t}\right) : i = 1,\dots,N\right\}.$$

Take a compact subset $A(t) \subset F(t)$. We denote by $A_N(s)$ the push-forward of the set $A(t)$ along the approximating rays $p_N(t,x)$. Then we get

$$\mathcal{L}^d(A_N(s)) \geq \left(\frac{s}{t}\right)^d \mathcal{L}^d(A(t)) \quad \text{for } s \leq t.$$

Up to a set of measure ϵ, we can assume that $p_N(t)$, $p(t)$ are continuous and $p_N(t) \to p(t)$ uniformly on $A(t)$. Then $A_N(s)$ is compact for $s \leq t$, and it converges to $A(s)$ in Hausdorff distance. Since $\mathcal{L}^d$ is upper semicontinuous with respect to the Hausdorff distance, this implies that

$$\mathcal{L}^d(A(s)) \geq \left(\frac{s}{t}\right)^d \mathcal{L}^d(A(t)) \quad \text{for } s \leq t.$$

By repeating the above approximation with finitely many points at $t = 1$, one obtains the corresponding estimate

$$\mathcal{L}^d(A(s)) \geq \left(\frac{1-s}{1-t}\right)^d \mathcal{L}^d(A(t)) \quad \text{for } s \geq t.$$

We thus obtain the following estimate for the push-forward of the Lebesgue measure.

Lemma 6.1 *Let*

$$\mu(s) = [z + (s-t)p]_\sharp \mathcal{L}^d.$$

Then

$$\mu(s) = c(s,t,z)\mathcal{L}^d|_{F(t)},$$

with

$$c(s,t,z) \in \left[\left(\frac{s}{t}\right)^d, \left(\frac{1-s}{1-t}\right)^d\right] \qquad \text{for } s \leq t,$$

$$c(s,t,z) \in \left[\left(\frac{1-t}{1-s}\right)^d, \left(\frac{t}{s}\right)^d\right] \qquad \text{for } t \leq s.$$

Proof. By the previous estimates, we have for $s \geq t$,

$$\left(\frac{1-t}{1-s}\right)^d \mathcal{L}^d(A(s)) \leq \mathcal{L}^d(A(t)) \leq \left(\frac{t}{s}\right)^d \mathcal{L}^d(A(s)).$$

By the definition of the image measure,

$$\mathcal{L}^d(A(t)) = \mu(s)(A(s)).$$

Thus the result follows.

The function $c(s,t,z)$ is the Jacobian of the transformation.

6.1 Disintegration of the Lebesgue Measure

Using Lemma 6.1, we now apply the Fubini-Tonelli theorem to a measurable set $A = \bigcup_t \{t\} \times A(t) \subset \bigcup_t \{t\} \times F(t)$ to obtain

$$\begin{aligned} \int_A \mathrm{d}t \times \mathcal{L}^d &= \int \mathrm{d}t \int_{A(t)} \mathcal{L}^d \\ &= \int \mathrm{d}t \int_{A(t,s)} c(t,s)\mathcal{L}^d \\ &= \int \mathcal{L}^d \int \mathrm{d}t\, c(t,s)\chi_{A(t,s)}, \end{aligned}$$

where $A(t,s)$ is the image of the set $A(t)$ by

$$A(t,s) = (z + (s-t)p(z))(A(t)).$$

Remark 6.2 *In the new coordinates,* $\mathrm{d}t\, c(t,s)$ *is concentrated on a single optimal ray.*

Since the rays do not intersect, we can disintegrate the Lebesgue measure along rays,

$$\mathfrak{L}^d \times \mathrm{d}t|_F = \int \mathrm{d}m(\alpha)\mu_\alpha.$$

We can parametrize the rays by the points of the plane $t = 1/2$, then the support of μ_α is the optimal ray passing through $\alpha \in F(1/2)$. Using the previous formula, we obtain the following theorem:

Theorem 6.3 *The disintegration of the Lebesgue measure on the set of optimal rays* F *is*

$$\int \mathrm{d}m(\alpha)\mu_\alpha,$$

with

$$m(\alpha) = \mathfrak{L}^d \int_0^1 c\left(t, \frac{1}{2}\right) \mathrm{d}t,$$

$$\mu_\alpha = \left(\int_0^1 c\left(t, \frac{1}{2}\right) \mathrm{d}t\right)^{-1} c\left(t, \frac{1}{2}\right) \mathrm{d}t,$$

where $c\left(t, \frac{1}{2}\right)$ *is the Jacobian along the ray* $\alpha + (t - 1/2)p(\alpha)$.

Remark 6.4 *By Fubini's Theorem,*

$$\int_0^1 c\left(t, \frac{1}{2}\right) \mathrm{d}t < +\infty \quad \mathfrak{L}^d\text{-a. e.},$$

therefore the formula makes sense.

In the following we denote $c(t,\alpha) = c(t, 1/2, \alpha)$.

6.2 Regularity of the Jacobian and Applications

Lemma 6.5 $c(t,\alpha) \in W^{1,1}_t$, *and there exists a* $K_d > 0$ *such that*

$$\int_0^1 \left| \frac{\mathrm{d}}{\mathrm{d}t} c(t,\alpha) \right| \mathrm{d}t \le K_d.$$

Proof. Since

$$\frac{\mathrm{d}}{\mathrm{d}t} c(t,\alpha) + \frac{d}{1-t} c(t,\alpha) \ge 0,$$

we can estimate

$$\begin{aligned}\int_0^{\frac{1}{2}} \left| \frac{\mathrm{d}}{\mathrm{d}t} c(t,\alpha) \right| \mathrm{d}t &\leq \int_0^{\frac{1}{2}} \frac{\mathrm{d}}{\mathrm{d}t} c(t,\alpha) + 2 \frac{d}{1-t} c(t,\alpha) \, \mathrm{d}t \\ &\leq c\left(\frac{1}{2}, \alpha\right) + 4d \int_0^{\frac{1}{2}} c(t,\alpha) \, \mathrm{d}t,\end{aligned}$$

and similarly

$$\int_{\frac{1}{2}}^1 \left| \frac{\mathrm{d}}{\mathrm{d}t} c(t,\alpha) \right| \mathrm{d}t \leq c\left(\frac{1}{2}, \alpha\right) + 4d \int_{\frac{1}{2}}^1 c(t,\alpha) \, \mathrm{d}t,$$

Hence

$$\mathrm{Tot.Var.}(c(\cdot,\alpha)) \leq 4d + 2c\left(\frac{1}{2}, \alpha\right). \tag{1}$$

In particular the limits

$$\lim_{t \to 0^+, 1^-} c(t,\alpha)$$

exist. From the normalization

$$\int_0^1 c(t,\alpha) \, \mathrm{d}t = 1$$

and the estimate

$$c(t,\alpha) \geq \min\left\{2^d |t|^d, 2^d |1-t|^d\right\} c\left(\frac{1}{2}, \alpha\right),$$

it follows that there is K'_d such that

$$c\left(\frac{1}{2}, \alpha\right) \leq K'_d,$$

so that by (1) there is K_d such that

$$\mathrm{Tot.Var.}(c(\cdot,\alpha)) \leq K_d$$

Corollary 6.6

$$\frac{1}{c} \left| \frac{\mathrm{d}}{\mathrm{d}t} c \right| \in L^1_{\mathrm{loc}}(\mathrm{d}t \, \mathrm{d}x).$$

6.3 Divergence Formulation

Proposition 6.7 *We have the following relation between c and the divergence of the vector field p:*

$$\mathrm{div}(1, p\chi_F) = \frac{1}{c}\frac{\mathrm{d}c}{\mathrm{d}t}\mathrm{d}t\,\mathrm{d}z\Big|_F .$$

From

$$\frac{1}{c}\frac{\mathrm{d}c}{\mathrm{d}t} \in \left(-\frac{d}{1-t}, \frac{d}{t}\right)$$

it follows that it is an absolutely continuous measure.

Proof. Take a test function $\phi \in C^1_c(F)$. Applying the disintegration along the rays, one obtains

$$\begin{aligned}
&\int_{\mathbb{R}^d}\int_0^1 \phi(t,z)\mathrm{div}(1, p_t\chi_{F(t)})\,\mathrm{d}t\,\mathrm{d}z \\
&\quad = -\int_{\mathbb{R}^d}\int_0^1 \chi_{F(t)}(z)\phi_t(t,z) + p_t(z)\cdot\nabla\phi(t,z)\,\mathrm{d}t\,\mathrm{d}z \\
&\quad = -\int \mathrm{d}m(\alpha)\int_0^1 \mathrm{d}t\,c(t,\alpha)\Big[\phi_t(t,(1-t)y+tx)+(x-y)\cdot\nabla\phi(t,(1-t)y+tx)\Big] \\
&\quad = -\int \mathrm{d}m(\alpha)\int_0^1 \mathrm{d}t\,c(t,\alpha)\frac{\mathrm{d}}{\mathrm{d}t}\phi(t,(1-t)y+tx) \\
&\quad = \int \mathrm{d}m(\alpha)\int_0^1 \mathrm{d}t\,\frac{\mathrm{d}c}{\mathrm{d}t}(t,\alpha)\phi(t,(1-t)y+tx) \\
&\quad = \int_{\mathbb{R}^d}\int_0^1 \left(\frac{1}{c}\frac{\mathrm{d}c}{\mathrm{d}t}\right)\phi(t,z)\mathrm{d}t\,\mathrm{d}z.
\end{aligned}$$

References

1. L. Ambrosio, N. Fusco, D. Pallara, Functions of bounded variation and free discontinuity problems (Oxford Mathematical Monographs, 2000)
2. F. Aubin, A. Cellina, *Differential inclusions, set-valued maps and viability theory* (Springer, New York, 1984)
3. S. Bianchini, C. De Lellis, R. Robyr, SBV regularity for Hamilton–Jacobi equations in $\mathbb{R}^d$, Arch. Rational Mech. Anal. **200**, 1003–1021 (2011) DOI: 10.1007/s00205-010-0381-z
4. S. Bianchini, M. Gloyer, On the Euler–Lagrange equation for a variational problem: the general case II. Math. Z. **265**, 889–923 (2010) DOI: 10.1007/s00209-009-0547-2
5. P. Cannarsa, A. Mennucci, C. Sinestrari, Regularity results for solutions of a class of Hamilton–Jacobi equations. Arch. Rational Mech. Anal. **140**, 197–223 (1997)
6. D.H. Fremlin, *Measure Theory*, vol. 3, *Measure Algebras.* (Torres Fremlin, 2004)
7. D.H. Fremlin, *Measure Theory*, vol. 4, *Topological Measure Spaces.* (Torres Fremlin, 2006)

Functional Inequalities and Dynamics

Eric A. Carlen

Abstract We give a survey of recent results on functional inequalities and evolution equations focusing on the relations between these two subjects. We discuss the use of evolution equations to prove sharp functional inequalities, as well as the use of sharp functional inequalities to prove precise results about the behavior of solutions of evolution equations. These note are based on a series of lectures given by the author at the C.I.M.E. Summer School in June 2008.

1 Introduction

Functional inequalities play a crucial role in the investigation of dynamical systems and evolution equations, and at the same time, many functional inequalities can be proved in their sharp form by employing a properly chosen dynamics to evolve an arbitrary trial function into an extremal function. Moreover, a dynamical approach often yields a very transparent analysis of the cases of equality.

In these lectures we will discuss a number of examples taken from recent research that provide a perspective on this interplay between research into functional inequalities and into dynamical systems and evolution equations.

A number of recent results will be proved, and a number of open problems will be posed. Although work with many of my collaborators will be presented, my long term collaboration with Michael Loss will feature particularly prominently, and much of the perspective that I present here has been developed in the context of that collaboration.

E.A. Carlen (✉)
Department of Mathematics, Hill Center, Rutgers University, 110 Frelinghuysen Road, Piscataway NJ 08854-8019, USA
e-mail: carlen@math.rutgers.edu

S. Bianchini et al., *Nonlinear PDE's and Applications*, Lecture Notes in Mathematics 2028, DOI 10.1007/978-3-642-21861-3_2,

We begin with some examples of the sorts of functional inequalities with which we shall be concerned. The examples in this section are all very closely related to one another: They are:

(1) The sharp Hardy-Littlewood-Sobolev inequality
(2) The sharp Sobolev inequality
(3) The sharp Logarithmic Hardy-Littlewood-Sobolev inequality
(4) The Euclidean Onofri inequality

Inequalities (1) and (2) are *dual* to one another, as are (3) and (4). Also inequality (3) is a *limiting case* of inequality (1), and (4) is a limiting case of (2).

On account of these relations, the cases of equality in all of these inequalities are closely related: Equality holds in each of them only for functions $f(x)$ that are up to multiples, translations and scalings, of the form $(1+|x|^2)^p$ for some power p (Actually, the power of $(1+|x|^2)$ becomes a logarithm of $(1+|x|^2)$ in case (4).) However, the connection runs even deeper than this: All of these inequalities are *conformally invariant*, as we shall explain later on. In fact, this conformal invariance shall be fundamental to the proofs we shall give of these inequalities. Interestingly, there are other classical inequalities whose sharp versions have powers of $(1+|x|^2)$ as their cases of equality, but which are *not* conformally invariant: Certain Gagliardo-Nirenberg inequalities, whose sharp form was established only recently by Dolbeault and Del Pino are our prime example here.

We shall see in the second section of these notes that these Gagliardo-Nirenberg inequalities are in fact closely connected to the inequalities (1) and (2) through certain *dynamical* considerations that yield another proof of at least certain cases of inequalities (1) and (2). Indeed, Sects. 2 and 3 present three examples of dynamical systems whose *equilibria* are the functions yielding equality in inequalities (1)–(4). The examples in Sect. 2 involve evolution equations (in the form of partial differential equations), while the example in Sect. 3 involves a discrete-time dynamical system. In Sect. 2 we shall discuss certain applications of the inequalities (1)–(4) to the study of the evolution equations introduced there, while in Sect. 3 we shall apply the discrete-time dynamics introduced there to prove the inequalities themselves. The remaining sections further develop the themes introduce in the first three, but in the context of other examples.

Our goal in the rest of this section is to introduce the inequalities (1)–(4), and to explain the close relations between them, going carefully into all of the details. The effort shall be rewarding; this is a subject in which the details are extremely beautiful.

1.1 The Sharp Hardy-Littlewood-Sobolev Inequality

For two non-negative measurable functions f and g on $\mathbb{R}^d$, $d \geq 1$, and a number λ with $0 \leq \lambda < d$, define the functional $I_\lambda(f, g)$ by

$$I_\lambda(f, g) = \int_{\mathbb{R}^d} \int_{\mathbb{R}^d} f(x) \frac{1}{|x-y|^\lambda} g(y) \mathrm{d}x\mathrm{d}y, \tag{1}$$

and for $1 \leq p < \infty$, let $\|f\|_p$ denote the L^p norm of f defined by

$$\|f\|_p = \left(\int_{\mathbb{R}^d} f^p \mathrm{d}x \right)^{1/p}. \tag{2}$$

The HLS inequality provides an upper bound on $I_\lambda(f, f)$ in terms of $\|f\|_{p(\lambda)}$ where

$$p(\lambda) = \frac{2d}{2d - \lambda}. \tag{3}$$

The value $p(\lambda) = \dfrac{2d}{2d-\lambda}$ is determined by the *scale invariance* of the functional $I_\lambda(f, f)$: For $s > 0$, define $f^{(s)}$ by

$$f^{(s)}(x) = f(x/s).$$

Then one computes

$$I_\lambda(f^{(s)}, f^{(s)}) = s^{2d-\lambda} I_\lambda(f, f) \qquad \text{and} \qquad \|f^{(s)}\|_p^2 = s^{2d/p} \|f\|_p^2. \tag{4}$$

An inequality of the form $I_\lambda(f, f) \leq C\|f\|_p^2$ with finite C is possible only if both sides scale the same way. Hence $2d - \lambda = 2d/p$ which yields the value of $p(\lambda)$ given in (3).

There is another way to view the computations we have just made. Define the functional Φ_λ by

$$\Phi_\lambda(f) := \frac{I_\lambda(f, f)}{\|f\|_{p(\lambda)}^2}. \tag{5}$$

Then by (4), $f \mapsto \Phi_\lambda(f)$ is invariant under the transformation

$$f \to f^{(s)}.$$

Invariance properties of functionals, also referred to as *symmetry properties of functionals* play a basic role in the theory of functional inequalities.

This invariance under scale transformations is the first of many symmetry transformations we shall encounter. Another one, quite obvious for the functional Φ_λ, is *translation invariance*,

$$f(x) \to f(x - x_0)$$

for some x_0 in $\mathbb{R}^d$. And of course, for any $c > 0$, $\Phi_\lambda(cf) = \Phi_\lambda(f)$. The remaining symmetries, or invariance properties, of Φ_λ are less obvious. However, they will be crucial to the story. We are now ready to state the sharp form of the HLS inequality:

Theorem 1.1 *[27] For all non-negative measurable functions f with $\|f\|_{p(\lambda)} < \infty$,*

$$\frac{I_\lambda(f,f)}{\|f\|^2_{p(\lambda)}} \le \frac{I_\lambda(h,h)}{\|h\|^2_{p(\lambda)}}$$

where

$$h(x) = \left(\frac{1}{1+|x|^2}\right)^{d/p} .$$

There is equality if and only if for some $x_0 \in \mathbb{R}^d$ and $s \in \mathbb{R}_+$, f is a non-zero multiple of

$$h(x/s - x_0).$$

This theorem is due to Elliott Lieb [27]. Hardy and Littlewood had treated the case $d = 1$ without proving the sharp constant, and Sobolev had treated $d > 1$, also without proving the sharp constant; see [27] for more background.

By Lieb's Theorem, the sharp constant is $C_{\lambda,d}$ where

$$C_{\lambda,d} := \frac{I_\lambda(h,h)}{\|h\|^2_{p(\lambda)}}, \tag{6}$$

so that we have the sharp inequality

$$I_\lambda(f,f) \le C_{\lambda,d}\|f\|^2_{p(\lambda)}. \tag{7}$$

The explicit value of the sharp constant $C_{\lambda,d}$ can be computed by directly computing the integrals $I_\lambda(h,h)$ and $\|h\|^{p(\lambda)}_{p(\lambda)}$. However, there is an easier and more instructive way that makes us of the Euler–Lagrange equation satisfied by h: Since by Lieb's Theorem, knowing that h maximizes Φ_λ, we may take any $g \in L^{p(\lambda)}(\mathbb{R}^d)$, and we have

$$\Phi_\lambda(h + \epsilon g) \le \Phi_\lambda(h).$$

Expanding the left hand side through first order in ϵ, we find:

$$\int_{\mathbb{R}^d} |x-y|^{-\lambda} h(y)\mathrm{d}y = \left[\frac{I_\lambda(h,h)}{\|h\|^2_{p(\lambda)}}\right] \|h\|^{2-p(\lambda)}_{p(\lambda)} h^{p(\lambda)-1}(x). \tag{8}$$

Since for large $|x|$,

$$\int_{\mathbb{R}^d} |x-y|^{-\lambda} h(y)\mathrm{d}y \sim \|h\|_1 |x|^{-\lambda} \qquad \text{and} \qquad h^{p(\lambda)-1}(x) \sim |x|^{-\lambda},$$

we deduce $C_{\lambda,d} = \|h\|_1 \|h\|_{p(\lambda)}^{p(\lambda)-2}$.

Now computing the two norms, one finds (since the integrals defining the norms readily convert to beta integrals, and hence Gamma functions):

$$C_{\lambda,d} = \pi^{\lambda/2} \frac{\Gamma(d/2-\lambda/2)}{\Gamma(d-\lambda/2)} \left[\frac{\Gamma(d/2)}{\Gamma(d)}\right]^{\lambda/2-1} . \tag{9}$$

1.2 Duality and Functional Inequalities

The functionals on both sides of (7) are continuous convex functions defined on all of $L^{p(\lambda)}(\mathbb{R}^d)$. (The continuity of the right hand side is obvious, and the continuity of the left hand side on $L^{p(\lambda)}(\mathbb{R}^d)$ is a consequence of the HLS inequality itself.)

In such circumstances, even with only lower semicontinuity, there is an equivalent *dual inequality* obtained by computing the Legendre transforms of both sides of the inequality.

Two real Banach spaces E and E^* are said to be a *dual pair* if there is a bilinear form $\langle \cdot, \cdot \rangle$ on $E^* \times E$ such that for each non-zero $f \in E$, $\langle \cdot, f\rangle$ is a non-zero continuous linear functional on E^*, and likewise, each non-zero $u \in E^*$, $\langle u, \cdot \rangle$ is a non-zero continuous linear functional on E. For example, consider $E = L^p(\mathbb{R}^d)$ and $E^* = L^q(\mathbb{R}^d)$ where $1/p + 1/q = 1$, and define

$$\langle u, f\rangle = 2\int_{\mathbb{R}^d} u(x) f(x)\mathrm{d}x, \tag{10}$$

which is easily seen to have the required properties.

Then, if φ and ψ are two lower semicontinuous convex functions defined on E, the *Legendre transforms* of φ and ψ are the functions φ^* and ψ^* on E^* defined by

$$\varphi^*(u) = \sup_{f\in E} \{\langle u, f\rangle - \varphi(f)\} \qquad \text{and} \qquad \psi^*(v) = \sup_{g\in E} \{\langle v, g\rangle - \varphi(g)\}, \tag{11}$$

and φ^* and ψ^* are automatically convex and lower semicontinuous on E^*, possibly taking on the value $+\infty$. On account of φ and ψ being lower semicontinuous, one has

$$\varphi(f) = \sup_{u\in E^*} \{\langle u, f\rangle - \varphi^*(u)\} \qquad \text{and} \qquad \psi(g) = \sup_{v\in E^*} \{\langle v, g\rangle - \varphi^*(v)\}. \tag{12}$$

Having recalled these facts concerning the Legendre transform, we now state the following theorem:

Theorem 1.2 *Let E and E^* be a dual pair of Banach spaces. Let φ and ψ be two lower semicontinuous convex functions on E, and let φ^* and ψ^* be their Legendre transforms on E^*. The the following two inequalities are equivalent:*

$$\varphi(f) \leq C\psi(f) \qquad \text{for all } f \in E \tag{13}$$

and

$$\varphi^*(u) \geq C\psi^*(u/C) \qquad \text{for all } \ u \in E^*. \tag{14}$$

in that if either one true for some positive constant C, then the other is also true for the same constant C. In particular, the least value of C for which (13) is true is the same as the least value of C for which (14)is true.

Proof: Suppose that (13) is true. Then, for all $f \in E$,

$$\langle u, f\rangle - \varphi(f) \geq \langle u, f\rangle - C\psi(f) \qquad \text{for all } \ u \in E^*,$$

and taking the supremum over f, we conclude that (14) is true. Indeed,

$$\langle u, f\rangle - C\psi(f) = C\left[\langle (u/C), f\rangle - \psi(x)\right],$$

from which the form of the right hand side of (14) readily follows.

Next, suppose that (14) is true. Then for all $f \in E$ and all $u \in E^*$,

$$\langle u, f\rangle - \varphi^*(u) \leq \langle u, f\rangle - C\psi^*(u/C).$$

Taking the supremum over u, we have from (12) that

$$\varphi(f) \leq C\psi(f) \qquad \text{for all } f \in E, \tag{15}$$

where we have used the fact that

$$\langle u, f\rangle - C\psi(u/C) = C\left[\langle (u/C), f\rangle - C\psi^*(u/C)\right].$$

□

To apply this result in the case at hand, we fix some p with $1 < p < \infty$, and choose $E = L^p(\mathbb{R}^d)$ and $E^* = L^q(\mathbb{R}^d)$ where $1/p + 1/q = 1$, and with the dual pairing given in (10).

Consider the continuous convex function ψ on E given by $\psi(f) = \|f\|_p^2$. To compute ψ^*, note that by Hölder's inequality,

$$\langle u, f\rangle - \psi(f) \leq 2\|f\|_p\|u\|_q - \|f\|_p^2 = \|u\|_q^2 - (\|f\|_p - \|u\|_q)^2 \leq \|u\|_q^2,$$

and by the conditions for equality in Hölder's inequality, there is equality if and only if

$$f = \frac{1}{\|u\|_q^{q-2}} |u|^{q-1} \mathrm{sgn}(u).$$

Thus,

$$\psi^*(u) = \|u\|_q^2. \tag{16}$$

Next, pick some λ with $0 < \lambda < d$, and specialize to the case $p = p(\lambda)$ consider the functional

$$f \mapsto I_\lambda(f).$$

A standard Fourier transform argument shows that the quadratic form $I_\lambda(f)$ is positive definite, and thus strictly convex. Moreover, the HLS inequality says that it is continuous on $E = L^{p(\lambda)}(\mathbb{R}^d)$.

In the interest of being as concrete as possible, we shall focus on the familiar case $\lambda = d - 2$. Recall that for $d \geq 3$, the kernel $K(x, y) = |x - y|^{-(d-2)}$ is, up to a constant, the kernel for $(-\Delta)^{-1}$. More precisely, with $|S^{d-1}|$ denoting the surface area of the $d - 1$ dimensional unit sphere in $\mathbb{R}^d$,

$$\langle f, (-\Delta)^{-1} f \rangle_{L^2(\mathbb{R}^d)} = \frac{1}{(d-2)|S^{d-1}|} \int_{\mathbb{R}^d} \int_{\mathbb{R}^d} f(x) \frac{1}{|x-y|^{d-2}} f(y) \mathrm{d}x \mathrm{d}y.$$

Since $(-\Delta)^{-1}$ is a positive operator, the left hand side is a convex function of f.

Thus, Lieb's Theorem gives the sharp constant in the inequality

$$\langle f, (-\Delta)^{-1} f \rangle_{L^2(\mathbb{R}^d)} \leq C_S \|f\|_{p(\lambda)}^2, \tag{17}$$

where

$$C_S := \frac{C_{\lambda,d}}{(d-2)|S^{d-1}|} = \frac{4}{d(d-2)} |S^d|^{-2/d}. \tag{18}$$

Therefore, for $d \geq 3$, let us consider the continuous convex functional φ on E given by

$$\varphi(f) = \frac{1}{|S^{d-1}|} I_{d-2}(f, f).$$

To compute φ^*, we complete the square (remembering the factor of 2 in (10)):

$$\begin{aligned} \langle u, f \rangle - \varphi(f) &= 2\langle (-\Delta)^{1/2} u, (-\Delta)^{-1/2} f \rangle_{L^2(\mathbb{R}^d)} - \|(-\Delta)^{-1/2} f\|_2^2 \\ &= \|(-\Delta)^{1/2} u\|_2^2 - \|(-\Delta)^{1/2} u - (-\Delta)^{-1/2} f\|_2^2 \\ &\leq \|(-\Delta)^{1/2} u\|_2^2 = \|\nabla u\|_2^2. \end{aligned} \tag{19}$$

There is equality if and only if $-\Delta u = f$, and so

$$\varphi^* = \|\nabla u\|_2^2. \tag{20}$$

Now, since (17) can be written as

$$\varphi(f) \leq C_S \psi(f)$$

for all f on E, we have from Theorem 1.2 that

$$\varphi^*(u) \geq C_S \psi^*(u/C_S).$$

By (16) and (20) this reduces to

$$C_S \|\nabla u\|_2^2 \geq \|u\|_{q(\lambda)}^2, \tag{21}$$

which is the classical Sobolev inequality. Moreover, by the equivalence asserted in Theorem 1.2, the constant C_S given in (18) is also best possible in (21).

Not only are the optimal constants in the two inequalities the same; there is also a complete correspondence among cases of equality. To see this, suppose the u is some function giving equality in (21); i.e., satisfying

$$\varphi^*(u) = C_S \psi^*(u/C_S). \tag{22}$$

As we have seen in the computation of ψ^*, there is a unique $f \in E$ so that

$$C_S \psi^*(u/C_S) = \langle u, f \rangle - C_S \psi(f), \tag{23}$$

and this f is a multiple of $u^{q(\lambda)-1} = u^{(d+2)/(d-2)}$. Then, with this choice of f and u, we have

$$\begin{aligned} \varphi^*(u) &\geq \langle u, f \rangle - \varphi(f) \\ &\geq \langle u, f \rangle - C_S \psi(f) \\ &= C_S \psi^*(u/C_S), \end{aligned} \tag{24}$$

where the final equality is (23). But then by (22), all of the inequalities in (24) must be equalities, and in particular, it must be the case that $\varphi(f) = C_S \psi(f)$. That is, f must be an optimizer for the HLS inequality, and since, up to a harmless constant, f is a multiple of $u^{q(\lambda)-1}$, we see that for u to be an optimizer for the Sobolev inequality (21), $u^{q(\lambda)-1}$ must be an optimizer for the HLS inequality for $\lambda = d - 2$. Similar reasoning applies in the other direction, and one concludes:

$$u \text{ saturates the Sobolev inequality}$$
$$\Longleftrightarrow$$
$$u^{(d+2)/(d-2)} \text{ saturates the HLS inequality for } \lambda = d-2$$

Thus, once we have determined the cases of equality in the HLS inequality for $\lambda = d-2$, we shall have determined the cases of equality in the Sobolev inequality for the square integral of the gradient. This has advantages as we shall see. However, the sharp value of the Sobolev constant, and all the cases of equality, were first worked out directly by Aubin [1] and Talenti [33].

1.3 A Limiting Case of the HLS Inequality

The case $\lambda = 0$ of the HLS Inequality is a triviality since $I_0(f,f) = \left(\int_{\mathbb{R}^d} f(x)\mathrm{d}x\right)^2$ and $p(0) = 1$. Thus, for $\lambda = 0$ we have the *identity*

$$I_0(f,f) = \|f\|_1^2.$$

This is not without interest: *Whenever a parameterized family of inequalities reduces to an identity, we get a new inequality by differentiating in the parameter.*

Starting from the sharp HLS inequality $I_\lambda(f,f) \le C_{\lambda,d}\|f\|_{p(\lambda)}^2$, and subtracting $I_0(f,f)$ from the left side, and $\|f\|_1^2$ from the right side, and dividing both sides by λ, one obtains

$$\frac{I_\lambda(f,f) - I_0(f,f)}{\lambda} \le \frac{C_{\lambda,d}\|f\|_{p(\lambda)}^2 - \|f\|_1^2}{\lambda}.$$

Formally taking limits under the integral sign,

$$\lim_{\lambda\to 0} \frac{I_\lambda(f,f) - I_0(f,f)}{\lambda} = -\int_{\mathbb{R}^d}\int_{\mathbb{R}^d} f(x)\ln(|x-y|)f(y)\mathrm{d}x\mathrm{d}y. \tag{25}$$

Also, since $p(0) = C(0) = 1$, and since

$$\left.\frac{\mathrm{d}}{\mathrm{d}\lambda}p(\lambda)\right|_{\lambda=0} = \frac{1}{2d},$$

$$\lim_{\lambda\to 0} \frac{C_{\lambda,d}\|f\|_{p(\lambda)}^2 - \|f\|_1^2}{\lambda} =$$

$$= \left[\left.\frac{\mathrm{d}}{\mathrm{d}\lambda}C_{\lambda,d}\right|_{\lambda=0}\right]\|f\|_1^2 + \frac{1}{d}\|f\|_1\left[\int_{\mathbb{R}^d} f\ln f\mathrm{d}x - \|f\|_1\ln\|f\|_1\right].$$

Thus defining

$$C'_d = \frac{\mathrm{d}}{\mathrm{d}\lambda} C_{\lambda,d}\bigg|_{\lambda=0}, \tag{26}$$

which we shall be able to explicitly compute using (9), we formally deduce the *Logarithmic HLS inequality*

$$\left[\int_{\mathbb{R}^d} f \ln f \mathrm{d}x - \|f\|_1 \ln \|f\|_1\right] + \frac{d}{\|f\|_1} \int_{\mathbb{R}^d}\int_{\mathbb{R}^d} f(x)\ln(|x-y|)f(y)\mathrm{d}x\mathrm{d}y \geq -dC'_d\|f\|_1 . \tag{27}$$

However, the limit and the integral in (25) require some care. The reason, of course, is that for f non-negative and measurable, the integrands in the integral defining $I_\lambda(f,f)$ are non-negative, and so the integral always yields a well defined, though possibly infinite, value. However, $\ln|x-y|$ is unbounded above and below, so non-negativity of f does not suffice to guarantee that the integrals in (27) makes sense. This problem is also present for the entropy functional, $\int_{\mathbb{R}^d} f \ln f \mathrm{d}x$: Mere integrability of $f \geq 0$ does not guarantee that either the positive or the negative parts of $f \ln f$ are integrable.

We therefore pause to make precise sense of the functionals involved in the inequality (27). We do this in full detail in the most interesting, and most subtle, case $d = 2$.

1.4 The Two Dimensional Energy Integral

It is easy to see that if the measurable function f is bounded and has compact support, then the integrals in

$$J(f,f) := -\frac{1}{4\pi}\int_{\mathbb{R}^d}\int_{\mathbb{R}^d} f(x)\ln(|x-y|)f(y)\mathrm{d}x\mathrm{d}y \tag{28}$$

converge, and so (28) defines a quadratic form on a dense subset of $L^1(\mathbb{R}^2)$.

The integral defining $J(f,f)$ is frequently referred to as the *energy integral* of f because of various physical interpretations. These are all based on the familiar fact that if $f \in C_0^\infty(\mathbb{R}^2)$, and u is defined by

$$u(x) = \frac{1}{2\pi}\int_{\mathbb{R}^2} \ln|x-y| f(y)\mathrm{d}y, \tag{29}$$

then

$$\Delta u(x) = f(x).$$

On this basis, one might be tempted to conclude that

$$-\frac{1}{4\pi}\int_{\mathbb{R}^2} f(x)\ln|x-y|f(y)\mathrm{d}x\mathrm{d}y = \frac{1}{2}\langle f, (-\Delta)^{-1} f\rangle_{L^2(\mathbb{R}^2)} \tag{30}$$

for all f in some dense subset of $L^1(\mathbb{R}^2)$. This would imply that $J(f,f)$ can be defined as a positive definite quadratic form on $L^1(\mathbb{R}^2)$, and hence convex. However, this line of argument would be erroneous.

To see why, suppose $f \in C_0^\infty(\mathbb{R}^2)$. Then with u defined in terms of f by (29), we have

$$J(f,f) = \frac{1}{2}\int_{\mathbb{R}^2} f(x)u(x)\mathrm{d}x = \frac{1}{2}\int_{\mathbb{R}^2}(-\Delta u(x))u(x)\mathrm{d}x.$$

So far, so good. Next, one might be tempted to integrate by parts, writing

$$\int_{\mathbb{R}^2}(-\Delta u(x))u(x)\mathrm{d}x = \int_{\mathbb{R}^2}|\nabla u(x)|^2\mathrm{d}x,$$

and the right hand side is another object frequently referred to as an energy integral.

The problem with this calculation is that for large $|x|$,

$$\nabla u(x) \sim \frac{1}{2\pi}\frac{x}{|x|^2}\left[\int_{\mathbb{R}^2} f(x)\mathrm{d}x + \mathcal{O}\left(\frac{1}{|x|}\right)\right],$$

and so ∇u is *not* square integrable unless $\int_{\mathbb{R}^2} f(x)\mathrm{d}x = 0$. In particular, for $f(x) \geq 0$, ∇u is not square integrable except in the trivial case that $f(x) = 0$ everywhere. Summarizing, while (30) is valid at least for $f \in C_0^\infty(\mathbb{R}^2)$ with $\int_{\mathbb{R}^2} f\mathrm{d}x = 0$, we are interested in the case that f is a probability density, for which (30) is not generally valid.

Therefore, let us back up and consider the integral defining $J(f,f)$ in (28) and ask: Under what conditions is this integral well defined?

Clearly we need some growth condition on f to tame the logarithmic growth of the negative part of the kernel $-\ln|x-y|$ for large $|x-y|$. Since

$$\ln|x-y| \leq \frac{1}{2}[\ln 2 + \ln(1+|x|^2) + \ln(1+|y|^2)], \tag{31}$$

the integral defining $J(f,f)$ is well defined, though possibly infinite, for f in the set of non-negative measurable functions with

$$\int_{\mathbb{R}^2} f(x)\ln(e+|x|^2)\mathrm{d}x < \infty.$$

Therefore, let E denote the Banach space $L^1(\mathbb{R}^2, \ln(e+|x|^2)\mathrm{d}x)$. If we define

$$J^+(f,f) = \frac{1}{4\pi}\int_{\mathbb{R}^2} f(x)\left(\ln|x-y|\right)_- f(y)\mathrm{d}x\mathrm{d}y \qquad \text{and}$$

$$J^-(f,f) = \frac{1}{4\pi}\int_{\mathbb{R}^2} f(x)\left(\ln|x-y|\right)_+ f(y)\mathrm{d}x\mathrm{d}y, \tag{32}$$

where $(\cdot)_-$ and $(\cdot)_+$ denote the negative and positive part, respectively, so that

$$J(f,f) = J^+(f,f) - J^-(f,f),$$

we have from (31) that $f \mapsto J^-(f,f)$ is strongly continuous on E. Moreover, if $\{f_n\}$ is a sequence of non-negative functions in E that is strongly convergent to f, then by considering an almost everywhere convergent subsequence, and then using Fatou's Lemma, one see that

$$J^+(f,f) \le \liminf_{n\to\infty} J^+(f_n,f_n).$$

Thus, $f \mapsto J^+(f,f)$ is lower semicontinuous on the positive cone in E. Putting the pieces together, we see that $f \mapsto J(f,f)$ is lower semicontinuous on the positive cone in E.

This takes care of the definitions and continuity properties of the functional J. To obtain convexity, we must further restrict the domain of J.

Definition 1.3 *for $E > 0$, define E_M to be the set of non-negative functions f in E with*

$$\int_{\mathbb{R}^2} f \mathrm{d}x = M.$$

Then on E define the functional $f \mapsto J_M(f,f)$ by

$$J_M(f,f) = \begin{cases} J(f,f) & \text{if } f \in E_M \\ \infty & \text{if } f \notin E_M \end{cases}. \tag{33}$$

The functional $f \mapsto J_M(f,f)$ is convex as well as lower semicontinuous. In fact, since we have already seen that it is lower semicontinuous, the convexity will follow once we show that for all $f_1, f_2 \in E_M$,

$$\frac{J_M(f_1,f_1) + J_M(f_2,f_2)}{2} \ge J_M\left(\frac{f_1+f_2}{2}, \frac{f_1+f_2}{2}\right).$$

In showing this, it obviously suffices to consider the case in which $J_M(f_1,f_1), J_M(f_2,f_2) < \infty$. A simple truncation and monotone convergence argument then shows that we need only consider the case in which f_1, f_2 are bounded with compact support. Then with $g := f_1 - f_2$,

$$\frac{J_M(f_1,f_1) + J_M(f_2,f_2)}{2} - J_M\left(\frac{f_1+f_2}{2}, \frac{f_1+f_2}{2}\right) = J(g,g).$$

Now, since $\int_{\mathbb{R}^2} g \mathrm{d}x = 0$, and since g has compact support, the identity in (29) applies, and we see that $J(g,g) \ge 0$. This proves that for each $M > 0$, the functional $f \mapsto J_M(f,f)$ is convex on E.

However, notice that to obtain the convexity, we have had to fix the total mass M, and have lost the homogeneity of $f \mapsto J(f,f)$: While for all $s > 0$, $J(sf,sf) = s^2 J(f,f)$, $J_M(sf,sf)$ can be finite for at most one value of s.

Finally, we turn to the other functional in (27), the entropy functional:

Definition 1.4 *For $M > 0$ the* entropy functional *is the functional $f \mapsto S_M(f)$ defined by*

$$S_M(f) = \begin{cases} \int_{\mathbb{R}^2} f \ln f \mathrm{d}x & \text{if } f \in E_M \\ \infty & \text{if } f \notin E_M \end{cases}. \tag{34}$$

In the special case $M = 1$, we have the standard Shannon *entropy functional on probability densities, and in this case will often write simply $S(f)$ in place of $S_1(f)$. Note that our sign convention for the entropy is the one frequently encountered in information theory, but* never *in statistical mechanics, where entropy would be given by $-\int_{\mathbb{R}^2} f \ln f \mathrm{d}x$, and would therefore be* concave.

It is easy to see that for each $M > 0$, $f \mapsto S_M(f)$ is well-defined, lower semicontinuous and convex on E: The key point is that to belong to the weighted L^1 space E, f must decay sufficiently fast at infinity that $f(\ln f)_-$ is automatically integrable. This follows from a standard *relative entropy argument*:

First, to prepare the way, we reduce to the consideration of densities $f \in E_M$ that are bounded. To do this, fix and $f \in E_M$, and define the function g by

$$g(x) = \begin{cases} f(x) & \text{if } f(x) < 1 \\ C & \text{if } f(x) \geq 1 \end{cases},$$

where C is a constant chosen so that $g \in E_M$. Clearly whenever the set $\{x : f(x) \geq 1\}$ has positive measure, C has a uniquely determined value, and $C \geq 1$. In the other cases, the value of C is irrelevant, and anyway, f is already bounded by 1. Either way, we have

$$g(\ln g)_- = f(\ln f)_- .$$

Therefore, it suffices to show that $g(\ln g)_-$ is integrable. Here is where the relative entropy comes in: For each $M > 0$, introduce the function h_M defined by

$$h_M(x) = \frac{M}{\pi} \left(\frac{1}{1+|x|^2} \right)^2 . \tag{35}$$

It is easy to see that for each $M > 0$, $h_M \in E_M$. Then by Jensen's inequality

$$0 \leq \int_{\mathbb{R}^2} \left(\frac{g}{h_M} \right) \ln \left(\frac{g}{h_M} \right) h_M \mathrm{d}x = \int_{\mathbb{R}^2} g[\ln g - \ln h_M] \mathrm{d}x. \tag{36}$$

The integral on the left side of (36) defines the *relative entropy of g with respect to h_M*.

By the definition of E, $g \ln h_M$ is integrable. By the definition of g, $g(\ln g)_+$ is integrable. But then by (36),

$$\int_{\mathbb{R}^2} g(\ln g)_- \mathrm{d}x \leq \int_{\mathbb{R}^2} g(\ln g)_+ \mathrm{d}x - \int_{\mathbb{R}^2} g \ln h \mathrm{d}x < \infty.$$

Thus, for $f \in E_M$, $f(\ln f)_-$ is integrable, and so $S_M(f)$ is always well defined, though possibly infinite (like $J_M(f,f)$). The convexity and lower semicontinuity of $f \mapsto S_M(f)$ now follow easily from the convexity and continuity of the function $t \mapsto f \ln t$ on $\mathbb{R}_+$.

The arguments we have made about J_M and S_M for $d = 2$ are easily extended to all $d > 2$, however, the applications we shall discuss here only concern the case $d = 2$.

1.5 The Logarithmic HLS Inequality for d = 2

We are now ready to give a precise statement of the sharp logarithmic Hardy-Littlewood-Sobolev (log HLS) inequality. The inequality itself is simply a restatement of (27) for $d = 2$, using the definitions and notation of the previous section, and an explicit calculation of C_2'.

Theorem 1.5 (Logarithmic Hardy-Littlewood-Sobolev inequality) *Let f be a non-negative function in $E = L^1(\mathbb{R}^2, \ln(e + |x|^2)\mathrm{d}x)$. Let $\int_{\mathbb{R}^2} f \mathrm{d}x = M$, so that $f \in E_M$ Then*

$$S_M(f) + C_M \geq \frac{8\pi}{M} J(f,f) \tag{37}$$

with

$$C_M := M(1 + \ln \pi - \ln M). \tag{38}$$

There is equality with both sides finite if and only if for some $x_0 \in \mathbb{R}^d$ and $s \in \mathbb{R}_+$,

$$f(x) = s^{-2} h_M(x/s - x_0).$$

One can prove the inequality by showing that the formal limits that led to (27) make sense when $f \in E_M$, and directly calculating the constant C_d' defined in (26).

However, there is an easier, more direct, way to calculate C_M, once we know that h_M minimizes the functional

$$f \mapsto S_M(f) - \frac{8\pi}{M} J(f,f), \tag{39}$$

which is what Theorem 1.5 asserts. Parts of this computation will be useful later on, in working out the Legendre transform of J_M, so we explain this now:

Knowing that h_M is a minimizer, the Euler–Lagrange equation for the minimization of the functional in (39) says that

$$\ln h_M(x) + \frac{4}{M}\int_{\mathbb{R}^2} \ln|x-y| h_M(y)\mathrm{d}y = K_M \tag{40}$$

for some constant K_M. Since for large $|x|$, up to constant order,

$$\ln h_M(x) \sim \ln M - \ln\pi - 4\ln|x| \quad \text{and} \quad \frac{4}{M}\int_{\mathbb{R}^2} \ln|x-y| h_M(y)\mathrm{d}y \sim 4\ln|x|,$$

we deduce $K_M = \ln M - \ln\pi$.

Now, multiplying both sides of (8) by h_M, and integrating, one obtains

$$\int_{\mathbb{R}^2} h_M(x)\ln h_M(x)\mathrm{d}x + \frac{4}{M}\int_{\mathbb{R}^2}\int_{\mathbb{R}^2} h_M(x)\ln|x-y| h_M(y)\mathrm{d}x\mathrm{d}y = M(\ln M - \ln\pi). \tag{41}$$

A direct calculation shows that

$$\int_{\mathbb{R}^2} h_M(x)\ln h_M(x)\mathrm{d}x = M(\ln M - \ln\pi - 2), \tag{42}$$

and so we conclude that

$$\frac{4}{M}\int_{\mathbb{R}^2}\int_{\mathbb{R}^2} h_M(x)\ln|x-y| h_M(y)\mathrm{d}x\mathrm{d}y = 2M, \quad \text{or equivalently,}$$

$$J_M(h_M, h_M) = -\frac{M^2}{8\pi}. \tag{43}$$

From here, we see that $C_M = M(1 + \ln\pi - \ln M)$, as claimed.

In summary, once we have shown that the only minimizers of the functional (39) are translates of scalings of h_M, we shall have proved all of Theorem 1.5.

It is also worth noting at this point that $J_M(h_M, h_M) = J(h_M, h_M)$ is *negative*, so $f \mapsto J(f, f)$ is manifestly *not* a positive quadratic form on any dense subset of E, as might be suggested by a naive argument using the two dimensional Green's function for the Laplacian.

So, how does one prove that the only minimizers of the functional (39) are translates of scalings of h_M? Given the derivation of the log HLS inequality as a limiting case of the HLS inequality, it is easy to guess and check that h_M yields equality in the log HLS inequality, and hence so would all of its translates and scalings. After all, we know from Lieb's Theorem that the HLS is saturated by, and only by, multiples, translates and scalings of

$$\left(\frac{1}{1+|x|^2}\right)^{d/p(\lambda)} .$$

Going through the limiting argument, one arrives at the conclusion, checkable by direct calculation, that h_M does saturate the log HLS inequality.

However, whenever one proves an inequality by such a limiting process, it is not *a priori* clear that additional cases of inequality do not "sneak in" through the limit. Thus, a separate proof is required for the statement that *there are not other non-trivial cases of equality, apart from translates and scaling of h_M*, where by a non-trivial case of equality, we mean one in which both sides are finite.

A proof of this was given Carlen and Loss [17], and at the same time by Beckner [5]. The proof in [17] is dynamical in nature, based on the *competing symmetries* method that shall be explained later in these lectures. In fact, this method give a simultaneous proof of all aspects of the theorem – the inequality itself, and the determination of all of the cases of equality. However, before we come to that, let us investigate the log HLS inequality from a dual point of view, using Theorem 1.2.

1.6 The Dual of the Sharp HLS Inequality for $\mathbb{R}^2$

The Logarithmic HLS Inequality is particularly interesting for $\mathbb{R}^2$, and this is why we have focused on this dimension. One reason is this: As we have seen for $d \geq 3$, the case $\lambda = d-2$ of the HLS inequality is equivalent by duality the sharp Sobolev inequality for $\mathbb{R}^d$:

$$\|u\|_{2d/(d-2)}^2 \leq \frac{C_{d-2,d}}{(d-2)|S^{d-1}|}\|\nabla u\|_2^2. \tag{44}$$

For $d=2$, $d-2=0$, and $2d/(d-2)=\infty$. But as we have seen,

$$I_0(f,f) \leq C_{0,d}\|f\|_1^2 \tag{45}$$

is a triviality. Even worse on the dual side, standard examples show that even when u has compact support in $\mathbb{R}^2$, $\|\nabla u\|_2 < \infty$ does not imply that $\|u\|_\infty < \infty$ – and so the $d=2$ case of (44) is not even trivial; it is just plain false.

The log HLS inequality in $\mathbb{R}^2$ gives us a nontrivial replacement (45), and its dual gives us a true replacement for (44), valid for functions on $\mathbb{R}^2$. To find this replacement, we simply have to calculate the Legendre transforms of S_M and J_M, and apply Theorem 1.2.

The result is:

Theorem 1.6 *For all functions η on $\mathbb{R}^2$ such that for some finite constant C, $|\eta(x)| \leq C\ln(e+|x|^2)$ almost everywhere, and such that the distributional gradient of η is locally square integrable.*

$$\left[M\ln\left(\int_{\mathbb{R}^2} e^{\eta}\mathrm{d}x\right)-M\ln M\right]+\int_{\mathbb{R}^2}(\ln h_M-\eta)h_M\mathrm{d}x\leq\frac{M}{16\pi}\int_{\mathbb{R}^2}|\nabla(\eta-\ln h_M)|^2\,\mathrm{d}x. \tag{46}$$

There is equality if and only if for some $s>0$ and some $x_0\in\mathbb{R}^2$,

$$\eta(x)=\ln h_M(x/s+x_0).$$

This inequality is a Euclidean version of *Onofri's inequality* for functions on the two dimensional sphere S^2 in $\mathbb{R}^3$. Onofri's Theorem [30] states that

$$\ln\left(\int_{S^2} e^u\mathrm{d}\sigma\right)-\int_{S^2}u\mathrm{d}\sigma\leq\frac{1}{4}\int_{S^2}|\nabla u|^2\mathrm{d}\sigma.$$

Here $\mathrm{d}\sigma$ denotes the uniform probability measure on S^2. For a proof of an equivalent version, see [16].

A fundamental fact discovered by Onofri is that the functional

$$u\mapsto\left[\frac{1}{4}\int_{S^2}|\nabla u|^2\mathrm{d}\sigma+\int_{S^2}u\mathrm{d}\sigma-\ln\left(\int_{S^2}e^u\mathrm{d}\sigma\right)\right]$$

is invariant under an action of the group of conformal transformation of S^2. We shall return to this point, and the relation between Onoffri's inequality and (46) later.

Before turning to the proof of Theorem 1.6, we first make an observation about a special case of (46).

Suppose that $\int_{\mathbb{R}^2}e^{\eta}\mathrm{d}x = M$ so that $g := e^{\eta}$ is a density of total mass M. Then

$$\int_{\mathbb{R}^2}(\eta-\ln h_M)\,h_M\mathrm{d}x=\int_{\mathbb{R}^2}\left(\frac{h_M}{g}\right)\ln\left(\frac{h_M}{g}\right)g\mathrm{d}x=:H(h_M|g),$$

the *relative entropy of h with respect to g*. Thus, the Euclidean Onofri inequality can be written as

$$\frac{1}{16\pi}\int_{\mathbb{R}^2}|\nabla\ln g-\nabla\ln h_M|^2\geq H(h_M|g),$$

which holds for all densities g on $\mathbb{R}^2$ with total mass M. This inequality can be viewed as an unusual sort of logarithmic Sobolev inequality – unusual in the sense that the roles of the reference density h_M and the variable density g are interchanged: A standard sort of logarithmic Sobolev inequality would give an upper bound on $H(g|h_M)$, not $H(h_M|g)$, in terms of an integral of a quadratic form in $\nabla \ln g$.

We shall explain certain applications of Theorem 1.6 later. Now, however, let us turn to its proof. For this, we must compute the Legendre transforms of S_M and J_M. We do this in the next section in some detail, as the results are of independent interest, and the calculations require some care.

1.7 The Legendre Transforms of S_M and J_M

The first step in the computation of the Legendre transform is to specify the dual pairing. Let E^* denote the Banach space of measurable functions φ on $\mathbb{R}^2$ with

$$\|\varphi\|_{E^*} := \text{esssup}_{x\in\mathbb{R}^2}\left\{\frac{|\varphi(x)|}{\ln(e+|x|^2)}\right\} < \infty.$$

(Notice that this time there is no factor of 2 included on the right hand side; it would be less convenient here.) Then it is easy to see that with the dual pairing

$$\langle \varphi, f\rangle = \int_{\mathbb{R}^2} \varphi(x) f(x)\mathrm{d}x,$$

E^* and E are a dual pair.

We begin with the entropy functional $S_M(f)$; the analysis here is easier and more standard. Recall that for $f \in E_M$, $f \ln h_M$ and $f(\ln f)_-$ are integrable, so that $S_M(f)$ is at least well defined.

Theorem 1.7 *For all $\varphi \in E^*$, and all $\alpha > 0$,*

$$\sup_{f\in E_M}\left\{\int_{\mathbb{R}^2} \varphi f \mathrm{d}x - \alpha S_M(f)\right\} = \alpha M \ln\left(\int_{\mathbb{R}^2} e^{\varphi/\alpha}\mathrm{d}x\right) - \alpha M \ln M. \quad (47)$$

In particular, the theorem says that the Legendre transform of the entropy function S_M is the functional S_M^* on E^* given by

$$S_M^*(\varphi) = M \ln\left(\int_{\mathbb{R}^2} e^{\varphi}\mathrm{d}x\right) - M \ln M.$$

It is a good exercise to check directly that in fact for all $f \in E$

$$S_M(f) = \sup_{\varphi\in E^*}\left\{\int_{\mathbb{R}^2} \varphi f \mathrm{d}x - S_M^*(\varphi)\right\},$$

and in particular that the right hand side is infinite if $f \notin E_M$. While this follows from the general properties of Legendre transforms, the exercise is instructive.

Proof: First of all, we need only consider $f \in E_M$, since for $f \notin E_M$, $S_M(f) = \infty$, and such an f is irrelevant to the computation of the supremum. Next, for any $\varphi \in E^*$ and any $f \in E_M$,

$$\int_{\mathbb{R}^2} \varphi f \mathrm{d}x - \int_{\mathbb{R}^2} f \ln f \mathrm{d}x = \int_{\mathbb{R}^2} f \ln\left(\frac{e^\varphi}{f}\right) \mathrm{d}x = M \int_{\{f>0\}} \frac{f}{M} \ln\left(\frac{e^\varphi}{f}\right) \mathrm{d}x.$$

By Jensen's inequality, the right hand side is less than or equal to

$$\begin{aligned} M \ln\left(\int_{\{f>0\}} \frac{f}{M}\frac{e^\phi}{f}\right) \mathrm{d}x &= M \ln\left(\int_{\mathbb{R}^2} e^\phi\right) \mathrm{d}x - M \ln M \\ &\leq M \ln\left(\int_{\mathbb{R}^2} e^\phi\right) \mathrm{d}x - M \ln M, \end{aligned} \tag{48}$$

with equality if and only if $\phi = \ln f + C$ for some constant C, and $f > 0$ almost everywhere. Altogether,

$$\int_{\mathbb{R}^2} \varphi f \mathrm{d}x - \int_{\mathbb{R}^2} f \ln f \mathrm{d}x \leq M \ln\left(\int_{\mathbb{R}^2} e^\phi\right) \mathrm{d}x - M \ln M. \tag{49}$$

Conversely, for each $n > 0$, define

$$f_n(x) = 1_{\{|x| \leq n\}} \frac{M e^{\varphi(x)}}{\int_{\{|y| \leq n\}} e^{\varphi(y)}\, dy},$$

and note that $f_n \in E_M$ for each n. Direct calculation now yields

$$\int_{\mathbb{R}^2} \varphi f_n \mathrm{d}x - \int_{\mathbb{R}^2} f_n \ln f_n \mathrm{d}x \leq M \ln\left(\int_{\{|x| \leq n\}} e^\phi \mathrm{d}x\right) - M \ln M. \tag{50}$$

From (49) together with (50) and the Dominated Convergence Theorem, we obtain

$$\sup_{f \in E_M} \left\{\int_{\mathbb{R}^2} \varphi f \mathrm{d}x - S_M(f)\right\} = M \ln\left(\int_{\mathbb{R}^2} e^\varphi \mathrm{d}x\right) - M \ln M. \tag{51}$$

Finally. for any $\alpha > 0$,

$$\int_{\mathbb{R}^2} \phi f \mathrm{d}x - \alpha S_M(f) = \alpha \left(\int_{\mathbb{R}^2} \left(\frac{\varphi}{\alpha}\right) f \mathrm{d}x\right).$$

From this and (51) we obtain (47). □

We now turn to the computation of the Legendre transform of $f \mapsto J_M(f,f)$. To efficiently state the main result, we first make some definitions.

First, let us use the convenient notation

$$-\frac{1}{2\pi}\int_{\mathbb{R}^2} \ln|x-y| g(y)\mathrm{d}y = G * g, \tag{52}$$

for non-negative $g \in E$, or more generally, whenever the integral is well defined. This is the *potential* of g.

Let $\mathcal{P}_M$ be the class of functions defined by

$$\mathcal{P}_M = \{\ G * g\ :\ g \in E_M\ \}.$$

For any function $\varphi \in E^*$, we define

$$D_M(\varphi) = \inf_{u\in\mathcal{P}_M}\left\{\frac{1}{2}\int_{\mathbb{R}^2}|\nabla\varphi - \nabla u|^2\mathrm{d}x\right\} \tag{53}$$

provided the distribution gradient of φ is locally square integrable, and we define $D_M(\varphi)$ to be infinite otherwise.

The Legendre transform of J_M will consist of two parts: One is D_M, and the other is the functional $\mathcal{E}_M$ defined as follows:

For all $\varphi \in E^*$,

$$\mathcal{E}_M(\varphi) = \frac{1}{2}\int_{\mathbb{R}^2}\left|\nabla\left(\varphi - \frac{M}{8\pi}\ln h_M\right)\right|^2\mathrm{d}x + \int_{\mathbb{R}^2}\varphi h_M\mathrm{d}x + \frac{M^2}{8\pi}, \tag{54}$$

provided the distribution gradient of φ is locally square integrable, and we define $\mathcal{E}_M(\varphi)$ to be infinite otherwise.

It is useful to observe that since $\nabla\ln h_M$ is independent of M,

$$\mathcal{E}^*_M(M\varphi) = M^2\mathcal{E}^*_1(\varphi). \tag{55}$$

We shall discuss the distinguished role of the log HLS optimizers h_M in this definition after we have proved the following theorem:

Theorem 1.8 *For all $\varphi \in E^*$,*

$$\sup_{f\in E_M}\left\{\int_{\mathbb{R}^2}\varphi f\mathrm{d}x - J_M(f,f)\ \right\} = \mathcal{E}_M(\varphi) + D_M(\varphi). \tag{56}$$

Moreover, for all $f \in E$ and $\varphi \in E^*$,

$$\int_{\mathbb{R}^2}\varphi f\mathrm{d}x \le J_M(f) + \mathcal{E}_M(\varphi)$$

with equality if and only if

$$\nabla\varphi = \nabla G * f. \tag{57}$$

Proof: For any $M > 0$, and f in the domain of J_M, note that

$$J_M(f,f) = J(f,f) = J(f - h_M, f - h_M) + 2J(f, h_M) - J(h_M, h_M). \tag{58}$$

The point of this is that since $\int_{\mathbb{R}^2}(f - h_M)\mathrm{d}x = 0$, $J(f - h_M, f - h_M)$ is a positive definite quadratic form in $f - h_M$, while $J(f, h_M)$ is linear in f, and $J(h_M, h_M)$ is independent of f.

In (43), we have already determined that

$$J(h_M, h_M) = -\frac{M^2}{8\pi}.$$

To deal with the linear term, note that by (8)

$$-\frac{1}{2\pi}\int_{\mathbb{R}^2} \ln|x-y| h_M(y)\mathrm{d}y = \frac{M}{8\pi}\ln h_M(x) + \frac{M(\ln\pi - \ln M)}{8\pi}. \tag{59}$$

Therefore,

$$J_M(f,f) = J(f-h_M, f-h_M) + \frac{M}{8\pi}\int_{\mathbb{R}^2} f(x)\ln h_M(x)\mathrm{d}x + \frac{M^2(\ln\pi + 1 - \ln M)}{8\pi},$$

and so

$$\int_{\mathbb{R}^2} \varphi f \mathrm{d}x - J_M(f) =$$
$$\int_{\mathbb{R}^2}\left[\varphi - \frac{M}{8\pi}\ln h_M\right] f(x)\mathrm{d}x - J\left(f - h_M, f - h_M\right) - \frac{M^2(\ln\pi + 1 - \ln M)}{8\pi}. \tag{60}$$

Now define

$$\psi = \left[\varphi - \frac{M}{8\pi}\ln h_M\right] \qquad \text{and} \qquad g = f - h_M.$$

Then since

$$\varphi f = \left[\psi + \frac{M}{8\pi}\ln h\right][g + h_M],$$

$$\begin{aligned}\int_{\mathbb{R}^2}\varphi f\mathrm{d}x - J_M(f,f) &= \int_{\mathbb{R}^2}\psi f\mathrm{d}x - J(g,g) - \frac{M^2(\ln\pi+1-\ln M)}{8\pi}\\ &= \int_{\mathbb{R}^2}\psi\,[g+h_M]\,\mathrm{d}x - J(g,g) - \frac{M^2(\ln\pi+1-\ln M)}{8\pi}\\ &= \int_{\mathbb{R}^2}\psi g\mathrm{d}x - J\,(g,g) + \int_{\mathbb{R}^2}\psi h_M\mathrm{d}x - \frac{M^2(\ln\pi+1-\ln M)}{8\pi}\end{aligned} \tag{61}$$

$$\int_{\mathbb{R}^2}\psi g\mathrm{d}x - J\,(g,g) = \frac{1}{2}\int_{\mathbb{R}^2}|\nabla\psi|^2\mathrm{d}x - \frac{1}{2}\int_{\mathbb{R}^2}|\nabla\psi - \nabla G * g|^2\mathrm{d}x,$$

where the integration by parts is justified since $\int_{\mathbb{R}^2} g(x)\mathrm{d}x = 0$.
Therefore,

$$\int_{\mathbb{R}^2}\varphi f\mathrm{d}x - J_M(f,f) \le \frac{1}{2}\int_{\mathbb{R}^2}|\nabla\psi|^2\mathrm{d}x + \int_{\mathbb{R}^2}\psi h_M\mathrm{d}x - \frac{M^2(\ln\pi - \ln M)}{8\pi} - \frac{M^2}{8\pi} \tag{62}$$

with equality if and only if

$$\nabla\psi = \nabla G * g. \tag{63}$$

Rewriting this in terms of φ, we have

$$\begin{aligned}&\int_{\mathbb{R}^2}\varphi f\mathrm{d}x - J_M(f) \le\\ &\left[\frac{1}{2}\int_{\mathbb{R}^2}\left|\nabla\left(\varphi - \frac{M}{8\pi}\ln h_M\right)\right|^2\mathrm{d}x + \int_{\mathbb{R}^2}\left(\varphi - \frac{M}{8\pi}\ln h_M\right)h_M\mathrm{d}x\right] -\\ &\frac{M^2(\ln\pi+1-\ln M)}{8\pi}.\end{aligned} \tag{64}$$

Now, a direct calculation shows that

$$\Delta\ln h_M(x) = -\frac{8\pi}{M}h_M(x) \tag{65}$$

The condition for equality (63) amounts to $\Delta\psi = g$, and so by (65), $\Delta\varphi = \Delta f$, and so the condition for equality (63) can be rewritten as

$$\nabla\varphi = \nabla G * f. \tag{66}$$

Also note that by (42) and (43)

$$\frac{M}{8\pi}\int_{\mathbb{R}^2} h_M\ln h_M\mathrm{d}x = -\frac{M^2(\ln\pi+2-M)}{8\pi},$$

and so (64) becomes

$$\int_{\mathbb{R}^2} \varphi f \mathrm{d}x - J_M(f,f) \leq \left[\frac{1}{2}\int_{\mathbb{R}^2}\left|\nabla\left(\varphi - \frac{M}{8\pi}\ln h_M\right)\right|^2 \mathrm{d}x + \int_{\mathbb{R}^2}\varphi h_M \mathrm{d}x + \frac{M^2}{8\pi}\right] \tag{67}$$

with equality if and only if (57) holds. □

We are now ready to prove Theorem 1.6:

Proof of Theorem 1.6: Let us write the log HLS inequality in the form

$$\frac{M}{8\pi}S_M(f) + \frac{MC_M}{8\pi} \geq J_M(f,f).$$

Then of course

$$\int_{\mathbb{R}^2} \varphi f \mathrm{d}x - \frac{M}{8\pi}S_M(f) - \frac{MC_M}{8\pi} \leq \int_{\mathbb{R}^2} \varphi f \mathrm{d}x - J_M(f,f).$$

Taking the supremum over $f \in E_M$ we then conclude, using (47) and (56),

$$\frac{M^2}{8\pi}\ln\left(\int_{\mathbb{R}^2} e^{8\pi\varphi/M}\mathrm{d}x\right) - \frac{M^2\ln M}{8\pi} - \frac{MC_M}{8\pi} \leq \left[\frac{1}{2}\int_{\mathbb{R}^2}\left|\nabla\left(\varphi - \frac{M}{8\pi}\ln h_M\right)\right|^2 \mathrm{d}x + \int_{\mathbb{R}^2}\varphi h_M \mathrm{d}x + \frac{M^2}{8\pi}\right] \tag{68}$$

To simplify this, use the explicit expression for C_M, and introduce the new variable

$$\eta = \frac{8\pi}{M}\varphi.$$

We obtain:

$$\frac{M^2}{8\pi}\ln\left(\int_{\mathbb{R}^2} e^{\eta}\mathrm{d}x\right) - \frac{M^2(1+\ln\pi)}{8\pi} \leq \left[\left(\frac{M}{8\pi}\right)^2 \frac{1}{2}\int_{\mathbb{R}^2}\left|\nabla(\eta - \ln h_M)\right|^2 \mathrm{d}x + \frac{M}{8\pi}\int_{\mathbb{R}^2}\eta h_M \mathrm{d}x + \frac{M^2}{8\pi}\right] \tag{69}$$

Next, using (42),

$$\frac{M}{8\pi}\int_{\mathbb{R}^2}\eta h_M \mathrm{d}x = \frac{M}{8\pi}\int_{\mathbb{R}^2}(\eta - \ln h_M)h_M \mathrm{d}x + \frac{M^2(\ln M - \ln\pi - 2)}{8\pi}.$$

Therefore, (69) can be rewritten as

$$\ln\left(\int_{\mathbb{R}^2} e^{\eta} \mathrm{d}x\right) \le \frac{1}{16\pi}\int_{\mathbb{R}^2} |\nabla(\eta - \ln h_M)|^2 \mathrm{d}x + \frac{1}{M}\int_{\mathbb{R}^2} (\eta - \ln h_M) h_M \mathrm{d}x + \ln M, \tag{70}$$

and this is equivalent to (46). □

A number of other functional inequalities shall come up later in these lectures. However, the ones that we have introduced here: the *HLS inequality*, the *Sobolev inequality*, the *log HLS inequality* and the *Euclidian Onofri inequality*, are our core examples. We have seen that they are closely related to one another, and that the sharp HLS is in some sense the "master inequality" of this set, as all of the others can be deduced from it. However, we have not yet proved any of these inequalities starting from first principles. We now turn to our examples of dynamical systems whose connections with these functional inequalities shall be investigated here.

2 Some Evolution Equations for Which These Sharp Functional Inequalities Are Relevant

2.1 The Keller-Segel Model

We now give some examples of the sorts of evolution equations and dynamical systems with which we shall be concerned. These are chosen to relate to the examples of sharp functional inequalities that we have given in the last section. The first example comes from the work [7] of Blanchet, Dolbeault and Perthame.

The Keller-Segel system is a model of *chemotaxis*, describing the evolution of the population density of a colony of bacteria that is diffusing across a two dimensional surface; see [7] for more references and discussion. As the bacteria move across the surface, they continually emit a chemical attractant, which diffuses across the surface. The bacteria themselves also diffuse across the surface, but with a *drift*: The bacteria tend to move towards higher concentrations of the attractant, and this induces a drift term tending to concentrate the population, and countering the spreading effects of the diffusion.

• *The key point of mathematical interest in this model is the competition between the concentrating effects of the drift induced by the chemical attractant and the spreading effects of the diffusion.*

If $\rho(x)$ denotes the population density on $\mathbb{R}^2$, and $c(x)$ denotes the concentration of the chemical attractant, also on $\mathbb{R}^2$ of course, the system of equations is

$$\begin{cases} \dfrac{\partial \rho}{\partial t}(t,x) = \mathrm{div}\,[\nabla \rho(t,x) - \rho(t,x)\nabla c(t,x)] & t > 0\,,\ x \in \mathbb{R}^2\,, \\ -\Delta c(t,x) = \rho(t,x)\,, & t > 0\,,\ x \in \mathbb{R}^2\,, \\ \rho(0,x) = \rho_0(x) \geq 0 & x \in \mathbb{R}^2\,. \end{cases}$$

Here we take $c(x)$ to be the particular solution of $-\Delta c(x) = \rho(x)$ given by

$$c(x) = -\frac{1}{2\pi}\int_{\mathbb{R}^2} \log|x-y|\rho(y)\mathrm{d}y =: G * \rho\,, \tag{71}$$

G being the green's function for $-\Delta$ on $\mathbb{R}^2$, as in (52).

The initial data will be assumed to satisfy $\rho_0 \in L^1(\mathbb{R}^2)$, and we define the *total mass* M by

$$\int_{\mathbb{R}^2} \rho_0(x)\mathrm{d}x = M\,.$$

Because of the divergence form structure of the system, the solutions conserve the total mass M.

Insight into the behavior of solutions to the system can be gained by rewriting it in *gradient flow form*: Introduce the *free energy functional* $\mathcal{F}$ on non-negative integrable functions ρ on $\mathbb{R}^2$ and such that $f(x)\ln f(x)$ and $f(x)\ln(1+|x|^2)$ belong to $L^1(\mathbb{R}^2)$:

$$\mathcal{F}(\rho) = \int_{\mathbb{R}^2} \rho(x)\ln\rho(x)\mathrm{d}x + \frac{1}{4\pi}\int_{\mathbb{R}^2}\int_{\mathbb{R}^2} \rho(x)\ln|x-y|\rho(y)\mathrm{d}x\mathrm{d}y. \tag{72}$$

(As we have seen in the first section $\mathcal{F}$ is well-defined on this set.)

At this point we remark that we have something very precise in mind when we refer to

Then,

$$\frac{\delta\mathcal{F}}{\delta\rho}(x) = (\ln\rho(x)+1) + \frac{1}{2\pi}\int_{\mathbb{R}^2} \ln|x-y|\rho(y)\mathrm{d}y.$$

Thus, solutions $\rho(x,t)$ of the KS system satisfy

$$\frac{\partial\rho}{\partial t}(t,x) = \mathrm{div}\left(\rho(x,t)\nabla\left(\frac{\delta\mathcal{F}}{\delta\rho}(x,t)\right)\right). \tag{73}$$

It follows, at least formally, that along solutions,

$$\frac{\mathrm{d}}{\mathrm{d}t}\mathcal{F}(\rho(\cdot,t)) = -\int_{\mathbb{R}^2} \rho(x,t)\left|\nabla\frac{\delta\mathcal{F}}{\delta\rho}(x,t)\right|^2 \mathrm{d}x.$$

In particular, $t \mapsto \mathcal{F}(\rho(\cdot,t))$ is monotone non-increasing.

At this point we remark that we have something very precise in mind when we refer to an equation of the form (73) as describing *gradient flow* for some functional in place of $\mathcal{F}$. The notion of gradient flow requires the specification of a metric, and (73) describes gradient flow for the functional $\mathcal{F}$ with respect to the 2-Wasserstein metric. Felix Otto [31] has shown that mass transportation arguments and other tools associated with the 2-Wasserstein metric can be very powerful tools for constructing and analyzing evolution equations of the general form (73), especially when the functional $\mathcal{F}$ has good convexity properties (which, alas, it lacks in the case at hand). We shall not go into this aspect of the study of (73) here, but much information can be found in [31,35]. However, whenever we use the term "gradient flow" in these notes, we mean an evolution equation of the form (73) for some functional in place of $\mathcal{F}$. We now return to the case at hand, with $\mathcal{F}$ given by (72).

Recall the Logarithmic Hardy-Littlewood-Sobolev Inequality: if f is a non-negative function in $L^1(\mathbb{R}^2)$ such that $f(x)\ln f(x)$ and $f(x)\ln(1+|x|^2)$ belong to $L^1(\mathbb{R}^2)$, and $M := \int_{\mathbb{R}^2} f\mathrm{d}x$, then

$$\int_{\mathbb{R}^2} f(x)\log f(x)\mathrm{d}x + \frac{2}{M}\iint_{\mathbb{R}^2\times\mathbb{R}^2} f(x)\log|x-y|f(y)\mathrm{d}x\mathrm{d}y \geq -\,C(M)\,,$$

with $C(M) := M(1+\ln\pi - \ln M)$. There is equality if and only if

$$f(x) = \frac{c}{(a+|x-x_0|^2)^2}$$

for some $a, c > 0$ and some $x_0 \in \mathbb{R}^2$.

Now observe that we may write

$$\begin{aligned}\mathcal{F}(\rho) &= \frac{M}{8\pi}\left(\int_{\mathbb{R}^2} f(x)\ln\rho(x)\mathrm{d}x + \frac{2}{M}\iint_{\mathbb{R}^2\times\mathbb{R}^2}\rho(x)\log|x-y|\rho(y)\mathrm{d}x\mathrm{d}y\right)\\ &\quad + \left(1-\frac{M}{8\pi}\right)\int_{\mathbb{R}^2}\rho(x)\ln\rho(x)\mathrm{d}x\\ &\geq -\frac{M}{8\pi}C(M) + \left(1-\frac{M}{8\pi}\right)\int_{\mathbb{R}^2}\rho(x)\ln\rho(x)\mathrm{d}x. \qquad (74)\end{aligned}$$

2.2 *Consequence of the Logarithmic HLS Inequality: No Blow-Up for $M < 8\pi$*

It follows from (74) and the monotonic decrease of $\mathcal{F}(\rho(\cdot,t))$ for solutions $\rho(x,t)$ of the KS system with $M < 8\pi$

$$\int_{\mathbb{R}^2}\rho(x,t)\log\rho(x,t)\mathrm{d}x \leq \frac{1}{8\pi}\frac{\mathcal{F}(\rho_0) - C(M)}{(8\pi - M)},$$

and thus, the entropy $\int_{\mathbb{R}^2} \rho(x,t)\log\rho(x,t)\mathrm{d}x$ stays bounded, uniformly in time. This precludes the the collapse of mass into a point mass for such initial data.

The use of the sharp logarithmic HLS inequality to show that for $M < 8\pi$ diffusion wins over the attractant effects, so that no concentration into a point mass occurs, is due to Blanchet, Dolbeault and Perthame. Previous work in this direction, by Luckhaus and Jäger used one of the Gagliardo–Nirenberg inequalities to show this, but their results only prove regularity for masses below a threshold that is strictly less than 8π. However, as explained next, Blow-up *does* occur for $M > 8\pi$.

2.3 Evolution of the Second Moment: Blow-Up for $M > 8\pi$

That 8π is the actual critical value at which diffusive and concentrating effects are balanced can be seen by computing moments. When the initial data has a finite second moment, and $M > 8\pi$, such collapse, or "blow up" does indeed occur in a finite time.

To see this, we first note a weak formulation of our the KS evolution equation. Let ψ be any test function; i.e., $\psi \in C_c^\infty(\mathbb{R}^2)$. Then

$$\frac{\mathrm{d}}{\mathrm{d}t}\int_{\mathbb{R}^2}\psi(x)\rho(x,t)\mathrm{d}x = \int_{\mathbb{R}^2}\Delta\psi(x)\rho(x,t)\mathrm{d}x$$
$$-\frac{1}{4\pi}\int_{\mathbb{R}^2}\int_{\mathbb{R}^2}(\nabla\psi(x)-\nabla\psi(y))\cdot\frac{x-y}{|x-y|^2}\rho(x,t)\rho(y,t)\mathrm{d}x\mathrm{d}y$$

Fixing any $a \in \mathbb{R}^2$ and taking $\psi(x) = a\cdot x$, we see that the center of mass is conserved:

$$\frac{\mathrm{d}}{\mathrm{d}t}\int_{\mathbb{R}^2} x\rho(x,t)\mathrm{d}x = 0.$$

More interestingly, taking $\psi(x) = |x|^2$, we find

$$\frac{\mathrm{d}}{\mathrm{d}t}\int_{\mathbb{R}^2}|x|^2\rho(x,t)\mathrm{d}x = 4M - \frac{1}{2\pi}M^2 = 4M\left(1-\frac{M}{8\pi}\right).$$

Thus, if $M > 8\pi$, the lifetime of a smooth solution whose initial data ρ_0 a finite second moment cannot exceed

$$\frac{1}{(M-8\pi)}\frac{2\pi}{M}\int_{\mathbb{R}^2}|x|^2\rho_0(x)\mathrm{d}x,$$

since after this time, the second moment would become negative, which of course is impossible.

2.4 Equilibrium Solutions for $M = 8\pi$

When $M = 8\pi$, the *critical mass*, (74) shows that the functional figuring in the log HLS inequality is nothing other than $\mathcal{F}$. By (73), it follows that the minimizers of the log HLS functional (for $M = 8\pi$) are stationary solutions of the KS system. By the sharp log HLS inequality, these translates of the densities

$$\rho_{\lambda,\infty}(x) := \frac{8\lambda}{(\lambda + |x|^2)^2} \qquad \lambda > 0. \tag{75}$$

It is natural to ask: Are these stationary solutions stable? Do they have basins of attraction, and if so what are they? These questions are not so easy to answer by a direct analysis of the rate at which the KS evolution dissipates $\mathcal{F}$, in part at least because $\mathcal{F}$ is *not* a convex functional: It is a difference of two convex functionals of different orders. (The entropy term involves no derivatives, while the energy term involves the inverse Laplacian.)

However, there is another evolution equation for which the $\rho_{\lambda,\infty}$ whose asymptotic behavior is much easier to analyze on account of the convexity properties of the functional that it dissipates. This is the *critical fast diffusion equation in* $\mathbb{R}^2$. It shall turn out that there is a surprisingly close connection between the KS system and the critical fast diffusion equation, and this connection can be used to answer the questions we have raised about the stability of the $\rho_{\lambda,\infty}$ for the KS system, and moreover, to give a proof of the sharp log HLS inequality itself. In the next section we introduce the critical fast diffusion equation.

2.5 The Critical Fast Diffusion Equation as Gradient Flow of an Entropy

The porous medium equation on $\mathbb{R}^2$ has the form

$$\frac{\partial}{\partial t} u(x,t) = \Delta u^m(x,t).$$

We are interested in the fast diffusion case, which corresponds to $1/2 \leq m < 1$, and especially in the *critical fast diffusion* case $m = 1/2$. (The critical nature of $m = 1/2$ shall be explained shortly.) We will also look at the rescaled version, with a restoring drift

$$\frac{\partial}{\partial t} u(x,t) = \Delta u^m(x,t) + \kappa \mathrm{div}(x\, u(x,t)), \tag{76}$$

where κ is a positive constant. If we introduce the entropy functional $\mathcal{H}^{(m)}(u)$

$$\mathcal{H}^{(m)}(u) = \int_{\mathbb{R}^2} \left(\frac{1}{m-1} u^m(x) + \kappa \frac{|x|^2}{2} u(x) \right) \mathrm{d}x, \tag{77}$$

we find that (76) can be written as

$$\frac{\partial}{\partial t} u(x,t) = \mathrm{div} \left(u(x,t) \nabla \frac{\delta \mathcal{H}^{(m)}}{\delta u} \right), \tag{78}$$

which shows that the evolution equation (76) is gradient flow for $\mathcal{H}^{(m)}$, and that for solutions of (76),

$$\frac{\partial}{\partial t} \mathcal{H}^{(m)}(u(\cdot,t)) = -\int_{\mathbb{R}^2} u(x,t) \left| \nabla \frac{\delta \mathcal{H}^{(m)}}{\delta u} \right|^2 \mathrm{d}x.$$

Hence the evolution drives the solutions towards minimizers of $\mathcal{H}^{(m)}$. Since $\mathcal{H}^{(m)}(u)$ is a strictly convex function of u, it is easy to determine these: The Euler equation is

$$\frac{m}{m-1} u^{m-1} + \kappa \frac{|x|^2}{2} + C = 0$$

where C is a Lagrange multiplier for $\int_{\mathbb{R}^2} u(x) \mathrm{d}x$, which is conserved. Solving for u, one finds that $C > 0$, and

$$u(x) = \left(\frac{1-m}{m} \left(C + \frac{\kappa}{2} |x|^2 \right) \right)^{-1/(1-m)}.$$

In the case $m = 1/2$, this reduces to

$$u(x) = \left(\frac{2}{\kappa} \right)^2 \left(\frac{2C}{\kappa} + |x|^2 \right)^{-2}.$$

For

$$C = \sqrt{\frac{\lambda}{8}} \qquad \text{and} \qquad \kappa = \frac{2}{\sqrt{8\lambda}},$$

we finally obtain

$$u(x) = \frac{8\lambda}{(\lambda + |x|^2)^2} = \rho_{\infty,\lambda}(x).$$

One readily checks that $u = \rho_{\infty,\lambda}$ is a steady state solution to (76) with $\kappa = 2/\sqrt{8\lambda}$. However, in the limiting case $m = 1/2$, the integrals in the definition of $\mathcal{H}^{(m)}(u)$ do not converge for $u = \rho_{\infty,\lambda}(x)$: The square root of

$\rho_{\infty,\lambda}(x)$ is not integrable, and $\rho_{\infty,\lambda}(x)$ does not have a second moment. (No such difficulty arises for $1/2 < m < 1$.)

The cure is a simple renormalisation of $\mathcal{H}^{(1/2)}$: The concavity of the square root implies that

$$2\sqrt{u} \leq 2\sqrt{\rho_{\infty,\lambda}} + \frac{1}{\sqrt{\rho_{\infty,\lambda}}}(u - \rho_{\infty,\lambda}),$$

and so

$$u \mapsto \left(\frac{1}{\sqrt{\rho_{\infty,\lambda}}}(u - \rho_{\infty,\lambda})\right) - 2\sqrt{\rho_{\infty,\lambda}} = \frac{\left(\sqrt{u} - \sqrt{\rho_{\infty,\lambda}}\right)^2}{\sqrt{\rho}_{\infty,\lambda}}$$

defines a strictly convex function minimised at $u = \rho_{\infty,\lambda}$. Integrating this gives us our re-normalised entropy functional. Indeed,

$$\frac{\left(\sqrt{u(x)} - \sqrt{\rho_{\infty,\lambda}(x)}\right)^2}{\sqrt{\rho}_{\infty,\lambda}(x)} = u(x)\frac{\lambda + |x|^2}{\sqrt{8\lambda}} - 2\sqrt{u(x)} + \frac{\sqrt{8\lambda}}{\lambda + |x|^2}$$

$$= \left[u(x)\frac{|x|^2}{\sqrt{8\lambda}} - 2\sqrt{u(x)}\right] + \frac{1}{\sqrt{8\lambda}}u(x) + \frac{\sqrt{8\lambda}}{\lambda + |x|^2} + \frac{1}{\sqrt{8\lambda}}u(x). \tag{79}$$

The first term in brackets on the right is the integrand in $\mathcal{H}^{(1/2)}$. The next term, when integrated over $\mathbb{R}^2$ yields the constant $M/\sqrt{8\lambda}$, which is unchanged under variations of u that fix the total mass M. Thus we may discard this term. The final term is not integrable on $\mathbb{R}^2$. However, it is independent of u, and therefore irrelevant to our variational problem. Thus, we may discard it too. This leaves us with the re-normalised entropy functional:

$$\mathcal{H}_\lambda[u] = \int_{\mathbb{R}^2} \frac{\left(\sqrt{u} - \sqrt{\rho_{\infty,\lambda}}\right)^2}{\sqrt{\rho}_{\infty,\lambda}}.$$

It is easy to check that for $m = 1/2$, (76) can be written

$$\frac{\partial}{\partial t}u(x,t) = \operatorname{div}\left(u(x,t)\nabla\frac{\delta\mathcal{H}_\lambda}{\delta u}\right), \tag{80}$$

and the solutions satisfy

$$\frac{\mathrm{d}}{\mathrm{d}t}\mathcal{H}_\lambda[u(t)] = -\int_{\mathbb{R}^2} u(x,t)\left|\nabla\frac{\delta\mathcal{H}_\lambda}{\delta u}\right|^2 \mathrm{d}x.$$

2.6 Monotonicity of the Logarithmic HLS Functional Under Critical Fast Diffusion

At this point, we have introduced two evolution equations, namely the Keller-Segel system and the critical fast diffusion equation. We now turn to the relation between them. First, note that both have the *gradient flow* form

$$\frac{\partial}{\partial t}u(x,t) = \mathrm{div}\left(u(x,t)\nabla\frac{\delta \mathcal{G}}{\delta u}\right), \tag{81}$$

but for two quite different specifications of the functional $\mathcal{G}$ that they dissipate. Also, as we have noted, they both have the $\rho_{\infty,\lambda}$ as equilibrium solutions.

The connection turns out to run much deeper: *Each of these evolution equations dissipates the other evolution equations driving functional, providing it with a second Lypunov functional.*

There is no obvious reason to expect this. *A priori*, all that the functionals have in common is their common set of minimizers – the $\rho_{\infty,\lambda}$. Moreover, as we have pointed out, the KS functional is not even convex, so that one might expect it to have other local minima besides the global minimizers described by Theorem 1.5. However, as we now show, this is not the case. What follows is fully developed in a paper by myself, Carrillo and Loss [11], and we shall only provide a sketch here.

Consider the functional $\mathcal{F}(u)$ figuring in the Log HLS inequality

$$F(u) = \frac{2}{M}\int_{\mathbb{R}^2}\int_{\mathbb{R}^2} u(x)\ln|x-y|u(y)\mathrm{d}x\mathrm{d}y + \int_{\mathbb{R}^2} u(x)\ln u(x)\mathrm{d}x$$

where

$$M = \int_{\mathbb{R}^2} u(x)\mathrm{d}x,$$

with the domain of definition specified in Theorem 1.5.

We shall now show that $F(u(\cdot,t))$ is monotone decreasing when $u(x,t)$ is a solution of

$$\frac{\partial}{\partial t}u(x,t) = \Delta\sqrt{u(x,t)}.$$

(Since the functional $u \mapsto F(u)$ is *scale invariant* the rescaling term $\kappa\nabla\cdot[xu(x,t)]$ is irrelevant in this context.)

We now formally compute $\frac{\mathrm{d}}{\mathrm{d}t}\mathcal{F}(u(\cdot,t))$. The result, with $u = u(\cdot,t)$ is

$$-\frac{8\pi}{M}\langle u, (-\Delta)^{-1}\Delta u^{1/2}\rangle_{L^2(\mathbb{R}^2)} + \int_{\mathbb{R}^2}\ln u\Delta u^{1/2}\mathrm{d}x.$$

This reduces to

$$\frac{8\pi}{M}\int_{\mathbb{R}^2} u^{3/2}\mathrm{d}x - \frac{1}{2}\int_{\mathbb{R}^2} \frac{|\nabla u|^2}{u^{3/2}}\mathrm{d}x. \tag{82}$$

Making the change of variables

$$f = u^{1/4},$$

one has

$$\int_{\mathbb{R}^2} u^{3/2}\mathrm{d}x = \int_{\mathbb{R}^2} f^g\mathrm{d}x \quad , \quad \int_{\mathbb{R}^2} f^4\mathrm{d}x = M \quad \text{and} \quad \int_{\mathbb{R}^2} \frac{|\nabla u|^2}{u^{3/2}}\mathrm{d}x = 16\int_{\mathbb{R}^2} |\nabla f|^2\mathrm{d}x.$$

This quantity is *strictly negative* by particular case of the Gagliardo-Nirenberg inequalities for which the sharp form was recently found by Del Pino and Dolbeault [23], unless, up to a multiple and a translation, f coincides with $\rho_{\infty,\lambda}^{1/4}$ for some $\lambda > 0$:

Theorem 2.1 *For all functions f on $\mathbb{R}^2$ with a square integrable distributional gradient ∇f,*

$$\int_{\mathbb{R}^2} |\nabla f|^2\mathrm{d}x \int_{\mathbb{R}^2} |f|^4\mathrm{d}x \geq \pi \int_{\mathbb{R}^2} |f|^6\mathrm{d}x, \tag{83}$$

and there is equality if and only if f is a multiple of a translate of $(\rho_{\infty,\lambda})^{1/4}$ for some $\lambda > 0$.

Because of the good convexity properties of $\mathcal{H}_\lambda$, one can show that for all reasonable initial data u_0 with $\mathcal{H}_\lambda(u_0) < \infty$, the solutions to (80) tend to $\rho_{\infty,\lambda}$, as was first done by Lederman and Markowich [25]. (An earlier analysis of the long time behavior of solutions of the porous medium equation by Otto [31], using different methods, can be extended to the critical case $m = 1/2$, but this was not done in Otto's paper.)

One concludes

$$\mathcal{F}(u_0) \geq \mathcal{F}(\rho_{\infty,\lambda}).$$

From here, simple approximation properties lead to a proof of the sharp Log HLS inequality, and proceeding with a little more care, one can prove the full Theorem 1.5, including the statement about cases of equality. See [11] for details.

Proving the Log HLS inequality in this way gives us interesting new information about the functional $\mathcal{F}$ – i.e., that is has no strict local minima despite being non-convex. However, since it relies on the sharp Gagliardo-Nirenberg inequality of Theorem 2.1, one might ask how that can be proved.

The original proof of Dolbeault and Del Pino [23] was a direct argument in the calculus of variations. However, it is possible to give a dynamical proof of Theorem 2.1 which reveals it as an entropy–entropy dissipation inequality for the critical fast diffusion equation. Such a proof is implicit in the work of Carrillo and Toscani [21] on the long time behavior of solutions of the porous medium equation in which they used an entropy–entropy dissipation argument generalizing the well known Bakry-Emery [2] argument for the Fokker-Planck equation; i.e., the case $m = 1$. For more details on this, we again refer to [11], where it is also shown that this approach yields not only the sharp Log HLS inequality, but all of the cases of HLS inequality with $\lambda = d - 2$. Since as explained in Sect. 1, these cases are dual to the corresponding Sobolev inequalities, this yields a dynamical proof of these Sobolev inequalities, However, it is not clear how to prove Lieb's theorem in full generality by such methods.

We close this section by commenting on a complementary result: *For each* $\lambda > 0$*, the functional* $\mathcal{H}_\lambda$ *is decreasing along solutions of the KS system,* (Of course this is only interesting if $\mathcal{H}_\lambda$ is finite for the initial data, and since finiteness of $\mathcal{H}_\lambda$ evidently entails specific behavior at infinity, this can be true for at most one value of λ. Hence what we are about to show gives one additional Lyapunov functions for the critical mass KS system, and not a whole family of them.)

To see why this is the case, we formally compute: Let $\rho(t, x)$ be a sufficiently nice solution of the PKS system. Then, with G denoting the Green's function for $-\Delta$ on $\mathbb{R}^2$, as in (71),

$$
\begin{aligned}
\frac{\partial}{\partial t}\mathcal{H}_\lambda[\rho(t,\cdot)] &= \int_{\mathbb{R}^2}\left[\frac{\delta\mathcal{H}_\lambda[\rho]}{\delta\rho}\right]\mathrm{div}\left(\rho(x,t)\nabla\left[\frac{\delta\mathcal{F}_{\mathrm{PKS}}[\rho]}{\delta\rho}\right]\right)\mathrm{d}x\\
&= -\int_{\mathbb{R}^2}\rho\nabla\left[\frac{\delta\mathcal{H}_\lambda[\rho]}{\delta\rho}\right]\cdot\nabla\left[\frac{\delta\mathcal{F}_{\mathrm{PKS}}[\rho]}{\delta\rho}\right]\mathrm{d}x\\
&= -\int_{\mathbb{R}^2}\rho\nabla\left[\frac{1}{\sqrt{\rho_{\infty,\lambda}}}-\frac{1}{\sqrt{\rho}}\right]\cdot\nabla\left[\log\rho - G*\rho\right]\mathrm{d}x\\
&= -\int_{\mathbb{R}^2}\left[\frac{1}{\sqrt{2\lambda}}x\rho+\nabla\sqrt{\rho}\right]\cdot\nabla\left[\log\rho - G*\rho\right]\mathrm{d}x \qquad (84)
\end{aligned}
$$

Now, integrating by parts once more on the term involving the Green's function,

$$
\begin{aligned}
\int_{\mathbb{R}^2}\nabla\sqrt{\rho}\cdot\nabla\left[\log\rho - G*\rho\right]\mathrm{d}x &= \frac{1}{2}\int_{\mathbb{R}^2}\frac{|\nabla\rho|^2}{\rho^{3/2}} + \int_{\mathbb{R}^2}\sqrt{\rho}\Delta G*\rho\\
&= \frac{1}{2}\int_{\mathbb{R}^2}\frac{|\nabla\rho|^2}{\rho^{3/2}} - \int_{\mathbb{R}^2}\rho^{3/2}\mathrm{d}x.
\end{aligned}
$$

Also,

$$
\int_{\mathbb{R}^2} x\cdot\nabla\rho\mathrm{d}x = -2M
$$

and symmetrizing in x and y,

$$\int_{\mathbb{R}^2} x \cdot \nabla G * \rho \mathrm{d}x = \frac{1}{4\pi} \int_{\mathbb{R}^2} \int_{\mathbb{R}^2} (x-y) \cdot \frac{x-y}{|x-y|^2} \rho(t,x)\rho(t,y) \mathrm{d}x \mathrm{d}y = \frac{M^2}{4\pi}.$$

Using the last three calculations in (84), we find

$$\frac{\partial}{\partial t} \mathcal{H}_\lambda[\rho(t,\cdot)] = -\frac{1}{2} \int_{\mathbb{R}^2} \frac{|\nabla \rho|^2}{\rho^{3/2}} + \int_{\mathbb{R}^2} \rho^{3/2} \mathrm{d}x + \frac{1}{\sqrt{2\lambda}} \left(\frac{2M - M^2}{4\pi} \right). \quad (85)$$

Notice that the constant term vanishes in critical mass case $M = 8\pi$. Thus, the following lemma tells us that the right hand side of (85) is non-positive for $M = 8\pi$.

Lemma 2.2 (Dissipation of $\mathcal{H}_\lambda$) *For all densities ρ mass $M = 8\pi$,*

$$-\frac{1}{2} \int_{\mathbb{R}^2} \frac{|\nabla \rho(x)|^2}{\rho^{3/2}(x)} \mathrm{d}x + \int_{\mathbb{R}^2} \rho^{3/2}(x) \mathrm{d}x \leq 0, \quad (86)$$

and moreover, there is equality if and only ρ is a translate of $\rho_{\infty,\lambda}$ for some $\lambda > 0$.

The proof is, of course, another application of Theorem 2.1. This second Lyapunov function may be used to prove the stability of the equilibrium solutions $\rho_{\infty,\lambda}$ of the critical mass KS system, and to prove the existence of basins of convergence for them. This is done in [6], making full use of the convexity properties on $\mathcal{H}_\lambda$. It turns out to require considerable effort to make the above formal calculations rigorous, but again this is a subject in which the details are beautiful, and worth the effort. For the details, we refer to [6].

Let us close this section with a brief summary: We have introduced two dissipative evolution equations with the same equilibrium solutions. We have then seen a surprising connection between them: *Each dissipates the functional associated to the other.* We have seen how this has two types of applications. On the one had, one may use this to give a dynamical proof of certain functional inequalities – in this case the Log HLS inequality, or the $\lambda = d - 2$ cases of the HLS inequality. On the other hand, one can use the second Lyapunov functional provided by this circumstance to prove result about the long time behavior of solutions of one of the equations. As we shall see, later in these notes, there are other interesting and natural examples of this phenomena, and a natural scenario in which it arises has been investigated by Mathes, McCann and Savare [29]. However, our example here does not fall into the scope of the [29].

As indicated above, there is a dynamical approach to proving certain special cases of the HLS inequality using the porous medium equation for fast diffusion. At present, it is not known how to give such a PDE based proof in general. However, going to a discrete time dynamics, there is a dynamical proof of the full sharp HLS inequality, as we explain in the next section.

3 Competing Symmetries: A Dynamical Proof of the HLS Inequality and More

The functional $\Phi_\lambda(f)$ given by

$$\Phi_\lambda(f) = \frac{\int_{\mathbb{R}^d}\int_{\mathbb{R}^d} f(x)|x-y|^{-\lambda} f(y)\mathrm{d}x\mathrm{d}y}{\|f\|^2_{p(\lambda)}}$$

with $p(\lambda) = 2d/(2d-\lambda)$ has a large symmetry group, the extent of which went undiscovered until Lieb's work of 1983.

As we have already noted, it is invariant under the scaling transformations

$$f(x) \mapsto f(x/s),$$

and in fact, this invariance determines the value $p = p(\lambda) = 2d/(2d-\lambda)$. Also, it is even more obviously invariant under translations

$$f(x) \mapsto f(x - x_0).$$

But much less obviously, it is invariant under the full conformal group. To see this, it suffices to check that Φ_λ is invariant under the inversion

$$f(x) \mapsto |x|^{-2d/p(\lambda)} f(x/|x|^2).$$

This is easy to check directly, but a somewhat roundabout route provides more insight.

Let $(s_1,\ldots,s_d,s_{d+1}) =: (s, s_{d+1})$ denote a generic point in S^d, the unit sphere in $\mathbb{R}^{d+1}$. The stereographic projection $\mathcal{S}$ from S^d to $\mathbb{R}^d \cup \{\infty\}$ is the bijection given by

$$\mathcal{S}[(s, s_{d+1})] = \frac{s}{1+s_{d+1}}.$$

The inverse is

$$\mathcal{S}^{-1}[x] = \frac{1}{1+|x|^2}\left(2x, 1-|x|^2\right).$$

In the case $d = 1$, these definitions reduce to

$$\mathcal{S}[(s_1, s_2)] = \frac{s_1}{1+s_2} \qquad \text{and} \qquad \mathcal{S}^{-1}[x] = \left(\frac{2x}{1+|x|^2}, \frac{1-|x|^2}{1+|x|^2}\right).$$

Let $\mathbf{s}$ denote (s, s_{d+1}), a generic point in S^d. For $\mathbf{s}$ and $\mathbf{t}$ in S^{d+1}, let x and y be given by

$$x = \mathcal{S}^{-1}(\mathbf{s}) \qquad \text{and} \qquad y = \mathcal{S}^{-1}(\mathbf{t}).$$

Then a straightforward calculation yields

$$|\mathbf{s} - \mathbf{t}|^2 = \frac{4}{(1+|x|^2)(1+|y|^2)}|x-y|^2.$$

The Jacobian of the stereographic projection is readily computed, and one has

$$\mathrm{d}S = \left(\frac{2}{1+|x|^2}\right)^d \mathrm{d}x$$

where $\mathrm{d}S$ is the uniform measure on S^d induced by the embedding of S^d in $\mathbb{R}^{d+1}$.

For any p with $1 < p < \infty$, the point transformation $\mathcal{S}$ induces an isometry between $L^p(\mathbb{R}^d)$ and $L^p(S^d)$ through the identification

$$F \leftrightarrow f \qquad \text{where} \qquad F(\mathbf{s}) = f(x(\mathbf{s}))\left(\frac{2}{1+|x(\mathbf{s})|^2}\right)^{-d/p}.$$

Indeed,

$$\int_{S^d} |F(\mathbf{s})|^p \mathrm{d}S = \int_{\mathbb{R}^d} |F(\mathbf{s}(x))|^p \left(\frac{2}{1+|x|^2}\right)^{d/p} \mathrm{d}x = \int_{\mathbb{R}^d} |f(x)|^p \mathrm{d}x.$$

Now define $w(x) := \left(\frac{2}{1+|x(\mathbf{s})|^2}\right)^d$ so that $F[\mathbf{s}(x)] = f(x)w^{-1/p}(x)$. Then:

$$\int_{\mathbb{R}^d}\int_{\mathbb{R}^d} f(x)|x-y|^{-\lambda} f(y)\mathrm{d}x\mathrm{d}y = \int_{\mathbb{R}^d}\int_{\mathbb{R}^d} [f(x)w(x)^{-1/p}]$$
$$\times \left(w^{-1/p'}(x)|x-y|^{-\lambda} w^{-1/p'}(y)\right)[f(y)w^{-1/p}(y)]w(x)\mathrm{d}x w(y)\mathrm{d}y$$

Now recalling

$$|\mathbf{s} - \mathbf{t}|^2 = \frac{4}{(1+|x|^2)(1+|y|^2)}|x-y|^2$$

we see that provided

$$1/p' = \lambda/2d,$$

we have

$$\int_{\mathbb{R}^d}\int_{\mathbb{R}^d} f(x)|x-y|^{-\lambda}f(y)\mathrm{d}x\mathrm{d}y = \int_{S^d}\int_{S^d} F(\mathbf{s})|\mathbf{s}-\mathbf{t}|^{-\lambda}F(\mathbf{t})\mathrm{d}S(\mathbf{s})\mathrm{d}S(\mathbf{t}).$$

Since with $p(\lambda) = \dfrac{2d}{2d-\lambda}$, $\dfrac{1}{p'(\lambda)} = \dfrac{\lambda}{2d}$, we have:

$$\frac{\int_{\mathbb{R}^d}\int_{\mathbb{R}^d} f(x)|x-y|^{-\lambda}f(y)\mathrm{d}x\mathrm{d}y}{\|f\|_{p(\lambda)^2}} = \frac{\int_{S^d}\int_{S^d} F(\mathbf{s})|\mathbf{s}-\mathbf{t}|^{-\lambda}F(\mathbf{t})\mathrm{d}S(\mathbf{s})\mathrm{d}S(\mathbf{t})}{\|F\|_{p(\lambda)^2}}.$$

This calculation, due to Lieb [27], reveals the conformal invariance of the HLS functional: In the from on the right, it is obviously invariant under rotations on the sphere, which are of course conformal.

While the obvious symmetries – translation and scaling – make it non trivial to prove existence of minimizers, the larger full conformal symmetry can be used to *both* prove existence of minimizers *and* identify them. This shall be done by a *competing symmetries argument*, due to Michael Loss and myself, which uses the symmetries to construct a dynamics that drives trial functions to the optimizers while increasing the functional. (We are seeking a sharp *upper* bound.) The original paper developing these ideas is [15]. Additional applications, including one to the Euclidean Onofri inequality, and hence by duality to the Log HLS inequality in on $\mathbb{R}^2$ were given in [16]. Further developments were made in [17] and [18], and it is this last reference that we shall follow here.

A key ingredient in the definition of the discrete dynamics is a rearrangement operation that we discuss next.

3.1 Rearrangement

For a non-negative measurable function f on $\mathbb{R}^d$ that vanishes at infinity, define f^* to be the radial decreasing function such that

$$\mathfrak{L}(\{x\ :\ f(x)\geq t\ \}) = \mathfrak{L}(\{x\ :\ f^*(x)\geq t\ \})$$

for all $t>0$, where $\mathfrak{L}$ denotes Lebesgue measure.

From the very definition, f and f^* have the same distribution function, and so for any continuous increasing function φ on $\mathbb{R}_+$ with $\Phi(0)=0$,

$$\int_{\mathbb{R}^d}\varphi[f(x)]\mathrm{d}x = -\int_{\mathbb{R}_+}\varphi(t)\mathrm{d}\mathfrak{L}(\{x\ :\ f(x)\geq t\ \}) = \int_{\mathbb{R}^d}\varphi[f^*(x)]\mathrm{d}x.$$

In particular, for all $1\leq p\leq\infty$,

$$\|f\|_p = \|f^*\|_p.$$

For f in $L^p(\mathbb{R}^d)$, define $Vf = |f|^*$. As we have just seen, $\|Vf\|_p = \|f\|_p$. Less obvious, but also true is that $f \mapsto Vf$ is a contraction on $L^p(\mathbb{R}^d)$. That is, for $f, g \in L^p(\mathbb{R}^d)$,

$$\|Vf - Vg\|_p \le \|f - g\|_p .$$

We shall need a little more than this, so let us look at why it is true first in the special case $p = 2$ and $f, g \ge 0$:

We can then write

$$f(x) = \int_0^\infty 1_{\{f(x)>t\}} \mathrm{d}t \quad \text{and} \qquad g(x) = \int_0^\infty 1_{\{g(x)>t\}} \mathrm{d}t .$$

Thus,

$$\int_{\mathbb{R}^d} f(x) g(x) \mathrm{d}x = \int_{\mathbb{R}^d} \left[\int_0^\infty \int_0^\infty 1_{\{f(x)>t\}} 1_{\{g(x)>s\}} \mathrm{d}t \mathrm{d}s \right] \mathrm{d}x .$$

Now use Fubini's Theorem and observe that

$$\begin{aligned} \int_{\mathbb{R}^d} 1_{\{f(x)>t\}} 1_{\{g(x)>s\}} \mathrm{d}x &= \mathcal{L}\left[\{ x \ : \ f(x) \ge t \ \} \cap \{ x \ : \ g(x) \ge s \ \}) \right] \\ &\le \min \Big\{ \mathcal{L}\left(\{ x \ : \ f(x) \ge t \ \} \right), \mathcal{L}\left(\{ x \ : \ g(x) \ge s \ \} \right) \Big\} , \end{aligned} \tag{87}$$

since, after all, for any two sets A and B, $A \cap B$ is contained in both A and B. Indeed, for this reason, unless either $A \subset B$ or $B \subset A$, there will be strict inequality in (87).

Replacing f and g by the rearranged functions f^* and g^*, we *will* have equality, since, both $\{x \ : \ f^*(x) \ge t \ \})$ and $\{x \ : \ g^*(x) \ge t \ \})$ are centered balls, so *one of them is contained in the other.* Thus,

$$\begin{aligned} \int_{\mathbb{R}^d} 1_{\{f^*(x)>t\}} 1_{\{g^*(x)>s\}} \mathrm{d}x &= \min \Big\{ \mathcal{L}(\{x \ : \ f^*(x) \ge t\}), \quad \mathcal{L}(\{x \ : \ g^*(x) \ge s\}) \Big\} \\ &= \min \Big\{ \mathcal{L}(\{x \ : \ f(x) \ge t\}), \quad \mathcal{L}(\{x \ : \ g(x) \ge s\}) \Big\} \\ &\ge \int_{\mathbb{R}^d} 1_{\{f(x)>t\}} 1_{\{g(x)>s\}} \mathrm{d}x , \end{aligned} \tag{88}$$

where the second equality is the equimeasurability.

Thus,

$$\int_{\mathbb{R}^d} 1_{\{f(x)>t\}} 1_{\{g(x)>s\}} \mathrm{d}x \le \int_{\mathbb{R}^d} 1_{\{f^*(x)>t\}} 1_{\{g^*(x)>s\}} \mathrm{d}x$$

and there is equality if and only if for almost every s and t, one the sets $\{x \,:\, f(x) \geq t\,\}$ and $\{x \,:\, g(x) \geq s\,\}$ is contained in the other.

Since $\|f-g\|_2^2 = \|f\|_2^2 + \|g\|_2^2 - \int_{\mathbb{R}^d} f(x)g(x)\mathrm{d}x$, we see that

$$\|f-g\|_2 \geq \|f^* - g^*\|_2$$

with the same condition on cases of equality as just above.

The condition for equality can be made explicit in case $g = g^*$ and for each $R > 0$, there is a $t > 0$ so that

$$\{x \,:\, g(x) \geq t\} = \{x \,:\, |x| \leq R\} =: B_R.$$

In this case

$$\|f-g\|_2 = \|f^* - g^*\|_2 \quad \Rightarrow \quad f^* = f$$

Indeed, under these conditions, the statement that for almost every s and t one the sets $\{x \,:\, f(x) \geq t\,\}$ and $\{x \,:\, g(x) \geq s\,\}$ is contained in the other becomes:

For almost every t and R, either

$$\{x \,:\, f(x) \geq t\,\} \subset B_R \quad \text{or} \quad B_R \quad \subset \{x \,:\, f(x) \geq t\,\}.$$

Evidently, this forces $\{x \,:\, f(x) \geq t\,\}$ to be a centered ball, so $f = f^*$.

This argument is easily extended to L^p norms (and beyond) and one has

Theorem 3.1 *For all* $1 < p < \infty$, *and all* f *and* g *in* $L^p(\mathbb{R}^d)$,

$$\|f^* - g^*\|_p \leq \|f-g\|_p.$$

If furthermore $g = g^*$ *and for each* $R > 0$, *there is a* $t > 0$ *so that*

$$\{x \,:\, g(x) \geq t\} = \{x \,:\, |x| \leq R\} =: B_R,$$

then

$$\|f-g\|_p = \|f^* - g^*\|_p \quad \Rightarrow \quad f^* = f$$

The inequality has been proved by a number of authors, including Crandal and Tartar, and Chiti. The statement about cases of equality is from [15], where the full Theorem is proved.

3.2 The Riesz Rearrangement Inequality

For three non-negative functions f, g and h on $\mathbb{R}^d$, the Riesz Rearrangement Inequality.

If f^* denotes the spherical decreasing rearrangement of f, then we have as a special case of the the *Riesz rearrangement inequality* say that

$$\int_{\mathbb{R}^d} f(x)g(x-y)h(y)\mathrm{d}x\mathrm{d}y \leq \int_{\mathbb{R}^d} f^*(x)g^*(x-y)h^*(y).$$

The cases of equality here are quite subtle, but have been completely determined by Almut Burchard [10]. However, in our application (and many others) the "middle function" g is already rearranged, and have every ball as a level set. In this case, there is a straightforward statement one can make about cases of equality, which is due to Lieb [26]:

Theorem 3.2 *Suppose that g is a non-negative function on $\mathbb{R}^d$ such that $g = g^*$ and for each $R > 0$, there is a $t > 0$ so that*

$$\{x \ : \ g(x) \geq t\} = \{x \ : \ |x| \leq R\} =: B_R,$$

then

$$\int_{\mathbb{R}^d} f(x)g(x-y)f(y)\mathrm{d}x\mathrm{d}y = \int_{\mathbb{R}^d} f^*(x)g(x-y)f^*(y)$$

implies that f^* is a translate of f; i.e., there exists an $x_0 \in \mathbb{R}^d$ so that $f^*(x) = f(x - x_0)$ almost everywhere.

3.3 The Main Ideas of Competing Symmetries as Applied to the HLS Inequality

Putting together what we have learned about rearrangements,

$$\frac{\int_{\mathbb{R}^d}\int_{\mathbb{R}^d} f^*(x)|x-y|^{-\lambda}f^*(y)\mathrm{d}x\mathrm{d}y}{\|f\|_{p(\lambda)^2}} \geq \frac{\int_{\mathbb{R}^d}\int_{\mathbb{R}^d} f(x)|x-y|^{-\lambda}f(y)\mathrm{d}x\mathrm{d}y}{\|f\|_{p(\lambda)^2}},$$

and there is equality if and only if f^* is a translate of f.

This tells us we need only look at trial functions f for which $f = f^*$. However, this reduction by itself does not help so much: There is no uniqueness result for the Euler–Lagrange equation

$$\int_{\mathbb{R}^d} |x-y|^{-\lambda}f(y)\mathrm{d}y = C_{\lambda,d}\|f\|_{p(\lambda)}^{2-p(\lambda)}f^{p(\lambda)-1}(x)$$

in the class of radial decreasing functions f. Indeed, there do exist spurious solution that are *almost* in $L^p(\mathbb{R}^d)$; see [27].

However, there is a way to get more out of the rearrangement using the identity

$$\Phi_\lambda(f) = \frac{\int_{\mathbb{R}^d}\int_{\mathbb{R}^d} f(x)|x-y|^{-\lambda} f(y)\mathrm{d}x\mathrm{d}y}{\|f\|_{p(\lambda)^2}} = \frac{\int_{S^d}\int_{S^d} F(\mathbf{s})|\mathbf{s}-\mathbf{t}|^{-\lambda} F(\mathbf{t})\mathrm{d}S(\mathbf{s})\mathrm{d}S(\mathbf{t})}{\|F\|_{p(\lambda)^2}}.$$

Consider an f_0 that is already rearranged (we done that). Then "lift" f up to the sphere using the stereographic projection, getting F_0.

Recalling

$$F_0(\mathbf{s}) =: f_0(x(\mathbf{s})) \left(\frac{2}{1+|x(\mathbf{s})|^2}\right)^{-d/p},$$

we see that since f_0 is rearranged, its level sets are centered balls. It follows that F_0 is constant in the slices of S^d consisting points that have the same distance from the "north pole" $(0,\ldots,0,1) \in S^d$.

Now rotate F_0, and then project it back down to $\mathbb{R}^d$. For $d > 1$, use the rotation that takes the north pole to somewhere on the equator. Call the result UF_0. Now define Uf_0 by

$$UF_1(\mathbf{s}) =: Uf_0(x(\mathbf{s})) \left(\frac{2}{1+|x(\mathbf{s})|^2}\right)^{-d/p},$$

Observe that all of the new rotated slices cut across several of the original slices, and the one at the rotated equator cuts across *all* of the original slices. It follows that unless F_0 is constant, it is impossible for Uf_0 to be rearranged (or even radial).

- *We can now further increase the HLS functional by rearranging Uf_0, so define*

$$f_1 = VUf_0.$$

Iterating this two-step procedure, we get the discrete time dynamics introduced in [15]. In more detail, it is this:

Starting from any trial function f and defining $f_0 = f^*$, we inductively define the sequence $\{f_n\}_{n\geq 0}$ through

$$f_n = VUf_{n-1},$$

and we have

$$\Phi_\lambda(f_n) \geq \Phi_\lambda(f_{n-1}).$$

Here is what this is good for. Suppose that this sequence converges strongly in $L^p(\mathbb{R}^d)$ – due to some compactness property. Then, as we shall show,

$$g := \lim_{n\to\infty} f_n$$

must be a fixed point of both U and V.

This allows us to identify g*:* Since g is fixed under V, g is rearranged. Since g is fixed under U, it follows that both g and Ug are rearranged. As we have observed before, this means that G is constant, where

$$G(\mathbf{s}) = g(x(\mathbf{s}))\left(\frac{2}{1+|x(\mathbf{s})|^2}\right)^{-d/p} .$$

Therefore, $g(x)$ is a constant multiple of

$$h(x) := \left(\frac{2}{1+|x|^2}\right)^{-d/p} .$$

From this, and the very nice convergence and uniformity properties of the sequence $\{f_n\}_{n\geq 0}$, we shall be able to show that

$$\Phi_\lambda(h) = \Phi_\lambda(g) = \lim_{n\to\infty} \Phi_\lambda(f_n) \geq \Phi_\lambda(f_0) \geq \Phi_\lambda(f),$$

which proves the inequality $\Phi_\lambda(h) \geq \Phi_\lambda(f)$ for arbitrary f.

3.4 A General Competing Symmetries Theorem

The application to the HLS inequality that we have indicated above is just one application of the competing symmetries scheme, and so we now formulate a general version. While it will be useful to keep in mind the specific example that we have just introduced, the proof of convergence will be simpler if we free it from this specific context.

We work in the following setting: Let B denote a Banach space of real–valued functions on a topological space M. Its norm is denoted by $\|\cdot\|$ and the cone of nonnegative functions by B_+; we suppose the norm is such that B_+ is closed. The ordering $g \geq f$ means that $g - f \in B_+$. The applications that we shall make concern the case in which B is $L^p(\mathbb{R}^n)$.

Our goal is to construct a "dynamics" on these spaces which carries any given trial function onto a small "attractor" on which a search for the extremals of some functional can be readily accomplished. *In particular, this is the case if the attractor consists of a single point, which as we have indicated, is the case in the application to HLS.*

We shall construct our dynamics using operators which share certain standard but useful properties; we refer to operators with these characteristics as *properly contractive.*

Definition 3.3 (Properly contractive) *An operator* A *on* B *is called* properly contractive *provided that:*

(i) *A is norm preserving on B_+, i.e.,* $\|Af\| = \|f\|$ for all $f \in B_+$,
(ii) *A is contractive on B_+, i.e., for all $f, g \in B_+$* $\|Af - Ag\| \leq \|f - g\|$,
(iii) *A is order preserving on B_+, i.e., $f \leq g \to Af \leq Ag$* for all $f, g \in B_+$,
(iv) *A is homogeneous of degree one on B_+, i.e., $A(\lambda f) = \lambda Af$* for all $\lambda \geq 0$ and $f \in B_+$.

Note that we do not require linearity. The examples of properly contractive operators relevant to our applications include rearrangements – which are not linear – and the isometries induced by certain point transformations – which are linear.

The next definition contains the essential conditions which will enable us to prove the convergence of our iteration procedure. In the definition, and what follows, $\mathcal{R}(T)$ denotes the range of the operator T.

Definition 3.4 (competitive operators) *Given a pair of properly contractive operators U and V, it is said that U* competes *with V if for $f \in B_+$*

$$f \in \mathcal{R}(V) \cap U\mathcal{R}(V) \quad \Rightarrow \quad Uf = f,$$

i.e., that f is a fixed point of U.Here R denotes the range.

The example to keep in mind here is the one we have explained above: V is the symmetric deceasing rearrangement, and U is the isometry induced by the conformal transformation that maps the "north pole" onto the some point on the "equator" of S^d, identified with $\mathbb{R}^d \cup \{\infty\}$ via the stereographic projection.

Theorem 3.5 (competing symmetries) *Suppose U and V are properly contractive operators, $V^2 = V$, and that U competes with V. Suppose further that there is a dense set $\tilde{B}_+ \subset B_+$ and sets K_N with $\bigcup_N K_N = \tilde{B}_+$ such that UK_N, $VK_N \subset K_N$ and $\overline{VK_N}$ is compact in B. Finally suppose that there exists some function $h \in B_+$ with $Uh = Vh = h$ and such that for $f \in B_+$*

$$\|Vf - h\| = \|f - h\| \Rightarrow Vf = f.$$

Then for any $f \in B_+$

$$Tf \equiv \lim_{n\to\infty} (VU)^n f$$

exists. Extended to all of B, T is properly contractive and its range is invariant under both U and V, i.e.,

$$UT = T, \quad VT = T$$

Furthermore $TK_N \subset \overline{TK_N}$ which is compact for all N.

Remark 3.6 *Since h is invariant under V and since V is contractive it follows automatically that $\|Vf - h\| \leq \|f - h\|$. The condition*

$$\|Vf - h\| = \|f - h\| \Rightarrow Vf = f.$$

says that if f is not invariant under V, then it moves strictly closer to h under V. This enables us to use the distance to h as a "Liapounov–function" for our dynamics given by $(VU)^n$.

Proof: First suppose that $f \in \tilde{B}_+$. Then $f \in K_N$ for some N and $f_n = (VU)^n f$ is a compact sequence in B_+. The next point to note is that the sequence $\{f_n\}$ moves simultaneously closer to all functions in B_+ invariant under both U and V. For $\omega \in B_+$ with $U\omega = \omega$ and $V\omega = \omega$ one has

$$\|f_{n+1} - \omega\| = \|VUf_n - VU\omega\| \leq \|Uf_n - U\omega\| \leq \|f_n - \omega\|$$

by "peeling off" the operators V and U. This has the consequence that if for a subsequence $\lim_{k\to\infty} f_{n_k} = \omega$ with ω invariant as above then the *whole* sequence converges to ω.

Next consider

$$D = \inf_n \|f_n - h\|$$

Since $\{f_n\}$ is compact there exists a subsequence f_{n_k} with $\lim_{k\to\infty} f_{n_k} = g$ and $D = \|g - h\|$. Since U and V are continuous $VUg = \lim_{k\to\infty} f_{n_k+1}$ and thus

$$D \leq \|VUg - h\| \leq \|Ug - h\| \leq \|g - h\| = D$$

– again, this a "peeling off" argument, using that $Uh = Vh = h$.

Thus

$$\|VUg - h\| = \|Ug - h\|$$

and by the strict contraction property of V; i.e., $\|V\varphi - h\| = \|\varphi - h\| \Rightarrow V\varphi = \varphi$, applied with $\varphi = Ug$,

$$VUg = Ug.$$

• *This shows that Ug is invariant under V, and hence $Ug \in \mathcal{R}(V)$.*

Next we shall show that $Ug \in U\mathcal{R}(V)$. Recall that since U and V compete,

$$Ug \in \mathcal{R}(V) \cap U\mathcal{R}(V) \quad \Rightarrow \quad U^2 g = Ug,$$

and hence this will have the consequence that

• *Ug is invariant under U.*

To see that $Ug \in U\mathcal{R}(V)$, observe that since $g = \lim_{k\to\infty} f_{n_k}$ and since V is continuous, $Vg = \lim_{k\to\infty} Vf_{n_k}$. But by construction, $f_n \in \mathcal{R}(V)$ for all n.

Then, since $V^2 = V$,

$$Vg = \lim_{k\to\infty} Vf_{n_k} = \lim_{k\to\infty} f_{n_k} = g.$$

Thus, $g \in \mathcal{R}(V)$. But then tautologically, $Ug \in U\mathcal{R}(V)$.

In summary so far,

- *Ug is invariant under both U and V.*

Now, by the initial observation made in the proof,

$$\lim_{k\to\infty} f_{n_k} = g \quad \Rightarrow \quad \lim_{n\to\infty} f_n = g.$$

Then since U and V are continuous,

$$g = \lim_{n\to\infty}(VU)^n f = \lim_{n\to\infty} VU(VU)^n f = VUg = Ug.$$

Thus,

- *g itself is invariant under both U and V.*

Now define $Tf = \lim_{n\to\infty}(VU)^n f$ for f in $\tilde{B}_+$, so that in the above notation, $Tf = g$. As we have seen g is invariant under U and V, so

$$UT = T \quad \text{and} \quad VT = T.$$

It is obvious that T thus defined is properly contractive and that $TK_N \subset \overline{K_N}$ since U and V map each K_N into itself. □

3.5 Application to the HLS Inequality

Let U and V be the transformation that we have introduced above in connection with the HLS inequality.

To apply the theorem to this pair U and V, we may choose for K_N the set of nonnegative functions f with $f(x) \leq Nh(x)$ with $h(x) = |S^n|^{-1/p}\left(\frac{2}{1+|x|^2}\right)^{n/p}$. Functions in $\bigcup_N K_N$ are dense in $L^p(\mathbb{R}^n)_+$. The desired compactness comes from the Helly selection principle, and the facts that $VK_N \subset K_N$ and $UK_N \subset K_N$ are clear.

Now we can apply the theorem and obtain

Theorem 3.7 (conformal rotations of $\mathbb{R}^n$ compete with rearrange– ments) For any $f \in L^p(\mathbb{R}^n)_+$, V, U as above we have that

$$\lim_{n\to\infty}(VU)^n Vf = \|f\|_p \cdot h$$

with the convergence taking place in $L^p(\mathbb{R}^n)$. If f is contained in K_N for some N, then so is each $(VU)^n Vf$.

To conclude, we must show that

$$\lim_{n\to\infty} \Phi_\lambda(f_n) = \Phi_\lambda(\|f\|h) = \Phi_\lambda(h).$$

This follows from the order preserving properties of U and V, and the Dominated Convergence Theorem since

$$f_n(x)|x-y|^{-\lambda} f_n(y) \leq N^2 h(x)|x-y|^{-\lambda} h(y),$$

and the right hand side is integrable.

3.6 Cases of Equality

By Lieb's Theorem [26] on the cases of equality in the Riesz rearrangement inequality, along the sequence $\{(VU)^n f_0\}_{n\geq 0}$ we have:

$$\frac{\int_{\mathbb{R}^d}\int_{\mathbb{R}^d} f_n(x)|x-y|^{-\lambda} f_n(y)\mathrm{d}x\mathrm{d}y}{\|f\|_{p(\lambda)^2}} \geq \frac{\int_{\mathbb{R}^d}\int_{\mathbb{R}^d} f_{n-1}(x)|x-y|^{-\lambda} f_{n-1}(y)\mathrm{d}x\mathrm{d}y}{\|f\|_{p(\lambda)^2}}$$

with *strict inequality* unless f_n is a translate of f_{n-1}.

But if for each n, V acts by translation – which is conformal – then, each f_n is a conformal transformation of f_0. Since $\{(VU)^n f_0\}_{n\geq 0}$ converges strongly to

$$\|f_0\|_p |S^d|^{-1}\left(\frac{2}{1+|x|^2}\right)^{d/p},$$

it must be the case that f_0 is a conformal transformation of this last function. That is, f_0 must have the form

$$c\left(a+|x-x_0|^2\right)^{-d/p}$$

for some $c, a > 0$, and some $x_0 \in \mathbb{R}^d$.

Indeed, it is easy to see that this class of functions is conformally invariant: Invariance under scalings and translations is obvious, and it is easy to check invariance under inversion:

$$f(x) \mapsto |x|^{-2d/p(\lambda)} f(x/|x|^2).$$

Since the operations generate the full (global) conformal group, our class of functions is invariant under it in general. Thus, this is all of what you get by acting on h by conformal transformations.

Evidently, if f_0 is one of these functions, i.e., one of the HLS optimizers, the sequence $\{(VU)^n f_0\}_{n\geq 0}$ generated by our dynamics simply moves across "submanifold" of optimizers, finally reaching the one we have singled out, that is, $\|f - 0\|_p h$.

This particular optimizer was singled out when we set up the stereographic projection: We picked an origin and a scale in doing so.

The following theorem may be used to settle all the cases of equality in various minimization problems. This result is applicable to a broader class of functionals that those to which one can apply Lieb's Theorem [26] on cases of equality in the Riesz rearrangement inequality. For example a theorem of Brothers and Ziemer [9] characterizes those functions f for which $\|\nabla f\|_2 = \|\nabla f^*\|_2$: The level sets of f must all be balls, but they need not be concentric.

Theorem 3.8 *Let f be an integrable non-negative function on $\mathbb{R}^d$, $d \geq 2$, such that for every conformal transformation U, all of the sets*

$$\{x : Uf(x) > \lambda\} \quad , \quad \lambda > 0,$$

are equivalent to balls. Then for some $a, c > 0$ and $x_0 \in \mathbb{R}^d$, f is of the form

$$f(x) = c(a^2 + |x - x_0|^2)^{-d}.$$

The proof, which is given in [18], is somewhat intricate, but the starting point is worth mentioning: Any non negative measurable function f all of whose level sets are balls has a natural lower semicontinuous version $\tilde{f}$

$$\tilde{f}(x) = \int_0^\infty \chi_\lambda(x) \mathrm{d}\lambda$$

where χ_λ is the characteristic function of the open ball to which $\{x : f(x) > \lambda\}$ is equivalent.

In fact, as is easy to see, $\tilde{f}$ is the *pointwise maximal* lower semicontinuous version of f. Note that since U is order preserving $\tilde{Uf} = U\tilde{f}$; i.e. the choice of this version is conformally invariant.

3.7 Other Applications of Competing Symmetries

While a number of other applications have been given in the papers [15–18], let us emphasize, with reference to the inequalities discussed here, that one can use this method to prove the Logarithmic HLS Inequality [17], and the Euclidean Onofri inequality [16]. This finally brings us to a proof of Theorem 1.5 – including the determination of all of the cases of equality. (Recall that the inequality itself could be deduced by a differentiation procedure for the HLS inequality.)

Another interesting example is a *multilinear* HLS inequality, due to Beckner, and applied by Morpurgo. The simplest form involves the functional

$$f \mapsto \int_{\mathbb{R}^d}\int_{\mathbb{R}^d}\int_{\mathbb{R}^d} f(x)|x-y|^{-\mu}f(y)|y-z|^{-\mu}f(z)\mathrm{d}x\mathrm{d}y\mathrm{d}z.$$

4 Generalized Brascamp-Lieb Inequalities

In this section, we shall give dynamical proofs by PDE methods of another class of functional inequalities: The Brascamp-Lieb inequality [8] is a generalization and a sharpening of the classical Young's inequality [36, 37], which is the following: For non-negative measurable functions f_1, f_2 and f_3 on $\mathbb{R}$, and

$$1/p_1 + 1/p_2 + 1/p_3 = 2,$$

$$\int_{\mathbb{R}^2} f_1(x)f_2(x-y)f_3(y)\mathrm{d}x\mathrm{d}y$$
$$\leq \left(\int_{\mathbb{R}} f_1^{p_1}(t)\mathrm{d}t\right)^{1/p_1}\left(\int_{\mathbb{R}} f_2^{p_2}(t)\mathrm{d}t\right)^{1/p_2}\left(\int_{\mathbb{R}} f_3^{p_3}(t)\mathrm{d}t\right)^{1/p_3}. \tag{89}$$

Define the vectors $\mathbf{a}_1$, $\mathbf{a}_2$ and $\mathbf{a}_3$ in $\mathbb{R}^2$ by

$$\mathbf{a}_1 = (1,0) \qquad \mathbf{a}_2 = (1,-1) \qquad \text{and} \quad \mathbf{a}_3 = (0,1).$$

Then (89) can be rewritten as

$$\int_{\mathbb{R}^2}\left(\prod_{j=1}^3 f_j(\mathbf{a}_j\cdot x)\right)\mathrm{d}^2x \leq \prod_{j=1}^3\left(\int_{\mathbb{R}} f_j^{p_j}(t)\mathrm{d}t\right)^{1/p_j}.$$

For a long time it was an open problem to find the sharp constant C in Young's inequality, where C is given by

$$C = \sup\left\{\frac{\int_{\mathbb{R}^2}\left(\prod_{j=1}^3 f_j(\mathbf{a}_j\cdot x)\right)\mathrm{d}^2x}{\prod_{j=1}^3\|f_j\|_{p_j}}\right\},$$

again with

$$1/p_1 + 1/p_2 + 1/p_3 = 2,$$

which must hold on account of scaling if C is to be finite.

Let q be the index dual to p_3, so that $1/q = 1 - 1/p_3$. Then

$$1/p_1 + 1/p_2 = 1 + 1/q,$$

and we then have the *convolution inequality*

$$\|f_1 * f_2\|_q \le C\|f_1\|_{p_1}\|f_2\|_{p_2}.$$

The problem of determining the sharp value of C was solved in 1975 simultaneously by Brascamp and Lieb [8] on the one hand, and Beckner [4] on the other hand, using very different methods. Brascamp and Lieb proved a more general inequality, concerning more than three functions, based on an analysis of the following optimization problem:

For any $N \ge M$, let

$$\mathbf{a}_1, \mathbf{a}_2, \ldots, \mathbf{a}_N$$

be a set of N non zero vectors spanning $\mathbb{R}^M$. Let $f_1, f_2, \ldots, f_N$ be a set of N non negative measurable functions on $\mathbb{R}$. Given numbers p_j with $1 \le p_j \le \infty$ for $j = 1, 2 \ldots, N$, form the vector

$$\mathbf{p} = (1/p_1, 1/p_2, \ldots, 1/p_N),$$

and define

$$D(\mathbf{p}) = \sup \left\{ \frac{\int_{\mathbb{R}^M} \prod_{j=1}^N f_j(\mathbf{a}_j \cdot x) \mathrm{d}^N x}{\prod_{j=1}^N \|f_j\|_{p_j}} \ : \ f_j \in L^{p_j}(\mathbb{R}) \qquad j = 1, 2, \ldots, N \right\}.$$

The Brascamp-Lieb Theorem reduces the computation of $D(\mathbf{p})$ to a *finite dimensional* variational problem: Let $\mathcal{G}$ denote the set of all centered Gaussian function functions on $\mathbb{R}$. That is, $g(x) \in \mathcal{G}$ if and only if

$$g(x) = ce^{-(sx)^2/2}$$

for some $s > 0$ and some constant c. Define $D_{\mathcal{G}}(\mathbf{p})$ by

$$D_{\mathcal{G}}(\mathbf{p}) = \sup \left\{ \frac{\int_{\mathbb{R}^M} \prod_{j=1}^N g_j(\mathbf{a}_j \cdot x) \mathrm{d}^N x}{\prod_{j=1}^N \|g_j\|_{p_j}} \ : \ g_j \in \mathcal{G} \qquad j = 1, 2, \ldots, N \right\}.$$

The Brascamp-Lieb theorem says that $D(\mathbf{p}) = D_{\mathcal{G}}(\mathbf{p})$, and hence

$$\int_{\mathbb{R}^M} \prod_{j=1}^N f_j(\mathbf{a}_j \cdot x) \mathrm{d}^N x \le D_{\mathcal{G}}(\mathbf{p}) \prod_{j=1}^N \|f_j\|_{p_j}.$$

This can be used to explicitly compute sharp constants in certain cases. For instance, when $M = 2$ and $N = 3$, $D_{\mathcal{G}}(\mathbf{p})$ may be evaluated, and this gives the sharp constant in the classical Young's inequality for convolutions. In fact, in this case one may return to $\|f_1 * f_2\|_q \le C\|f_1\|_{p_1}\|f_2\|_{p_2}$, and see what one gets using centered Gaussians where both sides are easily computed.

Although the computation of $D_{\mathcal{G}}(\mathbf{p})$ is a finite dimensional problem, for $N > 3$ it is still quite non-trivial: It is quite non-trivial even to determine the full set of vectors $\mathbf{p}$ for which this quantity is finite, and when a Gaussian optimizer exists. In the present context, this problem was solved only recently by Barthe [3].

Of course, there is the obvious scaling condition

$$\sum_{j=1}^{N} \frac{1}{p_j} = M.$$

This must hold for exactly the same reasons (3) must hold in the HLS inequality.

Brascamp and Lieb showed that whenever one also has that every set of M vectors chosen from the $\{\mathbf{a}_1, \ldots, \mathbf{a}_N\}$ are linearly independent, a Gaussian maximizer exists.

Lieb [28] later further generalized Young's inequality by replacing the maps

$$x \mapsto \mathbf{a}_j \cdot x$$

by

$$x \mapsto A_j x$$

where A_j is a linear transformation from $\mathbb{R}^M$ to $\mathbb{R}^{m_j}$, and of course then f_j is to be a non-negative measurable function on $\mathbb{R}^{m_j}$.

However, one if free to consider much wider generalizations: One is not restricted to linear transformations, or even Lebesgue measure on Euclidean spaces:

Definition 4.1 *Given measure spaces $(\Omega, \mathcal{S}, \mu)$ and $(M_j, \mathcal{M}_j, \nu_j)$, $j = 1, \ldots, N$, not necessarily distinct, together with measurable functions $\phi_j : \Omega \to M_j$ and numbers $p_1, \ldots, p_N$ with $1 \leq p_j \leq \infty$, $1 \leq j \leq N$, we say that* a generalized Brascamp-Lieb inequality holds for $\{\phi_1, \ldots, \phi_N\}$ and $\{p_1, \ldots, p_N\}$ *in case there is a finite constant C such that*

$$\int_{\Omega} \prod_{j=1}^{N} f_j \circ \phi_j \mathrm{d}\mu \leq C \prod_{j=1}^{N} \|f_j\|_{L^{p_j}(\nu_j)}$$

holds whenever f_j is non negative and measurable on M_j, $j = 1, \ldots, N$. In the next three section, we give three examples.

4.1 The Brascamp-Lieb Inequality with a Gaussian Reference Measure

Theorem 4.2 *Let $V_1, \ldots, V_N$ are N non zero subspaces of $\mathbb{R}^n$, and for each j, and let d_j denote the dimension of V_j. Define P_j to be the orthogonal* *projection of $\mathbb{R}^n$ onto V_j. Given the numbers p_j, $1 \leq p_j < \infty$ for $j = 1, \ldots, N$, there exists a finite constant C such that*

$$\int_{\mathbb{R}^n} \prod_{j=1}^{N} f \circ P_j(x) \gamma_n(x) \mathrm{d}x \leq C \prod_{j=1}^{N} \left(\int_{V_j} f_j^{p_j}(y) \gamma_{d_j}(y) \mathrm{d}y \right)^{1/p_j}$$

holds for all non negative f_j on V_j, $j = 1, \ldots, N$, if and only if

$$\sum_{j=1}^{n} \frac{1}{p_j} P_j \leq I$$

and in this case, $C = 1$.

This theorem is proved in [13].

4.2 A Brascamp-Lieb Inequality the Sphere

Now let $\{e_1, \ldots, e_N\}$ denote the standard orthonormal basis in $\mathbb{R}^N$, and define $\phi_j(x)$ on S^{N-1} by $\phi_j(x) = e_j \cdot x$. Then:

Theorem 4.3 *For all non negative functions f_j on $[-1, 1]$, $j = 1, \ldots, N$*

$$\int_{S^{N-1}} \prod_{j=1}^{n} f_j(e_j \cdot x) \mathrm{d}\mu \leq \prod_{j=1}^{n} \left(\int_{S^{N-1}} f_j^p(e_j \cdot x) \mathrm{d}\mu \right)^{1/p},$$

for all $p \geq 2$. Moreover, for any $p < 2$ it is possible for the left hand side to be infinite, while the right hand side is finite. Finally, for $p \geq 2$, and $N \geq 3$, there is equality if and only if each f_j is constant, or one of the f_j is identically zero.

This theorem is proved in [14].

4.3 A Brascamp-Lieb Inequality the Permutation Group

Let $\mathcal{S}^N$ denote the symmetric group on N letters; i.e., the group of all permutations σ of $\{1, \ldots, N\}$. Let μ denote the uniform probability measure on $\mathcal{S}^N$ so that if g is any function on $\mathcal{S}^N$,

$$\int_{\mathcal{S}^N} g(\sigma) \mathrm{d}\mu = \frac{1}{N!} \sum_{\sigma \in \mathcal{S}^N} g(\sigma).$$

We may identify vectors in $\mathbb{C}^N$ with complex valued functions on $\{1,\ldots,N\}$ as follows: If $f : \{1,\ldots,N\} \to \mathbb{C}$, let $\mathbf{f}$ be the vector in $\mathbb{C}^N$ whose jth entry is $f(j)$. Conversely, given a vector $\mathbf{f}$ in $\mathbb{C}^N$, define the function f by setting $f(j)$ equal to the jth component of $\mathbf{f}$.

For $1 \leq k \leq N$, define the function $\phi_k : \mathcal{S}^N \to \{1,\ldots,N\}$ by

$$\phi_k(\sigma) = \sigma(k).$$

If $f : \{1,\ldots,N\} \to \mathbb{C}$, then

$$f \circ \phi_k : \mathcal{S}^N \to \mathbb{C}.$$

Theorem 4.4 *For all $N \geq 2$, given non-negative measurable functions $f_1,\ldots,f_N$, on $\{1,\ldots,N\}$,*

$$\int_{\mathcal{S}^N} \left(\prod_{j=1}^{N} f_j \circ \phi_j \right) \mathrm{d}\mu \leq \prod_{j=1}^{N} \|f_j \circ \phi_j\|_{L^p(\mathcal{S}^N)}. \tag{90}$$

for all $p \geq 2$. For every $p \geq 2$ and $N \geq 2$, there is equality if and only if some function f_j vanishes identically, or else each f_j is constant.

This theorem is proved in [14]. It may be reformulated as an inequality about permanents: Let $\{f_1,\ldots,f_N\}$ be any N complex valued functions on $\{1,\ldots,N\}$. For $1 \leq j \leq N$, let $\mathbf{f}_j$ denote the corresponding vector in $\mathbb{C}^N$, and let F denote the $N \times N$ matrix whose jth column is $\mathbf{f}_j$. Then

$$\int_{\mathcal{S}^N} \prod_{j=1}^{N} (f_j \circ \phi_j)\mathrm{d}\mu = \frac{1}{N!}\mathrm{perm}(F).$$

Let $\|\cdot\|_p$ denote the L^p norm on $(\mathcal{S}^N, \mu)$, and note that

$$\sqrt{N}\|f_j \circ \phi_j\|_2 = |\mathbf{f}_j|,$$

where $|\mathbf{f}_j|^2$ denote the sum of the absolute squares of the entries of $\mathbf{f}_j$.

Therefore, we can rephrase the Brascamp-Lieb inequality on the permutation group as a statement relating the size of $\mathrm{perm}(F)$ to the size of the columns of F, analogous to Hadamard's inequality for determinants which states that $|\det(F)| \leq \prod_{j=1}^{N} |\mathbf{f}_j|$. For permanents, the result is:

Theorem 4.5 *For any vectors $\mathbf{f}_1,\ldots,\mathbf{f}_N$ in $\mathbb{C}^N$ we have the inequality*

$$|\mathrm{perm}(F)| \leq \frac{N!}{N^{N/2}} \prod_{j=1}^{N} |\mathbf{f}_j|. \tag{91}$$

For $N > 2$, there is equality in (91) if and only if at least one of the vectors $\mathbf{f}_j$ is zero, or else F is a rank one matrix and, moreover, each of the vectors $\mathbf{f}_j$ is a constant modulus vector; i.e., its entries all have the same absolute value.

Consideration of such bounds arises naturally in combinatorics, and Alex Samorodnitsky [32] has also obtained a proof of this, among other things, at about the same time.

Our interest in it as an example of a Brascamp-Lieb inequality is this: *Formally*, the BL inequality on $\mathcal{S}^N$ and the BL inequality on S^{N-1} are very similar, and may be proved in the same way, using a non linear heat flow – as we shall see.

However, for S^{N-1} there is no finite constant for $p < 2$, while obviously for $\mathcal{S}^N$ there must be. Finding the sharp constant as a function of p for $p < 2$ is a problem of considerable interest.

4.4 Brascamp-Lieb Inequalities and Entropy

The original motivation for proving the Brascamp-Lieb inequality on S^{N-1} was to better understand a result on the subadditivity of the entropy on S^{N-1}.

Given a probability density f on some measure space $(\Omega, \mathcal{B}, \mu)$, define the entropy $S(f)$ by

$$S(f) = \int_\Omega f \ln f \mathrm{d}\mu$$

provided $f \ln f$ is integrable.

Now let $\phi : \Omega \to \mathbb{R}$ be measurable. Let ν be a Borel measure on $\mathbb{R}$, and define $f_{(\phi)}$ to be the probability density on $(\mathbb{R}, \mathcal{B}, \nu)$ such that for all bounded continuous functions ψ on $\mathbb{R}$,

$$\int_\Omega \psi(\phi(x)) f(x) \mathrm{d}\mu(x) = \int_{\mathbb{R}} \psi(t) f_{(\phi)}(t) \mathrm{d}\nu(t). \tag{92}$$

In other words, the measure $f_{(\phi)}\mathrm{d}\nu$ is the "push–forward" of the measure $f\mathrm{d}\mu$ under ϕ:

$$\phi\#(f\mathrm{d}\mu) = f_{(\phi)}\mathrm{d}\nu,$$

and can be thought of as a "marginal". The entropy of $S(f_{(\phi)})$ is defined just as $S(f)$ was, except with $(\mathbb{R}, \mathcal{B}, \nu)$ replacing $(\Omega, \mathcal{S}, \mu)$.

Now take $\Omega = S^{N-1}$, and let μ be the uniform probability measure on S^{N-1}. Then take ν to be the probability measure on $\mathbb{R}$ that is the law of $u \cdot x$, where u is any unit vector in $\mathbb{R}^n$, so that for all continuous functions ϕ,

$$\int_{S^{N-1}} \psi(u \cdot x) \mathrm{d}\mu(x) = \int_{\mathbb{R}} \psi(t) \mathrm{d}\nu(t).$$

(By the rotational invariance of μ, this does not depend on the choice of u.) Now let $\{e_1, \ldots, e_n\}$ denote the standard orthonormal basis in $\mathbb{R}^n$, and define $\phi_j(x)$ on S^{n-1} by $\phi_j(x) = e_j \cdot x$. Then one has:

Theorem 4.6 *For all $N \geq 2$, given any probability density f on S^{N-1}, let $f_{(\phi_j)}$ be the jth marginal of f, as above, for $j = 1, 2, \ldots, N$. Then*

$$\sum_{j=1}^{n} S(f_{(\phi_j)}) \leq 2S(f)\,,$$

and the constant 2 is best possible.

This was proved in [14].

That the constant 2 is cannot be reduced can be seen by considering a density f strongly concentrated near the "north pole". To see that the inequality is true with the 2, we deduce it as a simple consequence of the BL inequality on the sphere. Here is a short argument that provides the passage from the BL inequality to subadditivity: Let f be any probability density on S^{N-1}, and let $f_{(1)}, f_{(2)}, \ldots, f_{(n)}$ be its n marginals, using the shorter notation $f_{(j)}$ in place of $f_{(\phi_j)}$. Then define another probability density g on S^{n-1} by

$$g(x) := \frac{1}{C}\prod_{j=1}^{n} f_{(j)}^{1/2}(e_j\cdot x) \qquad \text{where} \qquad C := \int_{S^{n-1}} \prod_{j=1}^{n} f_{(j)}^{1/2}(e_j\cdot x)\mathrm{d}\mu.$$

Then by positivity of the relative entropy (Jensen's inequality), we have

$$\begin{aligned} 0 &\leq \int_{S^{N-1}} \ln\left(\frac{f}{g}\right) f\mathrm{d}\mu = S(f) - \int_{S^{N-1}} \left(\sum_{j=1}^{n} \ln f_{(j)}^{1/2}\right) f\mathrm{d}\mu + \ln C \\ &= S(f) - \frac{1}{2}\int_{S^{N-1}} \left(\sum_{j=1}^{n} f_{(j)} \ln f_{(j)}\right) \mathrm{d}\mu + \ln C \\ &= S(f) - \frac{1}{2}\sum_{j=1}^{n} S(f_{(j)}) + \ln C. \end{aligned} \tag{93}$$

Finally, the BL inequality implies that

$$C = \int_{S^{N-1}} \prod_{j=1}^{n} f_{(j)}^{1/2}(e_j\cdot x)\mathrm{d}\mu \leq \prod_{j=1}^{n}\left(\int_{S^{N-1}} f_{(j)}(e_j\cdot x)\mathrm{d}\mu(x)\right)^{1/2} = 1$$

since each $f_{(j)}$ is a probability density. Thus, $\ln(C) \leq 0$. This argument may give the impression that the BL inequality is a "stronger" inequality than the subadditivity inequality, but as we shall see, this is not the case, as was shown in later work by myself and Dario Cordero-Erausquin.

The subadditivity inequality on S^{N-1} may be compared to the familiar *subadditivity of the entropy* inequality on $\mathbb{R}^N$: Let G be any probability density on $\mathbb{R}^N$ with respect to the unit Gaussian measure $\mathrm{d}\gamma_N$, and let g_j denote its j marginal, which is obtained by integrating out all of the variables except the jth. In this case, there is no relation among the coordinate functions. Hence

$$\int_{\mathbb{R}^N} \prod_{j=1}^N g_j \mathrm{d}\gamma_N = \prod_{j=1}^N \left(\int_{\mathbb{R}^N} g_j \mathrm{d}\gamma_N \right) = 1,$$

so that $H = \prod_{j=1}^N g_j$ is another probability density on $\mathbb{R}^N$. Then by Jensen's inequality,

$$\begin{aligned} 0 &\leq \int_{\mathbb{R}^N} \left(\frac{G}{H}\right) \ln \left(\frac{G}{H}\right) H \mathrm{d}\gamma_N \\ &= \int_{\mathbb{R}^N} G \ln G \mathrm{d}\gamma_N - \int_{\mathbb{R}^N} G \ln H \mathrm{d}\gamma_N \\ &= \int_{\mathbb{R}^N} G \ln G \mathrm{d}\gamma_N - \sum_{j=1}^N \int_{\mathbb{R}^N} G \ln g_j \mathrm{d}\gamma_N, \end{aligned} \tag{94}$$

and there is equality if and only if $G = H$.

Defining the entropy $S(G)$ of a density G relative to $\mathrm{d}\gamma_N$ by $S(G) = -\int_{\mathbb{R}^N} G \ln G \mathrm{d}\gamma_N$, this says

$$\sum_{j=1}^N S(g_j) \geq S(G)$$

with equality if and only if $G = H$.

Note the difference between the Gaussian case and the spherical case: The latter requires an extra factor of 2 on the right, independent of N. This is due to the dependence of the coordinate functions v_j resulting from the constraint $\sum_{j=1}^N v_j^2 = 1$.

The difference between Gaussian case and the spherical case is striking given the close relation between $\mathrm{d}\mu$ and $\mathrm{d}\gamma_N$. The inequality in Theorem 4.6 does not depend on the radius of the sphere, since the uniform measure is normalized, and so Theorem 4.6 says that there is a *dimension independent* departure from the equivalence of ensembles as measured by subadditivity of the entropy.

This dimension independence would not be guessed by linearizing the inequality in Theorem 4.6 about the $F = 1$; *it is a non-perturbative effect.* The natural perturbative calculation would suggest that the difference between the Gaussian case and the spherical case "washes out" with increasing N, as we now explain.

Consider a probability density F on S^{N-1} of the form

$$F = 1 + \varepsilon H$$

where H is bounded and orthogonal to 1 in $L^2(S^{N-1})$. Then $f_j = P_j F = 1 + \varepsilon h_j$ where $h_j = P_j H$, and P_j is the orthogonal projection onto functions depending only on the jth coordinate. (So $P_j f$ is the jth marginal of f, except that now an operator notation is advantageous.) Of course h_j is also orthogonal to 1 in $L^2(S^{N-1})$.

A simple and frequently encountered computation gives us

$$S(F) = \frac{\varepsilon^2}{2}\|H\|^2_{L^2(S^{N-1})} + \mathcal{O}(\varepsilon^3) \quad \text{and} \quad \sum_{j=1}^N S(f_j) = \frac{\varepsilon^2}{2}\|h_j\|^2_{L^2(S^{N-1})} + \mathcal{O}(\varepsilon^3).$$

Since

$$\|h_j\|^2_{L^2(S^{N-1})} = \langle P_j H, P_j H\rangle_{L^2(S^{N-1})} = \langle H, P_j H\rangle_{L^2(S^{N-1})},$$

if we define the operator

$$P = \frac{1}{N}\sum_{j=1}^N P_j,$$

we have that

$$\frac{\sum_{j=1}^N S(P_j F)}{S(F)} = \frac{\langle H, PH\rangle_{L^2(S^{N-1})}}{\|H\|^2_{L^2(S^{N-1})}} + \mathcal{O}(\varepsilon).$$

An optimist might then hope that the infimum of $\sum_{j=1}^N S(P_j F)/S(F)$ taken over all probability densities F would be given by C_N where

$$C_N = \inf\left\{ \frac{\langle H, PH\rangle_{L^2(S^{N-1})}}{\|H\|^2_{L^2(S^{N-1})}} \ : \ H \in L^2(S^{N-1}),\ \langle H, 1\rangle_{L^2(S^{N-1})} = 0 \right\}.$$

The computation of the infimum is an eigenvalue problem, and has been done by myself, Carvalho and Loss. The result is

$$C_N = 1 + \frac{3}{N+1}.$$

The surplus over 1, namely $3/(N+1)$, measures the "departure from independence" as a function of N. Thus, if one considers densities F that deviate only slightly from the uniform density, one gets a correction term to the constant 1 in the Gaussian entropy inequality that "remembers" the dependence of the coordinates on the sphere, but which vanishes as $N \to \infty$.

The precise size of this "departure from independence" as a function of N is crucial in some problems of non-equilibrium statistical mechanics. The computation of C_N was at the heart of recent progress in computing the rate of relaxation to equilibrium in kinetic theory by direct consideration of an N body system, as proposed long ago by Mark Kac.

The fact that for more general densities F, the correction term to the Gaussian entropy inequality does not vanish as $N \to \infty$ complicates the estimation of rates of relaxation in entropic terms for N body systems in kinetic theory.

4.5 Proof of the BL Inequality on the Sphere

Finally, we prove the BL Theorem for S^{N-1} using a non-linear heat semigroup. Related heat flow proof of the other examples of generalized BL inequalities may be found in the references cited above, and the case we treat here is quite representative.

Here is the strategy: For $1 \leq i, j \leq N$, $i \neq j$ let

$$L_{i,j} = v_i \frac{\partial}{\partial v_j} - v_j \frac{\partial}{\partial v_i}.$$

The Laplacian on S^{N-1} is the operator

$$\Delta = \sum_{i<j} L_{i,j}^2 = \frac{1}{2} \sum_{i \neq j} L_{i,j}^2.$$

The normalization of the gradient on S^{N-1} implicit in this is convenient; for smooth functions f and g, we write

$$\nabla f \cdot \nabla g = \sum_{i<j} L_{i,j} f L_{i,j} g = \frac{1}{2} \sum_{i \neq j} L_{i,j} f L_{i,j} g.$$

and $|\nabla f|^2 = \nabla f \cdot \nabla f$.

Now fix any $p \geq 1$. For any smooth, non negative function g in $L^p(S^{N-1})$, and any $t > 0$, define

$$g(\mathbf{v}, t) = \left(e^{t\Delta} g^p(\mathbf{v})\right)^{1/p}.$$

The first thing to observe is that $g(\cdot,t)$ will be smooth for all $t>0$, and the $L^p(S^{N-1})$ norm of g is conserved under this evolution:

$$\|g(\cdot,t)\|_{L^p(S^{N-1})} = \|g\|_{L^p(S^{N-1})}$$

for all $t \geq 0$.

The second thing to observe is that:

- *If g depends only on v_j for some j, so does $g(\cdot,t)$.*

The reason is that g depends only on v_j if and only if g is invariant under all rotations that fix the jth coordinate axis, and these rotations commute with the Laplacian. We write $g(v_j,t)$ to denote the evolution of such a function.

The third thing to observe is that the evolution, though non-linear, has the semigroup property: For all $s,t>0$,

$$g(\mathbf{v}, s+t) = \left(e^{s\Delta} g^p(\mathbf{v},t)\right)^{1/p}.$$

The fourth thing to observe is that

$$\lim_{t\to\infty} g(\mathbf{v},t) = \|g\|_{L^p(S^{N-1})}$$

uniformly in $\mathbf{v}$.

Finally, a simple computation shows that for any smooth, non negative function g on S^{N-1},

$$\left.\frac{\partial}{\partial t} g(v,t)\right|_{t=0} = \frac{1}{p} g^{1-p}\Delta g^p = \Delta g + (p-1)\frac{|\nabla g|^2}{g}.$$

Lemma 4.7 *Consider any N non negative functions $g_1, g_2, \ldots, g_N$ in $L^2([-1,1], \mathrm{d}\nu_N)$. Use $p=2$ define $g_j(v_j,t)$ as above. Then by the smoothing properties of the heat equation, the function $\phi(t)$ defined by*

$$\phi(t) = \int_{S^{N-1}} \prod_{j=1}^N g_j(v_j,t)\mathrm{d}\mu$$

is differentiable for all $t>0$, and is right continuous at $t=0$. Moreover, introducing the functions h_k and G defined by

$$h_j(v_j,t) = \ln g_j(v_j,t) \qquad k = 1,2\ldots,N \qquad \text{and} \qquad G = \prod_{j=1}^N g_j,$$

$$\frac{\mathrm{d}}{\mathrm{d}t}\phi(t) = \frac{1}{2}\int_{S^{N-1}} \sum_{i\neq k} \left[(L_{i,k}h_k) - (L_{i,k}h_i)\right]^2 G\mathrm{d}\mu.$$

Proof: The statements about smoothness and continuity require no justification.

Taking $p = 2$ and $g = g_k(v_k, t)$ in

$$\frac{\partial}{\partial t} g(v,t)\bigg|_{t=0} = \frac{1}{p} g^{1-p} \Delta g^p = \Delta g + (p-1)\frac{|\nabla g|^2}{g}$$

we have

$$\frac{\partial}{\partial t} g_k(v_k,t)\bigg|_{t=0} = \Delta g_k(v_k,0) + \frac{|\nabla g_k(v_k,0)|^2}{g_k(v_k,0)}.$$

Hence, suppressing the arguments on the right,

$$\frac{\mathrm{d}}{\mathrm{d}t}\left(\int_{S^{N-1}} \prod_{j=1}^N g_j(v_j,t)\mathrm{d}\mu\right)\bigg|_{t=0} = \sum_{k=1}^N \int_{S^{N-1}} \left(\Delta g_k + \frac{|\nabla g_k|^2}{g_k}\right) \prod_{\ell=1,\ell\neq k}^N g_\ell \mathrm{d}\mu.$$

The integral on the right can be written as

$$\int_{S^{N-1}} \sum_{k=1}^N (\Delta g_k) \prod_{\ell=1,\ell\neq k}^N g_\ell \mathrm{d}\mu + \int_{S^{N-1}} \sum_{k=1}^N \left(\frac{|\nabla g_k|^2}{g_k}\right) \prod_{\ell=1,\ell\neq k}^N g_\ell \mathrm{d}\mu.$$

Clearly, the second integral on the right is non negative. We therefore examine the first integral.

Observe that $L_{i,j} g_k = 0$ unless either $i = k$ or $j = k$. Therefore,

$$\int_{S^{N-1}} \sum_{k=1}^N (\Delta g_k) \prod_{\ell=1,\ell\neq k}^N g_\ell \mathrm{d}\mu =$$

$$\int_{S^{N-1}} \sum_{k=1}^N \left(\sum_{i<k} L_{i,k}^2 g_k + \sum_{j>k} L_{k,j}^2 g_k\right) \prod_{\ell=1,\ell\neq k}^N g_\ell \mathrm{d}\mu$$

Integrating by parts,

$$\int_{S^{N-1}} \sum_{k=1}^N \left(\sum_{i<k} L_{i,k}^2 g_k\right) \prod_{\ell=1,\ell\neq k}^N g_\ell \mathrm{d}\mu =$$

$$- \int_{S^{N-1}} \sum_{k=1}^N \left(\sum_{i<k} L_{i,k} g_k L_{i,k} g_i\right) \prod_{\ell=1,\ell\neq i,k}^N g_\ell \mathrm{d}\mu.$$

Using the notations

$$h_j(v_j,t) = \ln g_j(v_j,t) \qquad k = 1,2\dots,N \qquad \text{and} \qquad G = \prod_{j=1}^N g_j,$$

the integral on the right side of above is

$$-\int_{S^{N-1}} \sum_{k=1}^N \sum_{i<k} (L_{i,k}h_k L_{i,k}h_i)\, G\mathrm{d}\mu.$$

Doing the same integration by parts on the $j > k$ terms in the identity at the top, and substituting i for j, we have:

$$\int_{S^{N-1}} \sum_{k=1}^N (\Delta g_k) \prod_{\ell=1,\ell\neq k}^N g_\ell \mathrm{d}\mu = -\int_{S^{N-1}} \sum_{i\neq k} (L_{i,k}h_k L_{i,k}h_i)\, G\mathrm{d}\mu.$$

Going back to the positive term, and writing it in terms of the h_j and G, we obtain

$$\int_{S^{N-1}} \sum_{k=1}^N \left(\frac{|\nabla g_k|^2}{g_k}\right) \prod_{\ell=1,\ell\neq k}^N g_\ell \mathrm{d}\mu = \int_{S^{N-1}} \sum_{k=1}^N (\nabla h_k)^2 G\mathrm{d}\mu$$

$$= \int_{S^{N-1}} \sum_{i\neq k} (L_{i,k}h_k)^2 G\mathrm{d}\mu = \frac{1}{2}\int_{S^{N-1}} \sum_{i\neq k} \left[(L_{i,k}h_k)^2 + (L_{i,k}h_i)^2\right] G\mathrm{d}\mu.$$

Altogether, we have

$$\frac{\mathrm{d}}{\mathrm{d}t}\left(\int_{S^{N-1}} \prod_{j=1}^N g_j(v_j,t)\mathrm{d}\mu\right)\Bigg|_{t=0} = \frac{1}{2}\int_{S^{N-1}} \sum_{i\neq k} \left[(L_{i,k}h_k) - (L_{i,k}h_i)\right]^2 G\mathrm{d}\mu.$$

□

Proof of Theorem 4.3: By the lemma above, the difference between the right and left hand sides of the BL inequality for S^{N-1} is

$$\int_0^\infty \left(\frac{\mathrm{d}}{\mathrm{d}t}\int_{S^{N-1}} \prod_{j=1}^N g_j(v_j,t)\mathrm{d}\mu\right) \mathrm{d}t =$$

$$\frac{1}{2}\int_0^\infty \left(\int_{S^{N-1}} \sum_{i\neq k} \left[(L_{i,k}h_k(v_k,t)) - (L_{i,k}h_i(v_i,t))\right]^2 G\mathrm{d}\mu\right) \mathrm{d}t \geq 0.$$

This proves the inequality.

Also, it is now clear that for all $t > 0$, each h_k is smooth and bounded, and G is strictly positive, so that there is equality if and only if

$$[(L_{i,k}h_k) - (L_{i,k}h_i)]^2 = 0$$

for all $t > 0$, all $\mathbf{v}$ and all $i \neq k$.

Fixing t, i and k, this requires $v_i h_k'(v_k) = -v_k h_i'(v_i)$ for all v_i and v_k. Applying

$$\frac{\partial^2}{\partial v_i \partial v_k}$$

to both sides, we obtain $h_k''(v_k) = h_i''(v_i)$. From here, the analysis is easy. □

There is yet another dynamical approach to the proof of generalized BL inequalities due to coredero-Erausquin and myself [12]. This turns on a duality argument showing that generalized BL inequalities are equivalent by duality to certain generalized *subadditivity of entropy inequalities.* The later may then be proven by a direct *linear* heat flow argument. We refer to [12]. for details on this.

5 The Slow Diffusion Equation

The slow diffusion equation has the same form as the fast diffusion equation

$$\frac{\partial}{\partial t}u(x,t) = \Delta u^m(x,t) + \kappa \mathrm{div}(x\,u(x,t)), \tag{95}$$

expect now we take $m > 1$.

In Sect. 2, we examined a remarkable relation between the critical mass Keller-Segel system and the critical fast diffusion equation. This may have seemed very special, but in this section we shall discuss another interesting case of such a relation between a pair of evolution equations. In this case, the companion equation is the thin film equation, which we now introduce.

5.1 About the Thin Film Equation

The thin film equation

$$\frac{\partial}{\partial t}u = -\left(u^n u_{xxx}\right)_x, \qquad x \in \mathbb{R}, t > 0,$$

with

$$u(x,0) = u_0(x) \geq 0, \qquad x \in \mathbb{R},$$

and where we are using subscripts to denote partial derivatives on the right.

We shall be concerned with the $n = 1$ case. Let us first rewrite it in gradient flow terms. Define the energy functional E_0 by

$$E_0(u) = \frac{1}{2}\int_{\mathbb{R}} |u_x|^2(x)\mathrm{d}x,$$

and then the $n = 1$ case of the thin film equation becomes

$$\frac{\partial}{\partial t}u = \left(u\left(\frac{\delta E_0(u)}{\delta u}\right)_x\right)_x .$$

This has the consequence that for solutions $u(\cdot,t)$, $E_0(u(\cdot,t)$ is monotone decreasing in time. Moreover, the equation is also a conservation law: The total mass $M = \int_{\mathbb{R}} u(x,t)\mathrm{d}t$ is conserved.

Furthermore, this equation has a scale invariance, and self–similar solutions. If one introduces

$$v(x,t) = \alpha(t)u(\alpha(t)x, \beta(t)),$$

where

$$\alpha(t) = e^t \quad \text{and} \quad \beta(t) = \frac{e^{5t}-1}{5},$$

it becomes

$$\begin{aligned} v_t = (xv - v^n v_{xxx})_x\,, \qquad & x \in \mathbb{R}, t > 0,\\ v(x,0) = v_0(x), \qquad & x \in \mathbb{R}. \end{aligned}$$

This rescaled equation has a unique steady state, found by Smyth and Hill:

$$v^{(\infty)}(x) = \frac{1}{24}\left(R^2 - |x|^2\right)_+^2, \tag{96}$$

where g_+ indicates the positive part of g, and where R is determine through the conservation of mass

As the solutions $v(x,t)$ approach these steady states, the corresponding solutions $u(x,t)$ approach to the corresponding self–similar solutions.

For the investigation of the rates at which this takes place, it is important that the rescaled equation also describes a gradient flow: Introduce the energy functional $E(v)$ where

$$E(v) = \int_{\mathbb{R}} (|v_x|^2(x) + |x|^2 v(x))\mathrm{d}x.$$

Then, the rescaled equation can be rewritten as

$$v_t = \left(u\left(\frac{\delta E(v)}{\delta v}\right)_x\right)_x .$$

Formally then, for any solution $v(x,t)$ of the rescaled equation $E(v(\cdot,t))$ is non increasing in t. Define

$$E(v|v^{(\infty)}) = E(v) - E(v^{(\infty)})$$

where $v^{(\infty)}$ is the stationary solution with the same mass as v.

This *relative energy* controls the $H_1(\mathbb{R})$ distance from v to $v^{(\infty)}$, so a rate on its decay yields a rate on this convergence. Differentiating, along for strong solutions $v(x,t)$ of the rescaled equation,

$$\frac{\mathrm{d}}{\mathrm{d}t}E(v(\cdot,t)|v^{(\infty)}) = -D_E(v(\cdot,t))$$

where the *energy dissipation* $D_E(v)$ is given by

$$D_E(v) := \int_{\mathbb{R}} v\left(v_{xxx} - x\right)^2 dx \tag{97}$$

If one had a good lower bound for $D_E(v)$ in terms of $E(v(\cdot,t)|v^{(\infty)})$, one could estimate the rate of decrease quantitatively. This seems hard to do in the present case, though one can get such estimates for an analogous problem in which $\mathbb{R}$ is replaced by a compact interval. Results here have been obtained in [19] and also by Tudorascu [34]. However, the methods used there do not apply in the unbounded domain case.

The starting point for the developments in the rest of this section the discovery of Carrillo and Toscani [22] that the rescaled thin film equation has a second, unexpected, Liapunov functional:

This is given by

$$H(v) = \int_{\mathbb{R}} \left(|x|^2 v(x) + \sqrt{\frac{2}{3}} v^{3/2}(x) \right) \mathrm{d}x.$$

This is an entropy that had been introduced into the study of the porous medium equation by Otto [31]. As above, one defines the relative entropy

$$H(v|v^{(\infty)}) = H(v) - H(v^{(\infty)}).$$

Carrillo and Toscani [22] have proved that

$$\frac{\mathrm{d}}{\mathrm{d}t}H(v(\cdot,t)|v^{(\infty)}) \le -D_H(v(\cdot,t))$$

where the *partial entropy dissipation* $D_H(v)$ is given by

$$D_H(v) := \frac{1}{2}\int_{\mathbb{R}} v\left(\frac{|x|^2}{2} + \sqrt{6}v^{1/2}\right)|x|^2 \mathrm{d}x. \tag{98}$$

Here is where this comes from: For $m > 1$, the *slow diffusion* case of the porous medium equation

$$\frac{\partial}{\partial t}u = \Delta u^m ,$$

has scaling solutions that are of the general form

$$(C - a|x - x_0|^2)^r_+ .$$

In one dimension one can make the constants match

$$v^{(\infty)} = \frac{1}{24}(R^2 - |x|^2)^2_+$$

by choosing $m = 3/2$.

Then the rescaled porous medium equation with the Smythe-Hill steady state

$$v^{(\infty)} = \frac{1}{24}(R^2 - |x|^2)^2_+$$

is

$$\frac{\partial}{\partial t}u = -\left(u\left(\frac{|x|^2}{2} + \sqrt{6}u^{1/2}\right)_x\right)_x .$$

This is gradient flow the functional

$$H(u) = \int_{\mathbb{R}}\left(|x|^2 u(x) + \sqrt{\frac{2}{3}}u^{3/2}(x)\right) \mathrm{d}x.$$

Unlike the critical fast diffusion equation that came discussed in Sect. 2, this fast diffusion equation is particularly nice: Both terms in the entropy are "displacement convex" – convex along geodesics in the Wasserstein space – and the second moment term is *strictly* displacement convex. As Felix Otto showed [31], this leads to an entropy–entropy dissipation bound for solutions of this equation:

$$H(v|v^{(\infty)}) \leq \frac{1}{2}D_H(v).$$

This has the consequence that solution of the rescaled porous medium equation satisfy

$$H(v(\cdot,t)|v^{(\infty)}) \leq e^{-2t}H(v_0|v^{(\infty)}).$$

The remarkable result of Carrillo and Toscani is that this porous medium entropy is also monotone for the rescaled thin film equation, *and that the dissipation functional for the rescaled thin film equation dominates the one for the rescaled porous medium equation.* Thus we can replace u by v in the inequalities above.

There is also a Csiszar–Kullback type inequality for $H(v(\cdot,t)|v^{(\infty)})$, namely

$$\|v - v^{(\infty)}\|_{L^1(\mathbb{R})} \leq \left[\frac{16}{3}\int_{\mathbb{R}} \sqrt{v^{(\infty)}}\mathrm{d}x\right]^{1/2} \sqrt{H(v(\cdot,t)|v^{(\infty)})}.$$

In this way, they proved strong $L^1(\mathbb{R})$ convergence, at an exponential rate, to the Smythe–Hill steady state. They raised, but left open, the question of $H_1(\mathbb{R})$ convergence. This was addressed in [19].

5.2 Equipartition of Energy

The main idea is very simple: We can use a zeroth order entropy (no derivatives) to control a first order entropy (involving first derivatives) because of an *equipartition of energy phenomenon.*

To explain, consider any smooth solution with finite energy $E(v(\cdot,t))$. Define

$$\alpha(v) = \int_{\mathbb{R}} |x|^2 v(x)\mathrm{d}x \qquad \text{and} \qquad \beta(v) = \int |v_x|^2(x)\mathrm{d}x$$

so that

$$E(v) = \alpha(v) + \beta(v).$$

By a simple computation,

$$\frac{\mathrm{d}}{\mathrm{d}t}\alpha(v(\cdot,t)) = -2\alpha(v(\cdot,t)) + 3\beta(v(\cdot,t)).$$

It follows that

$$2\alpha(v^{(\infty)}) = 3\beta(v^{(\infty)}).$$

Define

$$\alpha(v|v^{(\infty)}) = \alpha(v(\cdot,t)) - \alpha(v^{(\infty)}) \quad \text{and} \quad \beta(v|v^{(\infty)}) = \beta(v(\cdot,t)) - \beta(v^{(\infty)}).$$

Then one also has

$$\frac{\mathrm{d}}{\mathrm{d}t}\alpha(v(\cdot,t)|v^{(\infty)}) = -2\alpha(v(\cdot,t)|v^{(\infty)}) + 3\beta(v(\cdot,t)|v^{(\infty)}).$$

It is shown in [19] that for classical solutions to the rescaled thin film equation,

$$\lim_{t\to\infty}(2\alpha(v(\cdot,t)) - 3\beta(v(\cdot,t))) = 0.$$

Using the results of Carrillo and Toscani, *together with moment estimates,* it is easy to show that consequently

$$\lim_{t\to\infty}[\alpha(v(\cdot,t)) - \alpha(v^{(\infty)})] = 0.$$

We then have that $\lim_{t\to\infty}[\beta(v(\cdot,t)) - \beta(v^{(\infty)})] = 0$ and hence that

$$\lim_{t\to\infty} E(v(\cdot,t)|v^{(\infty)}) = 0.$$

This argument explains the term *equipartition*, but there is an even more direct way to take advantage of

$$\frac{\mathrm{d}}{\mathrm{d}t}\alpha(v(\cdot,t)|v^{(\infty)}) = -2\alpha(v(\cdot,t)|v^{(\infty)}) + 3\beta(v(\cdot,t)|v^{(\infty)}) \ :$$

Since

$$E(v(\cdot,t)|v^{(\infty)}) = \alpha(v(\cdot,t)|v^{(\infty)}) + \beta(v(\cdot,t)|v^{(\infty)}),$$

this mounts to

$$\frac{\mathrm{d}}{\mathrm{d}t}\alpha(v(\cdot,t)|v^{(\infty)}) = -5\alpha(v(\cdot,t)|v^{(\infty)}) + 3E(v(\cdot,t)|v^{(\infty)}) \ :$$

Now for $T > 0$, integrate both sides from T to $2T$. Since $E(v(\cdot,t)|v^{(\infty)})$ *is monotone decreasing*, we obtain

$$\alpha(v(\cdot,2T)|v^{(\infty)}) - \alpha(v(\cdot,T)|v^{(\infty)}) + 5\int_T^{2T}\alpha(v(\cdot,t)|v^{(\infty)})\mathrm{d}t \geq 3TE(v(\cdot,2T)|v^{(\infty)}).$$

By the results of Carrillo and Toscani on $L^1(\mathbb{R})$ convergence, together with moment bounds from [19], one has that $\alpha(v(\cdot,t)|v^{(\infty)})$ tends to zero exponentially fast, Then

$$\alpha(v(\cdot,2T)|v^{(\infty)}) - \alpha(v(\cdot,T)|v^{(\infty)}) + 5\int_T^{2T}\alpha(v(\cdot,t)|v^{(\infty)})\mathrm{d}t \geq 3TE(v(\cdot,2T)|v^{(\infty)})$$

shows that $E(v(\cdot,2T)|v^{(\infty)})$ decays at essentially this same exponential rate, and thus we have the exponential $H_1(\mathbb{R})$ convergence.

Another approach, with some advantages, to the exponential convergence has been made recently by Mathes, McCann and Savare [29]. However, the equipartition mechanism described here is of interest in its own right as well.

5.3 *For Which Class of Solutions?*

The theory of the thin film equation is not yet in a well developed state. Basic issues of existence of strong solutions and uniqueness remain open. So for which class of solutions can all of these formal calculations be made precise and rigorous?

A very natural framework is the discrete time version of Wasserstein gradient flow due to Kinderlehrer, Jordan and Otto [24]: Pick a small time step $\tau > 0$, and then given v_0, define

$$v_1 = \text{argmin}\left\{\tau E(v|v^{(\infty)}) + W_2^2(v, v_0)\right\}.$$

In the same way, inductively define v_n in terms of v_{n-1} for all $n > 1$.

The Euler–Lagrange equation for this variational problem, as shown by Otto [31], is that the optimal map $\nabla\psi$ that pushes v_1 forward to v_0 satisfies

$$\psi_x(x) = x + \tau[x - (v_1(x))_{xxx}].$$

Now define ρ_t for $0 \le t \le 1$ by

$$\rho_t = ((1-t)Id + t\nabla\psi)\# v_1$$

so that $\rho_0 = v_1$ and $\rho_1 = v_0$.

Then for any displacement convex functional $\mathcal{H}(\rho)$,

$$\mathcal{H}(\rho_1) \ge \mathcal{H}(\rho_0) + \int_{\mathbb{R}} \left(\frac{\delta\mathcal{H}}{\delta\rho}\right)_x [\psi_x(x) - x]\rho_0(x)\mathrm{d}x.$$

But this is

$$\mathcal{H}(v_0) - \mathcal{H}(v_1) \ge \int_{\mathbb{R}} \left(\frac{\delta\mathcal{H}}{\delta v_1}\right)_x [\psi_x(x) - x]v_1(x)\mathrm{d}x.$$

This applies in particular for $\alpha(v(\cdot,t)|v^{(\infty)})$ and higher even moments of v. Moreover, by adding and subtracting

$$\widetilde{\psi}_x(x) = x + h(x + \sqrt{6}(\sqrt{v})_x)$$

which is the corresponding transportation plan for the porous medium equation, one easily derives the Carrillo-Toscani result in the discrete setting. In fact, the whole convergence argument may be developed in a transparent way along these lines. This JKO scheme is what is used in [20] to construct a class of solutions for which the equipartition argument can be rigorously proved.

We close this section by remarking that in [29], there is a very interesting geometrical explanation given for why the $m = 3/2$ slow diffusion equation and $n=1$ thin film equation dissipate each others driving functionals. However, this explanation does not apply to the cases of the Keller-Segal system and the critical fast diffusion equation. It would be interesting to find such an explanation for this "coincidence".

This concludes our survey of dynamics and functional inequalities, and although we have surveyed a number of aspects of the interplay between these two subjects, there is much, much more that could have been included in a longer survey – and there is much, much more waiting to be discovered.

Acknowledgements Work partially supported by U.S. National Science Foundation grant DMS 06-00037.

References

1. Th. Aubin, Problèmes isopérimétriques et espaces de Sobolev. J. Diff. Geom. **11** 573–598 (1976)
2. D. Bakry, M. Emery, Hypercontractivité des semi-groupes de diffusion. C.R. Acad. Sci. Paris, I **299**, 775–778 (1984)
3. F. Barthe, On a reverse form of the Brascamp-Lieb inequality. Invent. Math. **134**(2), 335–361 (1998)
4. W. Beckner, Inequalities in Fourier analysis. Ann. Math. **102**(6), 159–182 (1975)
5. W. Beckner, Sharp Sobolev inequalities on the sphere and the Moser-Trudin- ger inequality. Ann. Math. **138**, 213–242 (1993)
6. A. Blanchet, E.A. Carlen, J. Carrillo, Functional inequalities, thick tails and asymptotics for the critical mass Patlak-Keller-Segel model. To appear in Journal of Functional Analysis.
7. A. Blanchet, J. Dolbeault, B. Perthame, Two-dimensional Keller-Segel model: optimal critical mass and qualitative properties of the solutions. Electron. J. Differ. Equat. **2006**(44), 1–33 (2006)
8. H.J. Brascamp, E.H. Lieb, The best constant in Young's inequality and its generalization to more than three functions. J. Math. Pures Appl. **86**(2), 89–99 (2006)
9. J.E. Brothers, W.P. Ziemer, Minimal rearrangements of Sobolev functions. J. Reine Angew. Math. **384**, 153–179 (1988)
10. A. Burchard, Cases of equality in the Riesz rearrangement inequality. Ann. Math. **143**, 49–527 (1996)
11. E.A. Carlen, J. Carrillo, M. Loss, On a PDE proof of certain cases of the sharp Hardy-Littlewod-Sobolev Inequality
12. E.A. Carlen, J.A. Carrillo, M. Loss, Hardy-Littlewood-Sobolev Inequalities via Fast Diffusion Flows. PNAS **46**, 19696–19701 (2010)
13. E.A. Carlen, D. Cordero-Erausquin, Subadditivity of entropy and its relation to Brascamp-Lieb type inequalities. Geom. Funct. Anal. **19**, 373–405 (2009)
14. E.A. Carlen, E.H. Lieb, M. Loss, A sharp form of Young's inequality on S^N and related entropy inequalities. J. Geom. Anal. **14**, 487–520 (2004)
15. E. Carlen, M. Loss, Extremals of functionals with competing symmetries. J. Funct. Anal. **88**(2), 437–456 (1990)

16. E.A. Carlen, M. Loss, in *Competing symmetries of some functionals arising in mathematical physics*, ed. by S. Albeverio, et al. Stochastic Processes, Physics and Geometry (World Scientific, Singapore, 1990), pp. 277–288
17. E.A. Carlen, M. Loss, Competing symmetries, the logarithmic HLS inequality and Onofri's inequality on S^n. Geom. Funct. Anal. **2**, 90–104 (1992)
18. E.A. Carlen, M. Loss, On the minimization of symmetric functionals. Rev. Math. Phys. **6**(5a), 1011–1032 (1994)
19. E.A. Carlen, S. Ulusoy, Asymptotic equipartition and long time behavior of solutions of a thin-film equation. J. Differ. Equat. **241**(2), 279–292 (2007)
20. E.A. Carlen, S. Ulusoy, preprint
21. J.A. Carrillo, G. Toscani, Asymptotic L^1 decay of the porous medium equation to self-similarity. Indiana Univ Math J. **46**, 113–142 (2000)
22. J.A. Carrillo, G. Toscani, Long-time asymptotics for strong solutions of the thin film equation. Comm. Math. Phys. **225**, 551–571 (2002)
23. M. Del Pino, J. Dolbeault, Best constants for Gagliardo-Nirenberg inequalities and applications to nonlinear diffusions. J. Math. Pures Appl. **81**, 847–875 (2002)
24. R. Jordan, D. Kinderleher, F. Otto, The variatinal formulation of the Fokker–plack equation. SIAM J. Math Anal. **29**, 1–17 (1998)
25. C. Lederman, P.A. Markowich, On fast-diffusion equations with infinite equilibrium entropy and finite equilibrium mass. Comm. Part. Differ. Equat. **28**, 301–332 (2003)
26. E.H. Lieb, Existence and uniqueness of the minimizing solution of Choquard's non-linear equation. Stud. Appl. Math. **57**, 93–105 (1977)
27. E.H. Lieb, Sharp constants in the Hardy-Littlewood-Sobolev and related inequalities. Ann. Math. **118**, 349–374 (1983)
28. E.H. Lieb, Gaussian kernels have only Gaussian maximizers. Invent. Math. **102**, 179–208 (1980)
29. D. Matthes, R. McCann, G. Savaré, A family of nonlinear fourth order equations of gradient flow type. Comm. Part. Differ. Equat. **34**, 1352–1397 (2009)
30. E. Onofri, On the positivity of the effective action in a theory of random surfaces. Comm. Math. Phys. **86**(3), 321–326 (1982)
31. F. Otto, *Doubly degenerate diffusion equations as steepest descent* (University of Bonn, 1996), preprint
32. A. Samorodnitsky, An upper bound for permanents of nonnegative matrices. J. Combin. Theor. A **115**(2), 279–292 (2008)
33. G. Talenti, Best constant in Sobolev inequality. Ann. Mat. Pure Appl. **110**, 353–372 (1976)
34. A. Tudorascu, Lubrication approximation for thin viscous films: asymptotic behavior of nonnegative solutions. Comm. Part. Differ. Equat. 1532–4133, **32**(7), 1147–1172 (2007)
35. C. Villani, *Topics in optimal transportation*, vol. 58 of Graduate Studies in Mathematics (American Mathematical Society, Providence, RI, 2003)
36. W.H.. Young, On the multiplication of successions of Fourier constants. Proc. Royal Soc. A. **97**, 331–339 (1912)
37. W.H. Young, Sur la généraliation du théoreme du Parseval. Comptes rendus. **155**, 30–33 (1912)

Differential, Energetic, and Metric Formulations for Rate-Independent Processes

Alexander Mielke

1 Introduction

In these notes we want to give an overview of the recently developed theory for rate-independent systems. Such systems are used to model hysteresis, dry friction, elastoplasticity, magnetism, and phase transformation, and they are characterized by the fact that the changes of the state are driven solely by changes of the loading. More specifically, if the loading profile is applied with a factor α faster to the system, then rescaling the solution with the same factor α gives again a solution.

General energy-driven systems, also called generalized gradient systems, are characterized by a triple $(\mathcal{Z}, \mathcal{I}, \mathcal{R})$ where $\mathcal{Z}$ is the state space and $\mathcal{I} : [0,T]\times\mathcal{Z} \to \mathbb{R}_\infty \stackrel{\text{def}}{=} \mathbb{R} \cup \{\infty\}$ is the energy functional. We use $\mathcal{Z}$ to denote a general topological state space, but we use $\boldsymbol{Z}$ if it is a Banach space. For simplicity we restrict the introduction to the latter case. The dissipation potential $\mathcal{R} : \boldsymbol{Z}\times\boldsymbol{Z} \to [0,\infty]$ allows us to write the evolution equation in the form

$$0 \in \partial_{\dot{z}}\mathcal{R}(z,\dot{z}) + \overline{\partial}_z\mathcal{I}(t,z) \quad \subset \boldsymbol{Z}^*, \tag{1}$$

where $\overline{\partial}_z$ denotes a suitable subgradient of $\mathcal{I}(t,\cdot)$, while $\partial_{\dot{z}}\mathcal{R}(z,\cdot)$ denotes the convex subdifferential of $\mathcal{R}(z,\cdot)$. The generalized gradient system $(\boldsymbol{Z}, \mathcal{I}, \mathcal{R})$ is rate independent if $\mathcal{R}(z,\cdot)$ is positively homogeneous of degree 1, since this implies $\partial_v\mathcal{R}(z,\alpha v) = \partial_v\mathcal{R}(z,v)$ for all $\alpha > 0$. We then call $(\boldsymbol{Z}, \mathcal{I}, \mathcal{R})$ a rate-independent system, shortly RIS. Hence, system (1) is necessarily nonsmooth. In fact, the convex subdifferential $\partial_v\mathcal{R}(z,\cdot) : \boldsymbol{Z} \rightrightarrows \boldsymbol{Z}^*$ is not continuous and set-valued.

A. Mielke (✉)
Weierstraß Institut für Angewandte Analysis und Stochastik, Mohrenstraße 39, 10117 Berlin, Germany
e-mail: mielke@wias-berlin.de

S. Bianchini et al., *Nonlinear PDE's and Applications*, Lecture Notes in Mathematics 2028, DOI 10.1007/978-3-642-21861-3_3,

However, the main difference to the usually studied generalized gradient flows is that $\mathcal{R}(z,\cdot)$ has at most linear growth, and we cannot guarantee continuity of the solutions $z : [0,T] \to \boldsymbol{Z}$. Thus, there is a need to study the question under what conditions we can guarantee absolute continuity, in such a way that (1) makes sense, see Sect. 4.3. In fact, this is only true under strong convexity assumptions, and we mainly discuss the question, how the strong differential form should be weakened to allow for solutions with jumps.

To motivate the main structures of the different solution concepts for RIS, we start from the Fenchel equivalence

$$\eta \in \partial_v \mathcal{R}(z,v) \quad \Longleftrightarrow \quad \mathcal{R}(z,v) + \mathcal{R}^*(z,\eta) \leq \langle \eta, v \rangle,$$

where $\mathcal{R}^*(z,\cdot)$ is the Legendre-Fenchel transform of $\mathcal{R}(z,\cdot)$. While the statement on the left-hand side of this equivalence is a force balance, the statement on the fight-hand side is given in terms of energy rates. Using $-\eta = \xi(t) \in \overline{\partial}_z \mathcal{I}(t,z(t))$ and a chain rule, we find that (1) is equivalent to the scalar, upper energetic inequality

$$\begin{aligned} &\mathcal{I}(T,z(T)) + \textstyle\int_0^T \mathcal{R}(z(t),\dot{z}(t)) + \mathcal{R}^*(z(t),-\xi(t))\,\mathrm{d}t \\ &\leq \mathcal{I}(0,z(0)) + \textstyle\int_0^T \partial_t \mathcal{I}(t,z(t))\,\mathrm{d}t. \end{aligned} \tag{2}$$

The particularity of RIS is that $\mathcal{R}^*(z,-\xi)$ only takes the two values 0 and ∞, viz. $\mathcal{R}^*(z,-\xi) = 0$ if and only if $0 \in \partial_v \mathcal{R}(z,0) + \xi$. Thus, the energetic inequality (2) can be rewritten in terms of two conditions

local stability $\qquad 0 \in \partial_v \mathcal{R}(z(t),0) + \overline{\partial}_z \mathcal{I}(T,z(t))$ a.e. in $[0,T]$, (3a)

energy inequality
$$\begin{aligned} &\mathcal{I}(T,z(T)) + \mathrm{Diss}_{\mathcal{R}}(z,[0,T]) \\ &\leq \mathcal{I}(0,z(0)) + \textstyle\int_0^T \partial_t \mathcal{I}(t,z(t))\,\mathrm{d}t, \end{aligned} \tag{3b}$$

where $\mathrm{Diss}_{\mathcal{R}}(z,[r,t]) = \int_r^t \mathcal{R}(z(s),\dot{z}(s))\,\mathrm{d}s$ is the energy dissipated during the time interval $[r,t]$.

The local stability condition is a purely static concept and does not involve any time dependence, which shows that RIS are very close to static systems. In particular, if the loading does not change on a time interval $[t_1,t_2]$, then the solution may also be constant. Relation (3b) is a simple scalar energy inequality, which in fact should hold as an identity and also for all times $t \in [0,T]$ and not just for $t = T$. In all the different solution concepts discussed below we have these two different principles, namely (a) a static stability condition and (b) an energy inequality. However, a crucial point in the definitions of solutions to RIS is always that the stability condition and the energy inequality interact in such a way that the stability condition implies a lower energy estimate on all subintervals of $[0,T]$, which together with the upper energy estimate (3b) provides energy balance on all subintervals.

These arguments apply to all our notions of solutions except for the *local solutions*, which ask for local stability and an upper energy estimate like (3b) on each subinterval $[r,t] \subset [0,T]$. This notion was introduced in [69], and it turns out that all solutions considered here fall into this class. In fact, we distinguish two important concepts, namely *energetic solutions* and *BV solutions*. The former were introduced in [50, 62] and surveyed in [42]. This notion is essentially the same as the notion of *irreversible quasistatic evolution* introduced and studied in [17, 21, 24, 25] in the context of crack or damage evolution. Since these solutions allow for jumps, the infinitesimal dissipation potential $\mathcal{R}$ is replaced by the more general *dissipation distance* $\mathcal{D} : \boldsymbol{Z}\times\boldsymbol{Z} \to [0,\infty]$, which is obtained via

$$\mathcal{D}(z_0,z_1) \overset{\text{def}}{=} \inf\Big\{ \int_{r=0}^{1} \mathcal{R}(z(r),\dot z(r))\,\mathrm{d}r \;\Big|\; z \in \mathrm{W}^{1,1}([0,1];\boldsymbol{Z}),\; z(0)=z_0,\ z(1)=z_1 \Big\}.$$

The local stability condition (3a) is replaced by the global stability condition (S), and the energy balance (E) is obtained from (3b) by replacing the dissipation functional $\mathrm{Diss}_{\mathcal{R}}(z,[r,t])$ by

$$\mathrm{Diss}_{\mathcal{D}}(z,[r,t]) \overset{\text{def}}{=} \sup\Big\{ \sum_{j=1}^{N} \mathcal{D}(z(s_{j-1}),z(s_j)) \;\Big|\; N\in\mathbb{N},\; r \le s_0 < s_1 < \cdots < s_N \le t \Big\},$$

see Definition 3.1.

The notion of *BV solutions*, introduced in [59, 60], is quite different from energetic solutions, since BV solutions jump as late as possible while energetic solutions jump as soon as possible, cf. Example 2.3. The BV solutions are constructed via the so-called *vanishing-viscosity limit* by adding a small viscosity to (1), namely

$$0 \in \partial_{\dot z}\mathcal{R}(z^\varepsilon,\dot z^\varepsilon) + \varepsilon\mathbb{V}\dot z^\varepsilon + \overline{\partial}_z\mathcal{I}(t,z^\varepsilon) \quad \subset \boldsymbol{Z}^*, \tag{4}$$

and studying the limits of the *viscosity approximations* z^ε for $\varepsilon \to 0$, see Sects. 4.5, 4.6, and 5.3. Hence, BV solutions contain the set of *approximable solutions*, which are defined as all limit points of this procedure, see [19, 34, 37, 69]. While the notion of approximable solutions is simply defined by all possible limits, the set of BV solutions is characterized by the local stability condition (3a) and an energy estimate using a dissipation functional $\mathrm{Diss}_{\mathfrak{p},\mathcal{I}}$ that is supplemented by additional terms involving the viscous effects in jumps. The new structure is the *vanishing-viscosity contact potential*

$$\mathfrak{p}(z,v,\xi) \overset{\text{def}}{=} \inf\{ \mathcal{R}_\varepsilon(z,v) + \mathcal{R}_\varepsilon^*(z,\xi) \mid \varepsilon > 0 \},$$

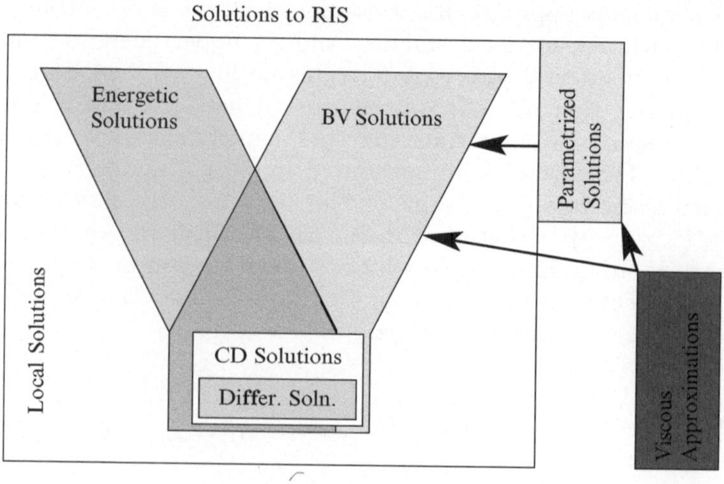

Fig. 1 Overview on the different solution types for RIS

where $\mathcal{R}_\varepsilon(z,v) = \mathcal{R}_0(z,v) + \frac{\varepsilon}{2}\langle \mathbb{V}v, v\rangle$ is the sum of the rate-independent dissipation potential $\mathcal{R}_0$ and the small viscosity term $\frac{\varepsilon}{2}\langle \mathbb{V}v, v\rangle$. The supplemented dissipation distance then reads

$$\Delta(t, z_0, z_1) \stackrel{\text{def}}{=} \inf\Big\{ \int_{r=0}^1 \mathfrak{p}(\widehat{z}(r), \dot{\widehat{z}}(r), -\mathrm{D}\mathcal{I}(t,\widehat{z}(r)))\,\mathrm{d}r \;\Big|\; \widehat{z} \in \mathrm{W}^{1,1}([0,1];\boldsymbol{Z}),\ \widehat{z}(0) = z_0,\ \widehat{z}(1) = z_1 \Big\}.$$

A useful tool for the understanding the vanishing-viscosity limit is the notion of *parametrized solutions*, which was studied in [22, 53]. Here the solutions $z^\varepsilon : [0,T] \to \boldsymbol{Z}$ are parametrized in the extended state space $\boldsymbol{Z}_T \stackrel{\text{def}}{=} [0,T]\times\boldsymbol{Z}$ such that $(t,z) = (\tau^\varepsilon(s), Z^\varepsilon(s))$ with $\dot\tau^\varepsilon(s) + \langle \mathbb{V}\dot Z^\varepsilon(s), \dot Z^\varepsilon(s)\rangle^{1/2} = 1$ for a.a. $s \in [0, S^\varepsilon]$. Under suitable conditions, see Sect. 4.4 and 5.2, it is then possible to show that S^ε stays bounded and that the parametrized curves converge to a limit $(\tau, Z) : [0,S] \to \boldsymbol{Z}_T$, namely the desired parametrized solution. Moreover, from this we obtain in a natural way BV solutions by taking any $z : [0,T] \to \boldsymbol{Z}$, such that for all t there exists $s \in [0,S]$ such that $(t, z(t)) = (\tau(s), Z(s))$.

Figure 1 summarizes the solution types we are discussing in this work. We emphasize that all these notions satisfy the natural conditions for multivalued evolutionary systems, namely the concatenation and restriction property. In these notes we concentrate on the main ideas and techniques for the different solution concepts for RIS. Thus, we refrain from giving an overview of the whole theory and application of RIS, which can be found in the forthcoming monography [48].

In particular, we refer to [18, 19, 23, 27, 41] for RIS describing the evolution of microstructure via Young-measure valued internal variables. For variational characterizations of solutions for RIS we refer to [14, 45, 49, 66].

2 Basics of Rate-Independent Systems

2.1 Definition of Rate Independence

Since we develop quite abstract notions of solutions, we give a definition of rate independence that does not use differential equations. RIS occur as limit problems in many physical and mechanical systems, if the interesting time scales are much longer than the intrinsic time scales of the system. RIS are sometimes also called quasi-static systems, however, the term "quasistatic" is often used in a more general sense, namely if the inertial terms in a system are neglected but viscous effects might still be present.

This survey only considers systems which satisfy the following exact definition of rate independence. The definition is formulated in terms of input functions $\ell : [t_1, t_2] \to \mathcal{X}$ and the set $\mathcal{O}([t_1, t_2], q_0, \ell)$ of possible output functions $q : [t_1, t_2] \to \mathcal{Q}$ with $q(t_1) = q_0$. The usage of input and output functions is necessary, since RIS have no own dynamics; they rather respond to changes in the input.

Definition 2.1. *A input-output $\mathcal{H}$ is called a* rate-independent system *with input data $q_0 \in \mathcal{Q}$ and $\ell \in \mathrm{F}_0([t_1, t_2]; \mathcal{X})$, if the output set $\mathcal{O}([t_1, t_2], q_0, \ell) \subset \mathrm{F}_1([t_1, t_2]; \mathcal{Q}) \cap \{q(t_1) = q_0\}$ (where F_0 and F_1 denote suitable function spaces) satisfies, for all strictly monotone and continuous time reparametrizations $\alpha : [t_1, t_2] \to [t_1^*, t_2^*]$ with $\alpha(t_1) = t_1^*$ and $\alpha(t_2) = t_2^*$, the relation*

$$q \in \mathcal{O}([t_1, t_2], q_0, \ell) \quad \Longleftrightarrow \quad q \circ \alpha \in \mathcal{O}([t_1^*, t_2^*], q_0, \ell \circ \alpha).$$

We call the system a multi-valued evolutionary system, *if the following additional conditions hold:*

$$\textit{Concatenation: } \widehat{q} \in \mathcal{O}([t_1, t_2], q_1, \ell),\ \widetilde{q} \in \mathcal{O}([t_2, t_3], \widehat{q}(t_2), \ell), \\ \Longrightarrow\ q \in \mathcal{O}([t_1, t_3], q_1, \ell) \text{ with } q(t) = \begin{cases} \widehat{q}(t) \text{ for } t \in [t_1, t_2], \\ \widetilde{q}(t) \text{ for } t \in [t_2, t_3], \end{cases}$$

$$\textit{Restriction: } \quad t_1 \le t_2 < t_3 \le t_4 \text{ and } q \in \mathcal{O}([t_1, t_4], q_1, \ell) \\ \Longrightarrow\ q|_{[t_2, t_3]} \in \mathcal{O}([t_2, t_3], q(t_2), \ell).$$

To compare this notion with the RIS $(\boldsymbol{Z}, \mathcal{I}, \mathcal{R})$ discussed previously, we let $\mathcal{I}(t, z) = \mathcal{U}(z) - \langle \ell(t), z \rangle$ and find the equation

$$0 \in \partial_v \mathcal{R}(z(t), \dot{z}(t)) + \overline{\partial} \mathcal{U}(z(t)) - \ell(t),$$

where we explicitly see the input ℓ. Assuming an existence result for this differential inclusion, it is then clear that the concatenation and restriction properties hold.

RIS are used to model hysteresis, which is often associated with memory effects (cf. [11, 35, 70]). Here we take a different approach using suitable internal variables that carry all memory information. This fact is the content of the concatenation property. If on $\mathcal{Q} = \mathcal{Y}\times\mathcal{Z}$ we consider a RIS we can easily obtain a system with memory by association with (q_0, ℓ) the output set $\mathcal{O}_\mathcal{Y}([t_1,t_2], q_0, \ell)$ given by

$$\{ y \in \mathrm{F}_1([t_1,t_2];\mathcal{Y}) \mid \exists\, z \in \mathrm{F}_1([t_1,t_2];\mathcal{Z})\colon (y,z) \in \mathcal{O}([t_1,t_2], q_0, \ell) \}.$$

Clearly, the concatenation and the restriction properties are then lost.

2.2 Differentiable Formulations and the Decomposition into Elastic and Dissipative Parts

The typical situations, which are the basis of this work, are differential inclusions of the form

$$0 \in \partial_{\dot q}\mathcal{R}(q,\dot q) + \mathrm{D}_q\mathcal{E}(t,q), \tag{5}$$

where the state q lies in a Banach space $\boldsymbol{Q}$. The functional $\mathcal{R} : \boldsymbol{Q}\times\boldsymbol{Q} \to [0,\infty]$ is called the dissipation potential and $\mathcal{E} : \boldsymbol{Q}_T \overset{\text{def}}{=} [0,T]\times\boldsymbol{Q} \to \mathbb{R}_\infty$ the energy functional. Thus, (5) can be interpreted as a force balance in $\boldsymbol{Q}^*$ where the dissipative force $\partial_{\dot q}\mathcal{R}(q,\dot q)$ must equilibrate the potential restoring force $-\mathrm{D}_q\mathcal{E}(t,q)$. Such systems are generalizations of the so-called doubly nonlinear systems considered in [13,15], where the special case $\mathcal{R}(q,\dot q) = \Psi(\dot q)$ is studied.

In many applications the state space $\boldsymbol{Q}$ decomposes into two parts, namely an elastic part $\boldsymbol{Y}$ and a dissipative part $\boldsymbol{Z}$, i.e. we have $q = (y,z)$ and $\boldsymbol{Q} = \boldsymbol{Y}\times\boldsymbol{Z}$. The distinction comes about because the functional $\mathcal{R}$ depends only on the z-component as follows:

$$\mathcal{R}(q,\dot q) = \mathcal{R}(z,\dot z) \quad \text{and} \quad \mathcal{R}(z,\dot z) = 0 \;\Rightarrow\; \dot z = 0. \tag{6}$$

In that case (5) takes the form of a coupled system, namely

$$0 = \mathrm{D}_y\mathcal{E}(t,y,z), \qquad 0 \in \partial_{\dot z}\mathcal{R}(z,\dot z) + \mathrm{D}_z\mathcal{E}(t,y,z). \tag{7}$$

Hence, the two components y and z need to be treated differently. In particular, often we study the reduced problem by minimizing with respect to y, viz.

$$\mathcal{I}(t,z) = \min\{\, \mathcal{E}(t,y,z) \mid y \in \boldsymbol{Y} \,\}. \tag{8}$$

Since this means that we have satisfied the first relation in (7), we are left with the reduced problem

$$0 \in \partial_{\dot{z}} \mathcal{R}(z, \dot{z}) + \mathrm{D}_z \mathcal{I}(t, z). \tag{9}$$

Moreover, we can go backward from (9). If $z : [0, T] \to \boldsymbol{Z}$ solves (9), we may choose $y : [0, T] \to \boldsymbol{Y}$ such that $y(t) \in \operatorname{Arg\,min} \mathcal{E}(t, \cdot, z(t))$, then $q : t \mapsto (y(t), z(t))$ solves (7).

These systems also include classical gradient flows of the form $G(q)\dot{q} = -\mathrm{D}_q \mathcal{E}(t, q)$, if we choose $\mathcal{R}(q, v) = \frac{1}{2}\langle G(q)v, v\rangle$. However, rate independence now means that the mapping $v \mapsto \partial_{\dot{q}} \mathcal{R}(q, v)$ is positively homogeneous of degree 0, i.e. $\partial_{\dot{q}} \mathcal{R}(q, \gamma v) = \partial_{\dot{q}} \mathcal{R}(q, v)$ for all $\gamma > 0$. We generally say that a mapping $f : \boldsymbol{X} \to Y$ is *p-homogeneous*, if it is positively homogeneous of degree p, i.e. $f(\lambda x) = \lambda^p f(x)$ for all $x \in \boldsymbol{X}$ and all $\lambda > 0$. Thus, $\mathcal{R}(q, \cdot) : \boldsymbol{Q} \to \mathbb{R}_\infty$ has to be 1-homogeneous, which implies that $\mathcal{R}(q, \cdot)$ either is identically 0 or it is not differentiable at $v = 0$.

Thus, from now on all dissipation potentials are assumed to satisfy

$$\begin{aligned} &\mathcal{R}(z, \cdot) : \boldsymbol{Z} \to [0, \infty] \text{ is convex, lower semicontinuous} \\ &\text{and satisfies } \mathcal{R}(z, 0) = 0. \end{aligned} \tag{10}$$

The derivative of $\mathcal{R}$ with respect to v is the set-valued convex subdifferential

$$\partial_{\dot{z}} \mathcal{R}(z, v) = \{\, \eta \in \boldsymbol{Z}^* \mid \forall\, w \in \boldsymbol{Z} : \ \mathcal{R}(z, w) \geq \mathcal{R}(z, v) + \langle \eta, w - v\rangle \,\}.$$

If $\mathcal{R}$ is 1-homogeneous, then the triple $(\boldsymbol{Q}, \mathcal{E}, \mathcal{R})$ is called a *rate-independent system* (RIS).

Very often we also look at rate-dependent versions of (5), i.e. we consider a potential $\mathcal{R}_{\mathrm{sl}}$ such that $\mathcal{R}_{\mathrm{sl}}(q, \cdot)$ is superlinear (whence not 1-homogeneous) and still assume (10). Note that the rate-independent case can be recovered by slowing down the loading rate. In fact, if we replace t in (5) by $\varepsilon\tau$ and let $\widetilde{z}(\tau) = z(\varepsilon t)$, then $\widetilde{z}$ solves the equation

$$0 \in \partial_{\widetilde{z}'} \widetilde{\mathcal{R}}_\varepsilon(\widetilde{z}, \widetilde{z}') + \mathrm{D}_{\widetilde{z}} \mathcal{I}(\tau, \widetilde{z}), \ \text{ where } \widetilde{\mathcal{R}}_\varepsilon(z, v) = \frac{1}{\varepsilon} \mathcal{R}_{\mathrm{sl}}(z, \varepsilon v).$$

By classical convexity arguments we have $\widetilde{\mathcal{R}}_\varepsilon(z, v) \searrow \mathcal{R}_0(z, v)$ for $\varepsilon \to 0$, where $\mathcal{R}_0$ is 1-homogeneous again. The limit passage $\varepsilon \to 0$ for the corresponding solutions z^ε is called the vanishing-viscosity limit and will be discussed in some detail in Sects. 4 and 5.

2.3 Some Canonical Examples

(1) The simplest example is obtained in the scalar case $z \in \boldsymbol{Z} = \mathbb{R}$ with the dissipation potential $\mathcal{R}(z, v) = |v|$ and the energy functional $\mathcal{I}(t, z) =$

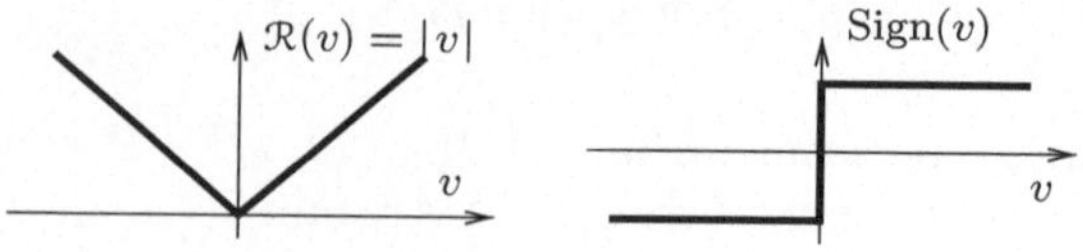

Fig. 2 Multivalued signum function $\mathrm{Sign} = \partial|\cdot|$

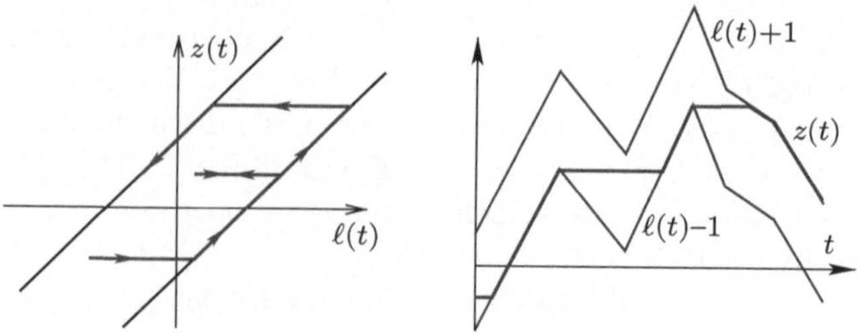

Fig. 3 The play operator associated with (11)

$\frac{1}{2}z^2 - \ell(t)z$. We obtain the equation

$$0 \in \mathrm{Sign}(\dot z) + z - \ell(t), \tag{11}$$

where Sign is the multi-valued Signum function depicted in Fig. 2. We observe that we always have $|z(t) - \ell(t)| \le 1$. Moreover, $|z(t) - \ell(t)| < 1$ implies $\dot z(t) = 0$, whereas $\pm \dot z(t) > 0$ implies $z(t) = \ell(t) \mp 1$. We obtain the so-called play operator, where q follows ℓ with a play of size 1, see Fig. 3.

(2) An infinite dimensional generalization leads to the most classical example of a rate-independent process formulated in a Hilbert space $\boldsymbol{Q}$ with a quadratic energy

$$\mathcal{E}(t, q) = \tfrac{1}{2}\langle \mathbb{A}q, q\rangle - \langle \ell(t), q\rangle$$

and a dissipation potential $\Psi : \boldsymbol{Q} \to [0, \infty]$. This situation is studied under the name *sweeping process*, see e.g. [36, 56]. The differential form reads $0 \in \partial\Psi(\dot q) + \mathbb{A}q - \ell(t)$. The input ℓ is considered as the center of the moving set $C(t) = \ell(t) - \partial\Psi(0)$ and the solution needs to satisfy $\mathbb{A}q(t) \in C(t)$. Nowadays the name *play operator* is used for this process, cf. [11, 35, 70].

(3) The motivation of the above sweeping process was the classical problem of *linearized elastoplasticity*, see [30, 55]. For a body $\Omega \subset \mathbb{R}^d$ the state q consists of the displacement $u \in \boldsymbol{Y} = \mathrm{H}^1_\Gamma(\Omega; \mathbb{R}^d) = \{\, u \in \mathrm{H}^1(\Omega; \mathbb{R}^d) \mid u|_\Gamma = 0 \,\}$ and the plastic strain tensor $z \in \mathrm{L}^2(\Omega; Z)$ with $Z = \{\, z \in \mathbb{R}^{d\times d} \mid z = z^\mathsf{T},\ \mathrm{tr}\, z = 0 \,\}$. The total energy contains the elastic energy, the hardening energy and the external loading:

$$\mathcal{E}(t, u, z) = \int_\Omega \tfrac{1}{2}(\mathrm{e}(u) - z):\mathbb{C}:(\mathrm{e}(u) - z) + \tfrac{1}{2} z:\mathbb{H}:z - u\cdot f_{\mathrm{ext}}(t)\,\mathrm{d}x,$$

where $\mathrm{e}(u) = \frac{1}{2}(\nabla u+\nabla u^{\mathsf{T}})$ is the infinitesimal strain tensor, and $\mathbb{C}$ and $\mathbb{H}$ are positive definite fourth-order tensors for elasticity and hardening, respectively. The dissipation potential reads $\mathcal{R}(z,\dot z) = \int_\Omega \sigma_{\mathrm{yield}}|\dot z(x)|\,\mathrm{d}x$. The subdifferential formulation then reads

$$-\operatorname{div}\big(\mathbb{C}:(\mathrm{e}(u)-z)\big) = f_{\mathrm{ext}}, \quad 0 \in \sigma_{\mathrm{yield}}\operatorname{Sign}(\dot z) + \mathbb{C}:(z-\mathrm{e}(u)) + \mathbb{H}:z.$$

For more details and general small-strain models we refer to [2,31].

(4) Elastoplastic models with finite strain lead to highly nonlinear rate-independent models. We refer to [40] for a discussion of the mathematical and mechanical background involving the associated Lie groups and to [39] for existence results for energetic solutions in the PDE context. Here we consider the simplified material-point mechanics, which applies to bodies that are deformed homogeneously.

The deformation gradient $F=\nabla\phi$ is treated as an element of the general linear group $\mathcal{Y} = \mathrm{GL}^+(\mathbb{R}^d) = \{\,F \in \mathbb{R}^{d\times d} \mid \det F>0\,\}$, and the plastic tensor P is taken from the special linear group $\mathcal{Z}=\mathrm{SL}(\mathbb{R}^d)=\{\,F \in \mathbb{R}^{d\times d} \mid \det F=1\,\}$. The energy takes the form

$$\mathcal{E}(t,F,P) = W_{\mathrm{elast}}(F\,P^{-1}) + W_{\mathrm{hard}}(P) - \Sigma(t):F,$$

where the multiplicative decomposition of the strain tensor $F = F_{\mathrm{elast}}P$ gives rise to the geometric nonlinearity $F_{\mathrm{elast}} = F\,P^{-1}$ appearing in W_{elast}. Here $\Sigma(t) \in \mathbb{R}^{d\times d}$ is the applied stress. Because of plastic invariance the dissipation potential takes the form

$$\mathcal{R}(P,\dot P) = \psi(\dot P\,P^{-1}) \;\text{ with } \psi(\eta) = \sigma_{\mathrm{yield}}|\eta|.$$

Thus, the dissipation potential depends intrinsically on the internal state. This and the strong geometric nonlinearity give rise to solutions with jumps, cf. [39,40].

2.4 The Basic a Priori Estimates

To understand the main difficulties in modeling RIS $(\boldsymbol{Q},\mathcal{E},\mathcal{R})$, we give the basic estimates, which follow from the differential formulation (5). This part is formal and needs proper justification for the construction of solutions. At the moment we just motivate some basic concepts.

We first provide a basic property of subdifferentials of 1-homogeneous functions.

Lemma 2.2. *Let $\Psi : \boldsymbol{Q} \to [0,\infty]$ be lower semicontinuous, convex and 1-homogeneous. Then, we have*

$$\partial\Psi(v) = \{\,\eta \in K \mid \Psi(v) = \langle \eta, v\rangle\,\}, \quad \textit{where } K = \partial\Psi(0).$$

Moreover, we have the characterization

$$0 \in \partial\Psi(v) + g \quad \Longleftrightarrow \quad 0 = \Psi(v) + \langle g, v\rangle \leq \Psi(w) + \langle g, w\rangle \ \textit{for all } w \in \boldsymbol{Q}.$$

Using the characterization of $\partial_{\dot q}\mathcal{R}(q,\cdot)$ we see that (5) is equivalent to

$$\forall\, v \in Q: \ \langle \mathrm{D}_q\mathcal{E}(t,q), v\rangle + \mathcal{R}(q,v) \geq 0, \tag{12a}$$

$$\langle \mathrm{D}_q\mathcal{E}(t,q), \dot q\rangle + \mathcal{R}(q,\dot q) = 0. \tag{12b}$$

Differentiation $\mathcal{E}(t,q(t))$ we see that (12b) is equivalent to the energy balance

$$\mathcal{E}(t,q(t)) + \int_r^t \mathcal{R}(q(s),\dot q(s))\,\mathrm{d}s = \mathcal{E}(r,q(r)) + \int_r^t \partial_s\mathcal{E}(s,q(s))\,\mathrm{d}s, \tag{13}$$

where $\partial_t\mathcal{E}$ denotes the usual partial derivative with respect to time, which has the physical meaning of the power induced by the temporal changes in the system. This identity now holds for all r,t with $0 \leq r < t \leq T$.

Throughout we assume that the power $\partial_t\mathcal{E}$ can be controlled by the energy itself, namely

$$\begin{array}{l} \exists\,\lambda_{\mathcal{E}} > 0\ \forall\,(s,q) \in \mathcal{Q}_T \text{ with } \mathcal{E}(s,q) < \infty: \\ \mathcal{E}(\cdot,q) \in \mathrm{C}^1([0,T]) \text{ and } |\partial_t\mathcal{E}(t,q)| \leq \lambda_{\mathcal{E}}\,\mathcal{E}(t,q) \text{ for all } t \in [0,T]. \end{array} \tag{14}$$

Using (13), (14), $\mathcal{R} \geq 0$, a Gronwall estimate gives the basic estimates

$$\mathcal{E}(t,q(t)) \leq \mathrm{e}^{\lambda_{\mathcal{E}} t}\mathcal{E}(0,q(0)) \quad \text{and} \quad \int_0^t \mathcal{R}(q(s),\dot q(s))\,\mathrm{d}s \leq \mathrm{e}^{\lambda_{\mathcal{E}} t}\mathcal{E}(0,q(0)).$$

The first estimate is useful, since we always assume that $\mathcal{E}(t,\cdot)$ is coercive. The second estimate controls the temporal behavior. First, we must take into account that $\mathcal{R}$ only controls the dissipative part z of $q = (y,z) \in \boldsymbol{Y}\times\boldsymbol{Z}$ and that this control may only be valid in a weaker norm, namely

$$\mathcal{R}(y,z,\dot y,\dot z) = \mathcal{R}(z,\dot z) \geq c_R\|\dot z\|_{\boldsymbol{X}}.$$

Second, the 1-homogeneity only provides a bound in $\mathrm{W}^{1,1}([0,T],\boldsymbol{X})$. However, since in this space the unit ball is not weakly closed, we have to work with the space $\mathrm{BV}([0,T],\boldsymbol{X})$, i.e., we have

$$\mathrm{Var}_{\boldsymbol{X}}(z,[0,T]) \leq \frac{1}{c_R}\int_0^T \mathcal{R}(q(t),\dot q(t))\,\mathrm{d}t \leq \frac{1}{c_R}\mathrm{e}^{\lambda_{\mathcal{E}} T}\mathcal{E}(0,q(0)).$$

This a priori estimate is essential to obtain temporal compactness and allows us to use a suitable version of Helly's selection principle for z. Yet, this estimate does not give control over the temporal behavior of y. Moreover, we have to be aware of the possibility of jumps, which occur in limit procedures.

The above estimates can be improved under suitable convexity assumptions. For this we use (12) as follows. We fix $\tau \in \,]0,T[$ and consider $\gamma(t) = \langle \mathrm{D}_q\mathcal{E}(t,q(t)),\dot{q}(\tau)\rangle + \mathcal{R}(q(t),\dot{q}(\tau))$. We have $\gamma(t) \geq 0$ and $\gamma(\tau) = 0$ from (12a) and (12b), respectively. Thus we conclude $\dot{\gamma}(\tau) = 0$, which gives

$$\langle \mathrm{D}_q^2\mathcal{E}(\tau,q(\tau))\dot{q}(\tau),\dot{q}(\tau)\rangle + \mathrm{D}_q\mathcal{R}(q(\tau),\dot{q}(\tau))[\dot{q}(\tau)] = -\langle \mathrm{D}_q\partial_t\mathcal{E}(\tau,q(\tau)),\dot{q}(\tau)\rangle.$$

Thus, assuming uniform convexity of $\mathcal{E}(t,\cdot) : \boldsymbol{Q} \to \mathbb{R}_\infty$ and that $\mathrm{D}_q\mathcal{R}$ is sufficiently small, we obtain a bound of the type

$$\|\dot{q}(\tau)\|_{\boldsymbol{Q}} \leq \frac{1}{\kappa-\rho}\|\partial_t\mathrm{D}_q\mathcal{E}(\tau,q(\tau))\|_{\boldsymbol{Q}^*},$$

if the *joint-convexity condition*

$$\kappa > \rho \tag{15}$$

holds, where $\kappa > 0$ and $\rho \in [0,\kappa[$ are such that

$$\langle \mathrm{D}_q^2\mathcal{E}(\tau,q(\tau))v,v\rangle \geq \alpha\|v\|_{\boldsymbol{Q}}^2 \quad \text{and} \quad |\mathrm{D}_q\mathcal{R}(q,v)[v]| \leq \rho\|v\|_{\boldsymbol{Q}}^2.$$

In Sect. 4.3 we discuss how these estimates can be used to prove the existence of differentiable solutions in the convex case.

2.5 Energetic Formulation of Generalized Gradient Flows

The starting point of the modeling are generalized gradient systems $(\boldsymbol{V},\mathcal{I},\mathcal{R})$, which are not necessarily rate independent. The state space $\boldsymbol{V}$ is a Hilbert space, $\mathcal{I} : \boldsymbol{V}_T \to \mathbb{R}_\infty$ is the energy functional, and the dissipation potential $\mathcal{R} : \boldsymbol{V}\times\boldsymbol{V} \to [0,\infty]$ satisfies (10). The evolution equation is given in the form

$$0 \in \partial_{\dot{z}}\mathcal{R}(z,\dot{z}) + \mathrm{D}_z\mathcal{I}(t,z).$$

The classical gradient flow is obtained, if $\mathcal{R}$ is given in terms of a Riemannian tensor $G(z) : \boldsymbol{V} \to \boldsymbol{V}^*$, which is symmetric and positive definite, viz. $\mathcal{R}(z,v) = \frac{1}{2}\langle G(z)v,v\rangle$. Then, we have

$$0 = G(z)\dot{z} + \mathrm{D}_z\mathcal{I}(t,z) \;\Leftrightarrow\; \dot{z} = -\nabla\mathcal{I}(t,z) = -G(z)^{-1}\mathrm{D}_z\mathcal{I}(t,z).$$

Rate-independent systems are special generalized gradient flows, namely those for which $\mathcal{R}(z,\cdot)$ is 1-homogeneous, i.e. $\mathcal{R}(z,\alpha v) = \alpha\mathcal{R}(z,v)$ for $\alpha > 0$.

Before going into more detail we discuss some equivalent formulations of generalized gradient flows.

Let $\boldsymbol{V}$ be a Hilbert space and $F : \boldsymbol{V} \to \mathbb{R}_\infty$ a proper, lower semicontinuous, and convex function. Its Legendre-Fenchel transform $F^* : \boldsymbol{V}^* \to \mathbb{R}_\infty$ is defined via $F^*(\xi) = \sup_{v \in \boldsymbol{V}} \langle \xi, v \rangle - F(v)$. The Fenchel equivalences for subdifferentials read

$$\xi \in \partial F(v) \quad \Leftrightarrow \quad v \in \partial F^*(\xi) \quad \Leftrightarrow \quad F(v) + F^*(\xi) = \langle \xi, v \rangle. \tag{16}$$

Moreover, by the definition of F^* we always have the lower bound

$$F(v) + F^*(\xi) \geq \langle \xi, v \rangle \quad \text{for all } v \in \boldsymbol{V} \text{ and } \xi \in \boldsymbol{V}^*. \tag{17}$$

Using the equivalences in (16), our subdifferential equations can be written in three equivalent ways:

Force balance $\quad 0 \in \partial_{\dot{z}} \mathcal{R}(z(t), \dot{z}(t)) + \xi(t) \subset \boldsymbol{V}^*;$ (18a)

Rate equation $\quad \dot{z}(t) \in \partial_\xi \mathcal{R}^*(z(t), -\xi(t)) \subset \boldsymbol{V};$ (18b)

Energy balance $\quad \mathcal{R}(z(t), \dot{z}(t)) + \mathcal{R}^*(z(t), -\xi(t)) = \langle -\xi(t), \dot{z}(t) \rangle \in \mathbb{R};$ (18c)

where in all three formulations the additional condition $\xi(t) \in \overline{\partial}_z \mathcal{I}(t, z(t)) \subset \boldsymbol{V}^*$ is imposed. The last relation can be combined with a chain rule, namely

$$\frac{\mathrm{d}}{\mathrm{d}t} \mathcal{I}(t, z(t)) = \langle \xi(t), \dot{z}(t) \rangle + \partial_t \mathcal{I}(t, z(t)). \tag{19}$$

Applying the chain rule and integrating (18c) over the time interval $[0, T]$ we obtain the *integral inequality*

$$\begin{aligned} &\mathcal{I}(T, z(T)) + \textstyle\int_0^T \mathcal{R}(z(t), \dot{z}(t)) + \mathcal{R}^*(z(t), -\xi(t))\,\mathrm{d}t \\ &\leq \mathcal{I}(0, z(0)) + \textstyle\int_0^T \partial_t \mathcal{I}(t, z(t))\,\mathrm{d}t \quad \text{with } \xi(t) \in \overline{\partial}_z \mathcal{I}(t, z(t)), \end{aligned} \tag{20}$$

which holds in fact as an equality. We call (20) the *energetic formulation* of the problem defined via (18). The point of importance is that already the integral inequality (20) is equivalent to the fact that the three formulations in (18) hold a.e. in $[0, T]$. To see this we use the lower estimate (17) and the chain rule (19) to obtain

$$\begin{aligned} &\langle -\xi(t), \dot{z}(t) \rangle \leq \mathcal{R}(z(t), \dot{z}(t)) + \mathcal{R}^*(z(t), -\xi(t)) \text{ a.e. on } [0, T], \\ &\textstyle\int_0^T \mathcal{R}(z(t), \dot{z}(t)) + \mathcal{R}^*(z(t), -\xi(t))\,\mathrm{d}t \leq \int_0^T \langle -\xi(t), \dot{z}(t) \rangle\,\mathrm{d}t, \end{aligned}$$

which immediately implies (18c) a.e.

A major advantage of the formulation as an integral inequality is seen for parameter-dependent dissipation potentials. Then, defining the functional

$$\mathcal{M}_\varepsilon(z,v,\eta) = \int_0^T \mathcal{R}_\varepsilon(z(t),\dot z(t)) + \mathcal{R}^*_\varepsilon(z(t),\eta(t))\,\mathrm{d}t,$$

we want to pass to the limit $\varepsilon \to 0$ in equations of the type

$$\mathcal{I}(T,z^\varepsilon(T)) + \mathcal{M}_\varepsilon(z^\varepsilon,\dot z^\varepsilon,-\xi^\varepsilon) \le \mathcal{I}(0,z^\varepsilon(0)) + \int_0^T \partial_t\mathcal{I}(t,z^\varepsilon(t))\,\mathrm{d}t.$$

If the solutions z^ε converge to a limit z, we can use then Γ-limit arguments to find a limit version of an integral inequality for z in the form

$$\mathcal{I}(T,z(T)) + \mathcal{M}_0(z,\dot z,-\xi) \le \mathcal{I}(0,z(0)) + \int_0^T \partial_t\mathcal{I}(t,z(t))\,\mathrm{d}t,$$

In this limit passage it is important to maintain the subdifferential property, i.e. we need a closedness of the subdifferential in the following form:

$$\xi^\varepsilon(t) \in \overline{\partial}_z\mathcal{I}(t,z^\varepsilon(t)),\;\; z^\varepsilon \rightsquigarrow z,\;\; \xi^\varepsilon \rightsquigarrow \xi \quad\Rightarrow\quad \xi(t) \in \overline{\partial}_z\mathcal{I}(t,z(t)).$$

Moreover, the limit functional $\mathcal{M}_0$ has to be such that it still interacts properly with a suitable chain rule, which allows us then to obtain the opposite inequality.

If $\mathcal{R}(z,\cdot)$ is 1-homogeneous, the generalized gradient system $(\boldsymbol{V},\mathcal{I},\mathcal{R})$ is a RIS. Then, $\mathcal{R}^*$ has a very specific form, namely

$$\mathcal{R}^*(z,\xi) = \begin{cases} 0 & \text{for } \xi \in K(z), \\ \infty & \text{otherwise,} \end{cases} \qquad \text{where } K(z) = \partial_v\mathcal{R}(z,0) \subset \boldsymbol{V}^*.$$

Thus, the term $\int_0^T \mathcal{R}^*(z(t),-\xi(t))\,\mathrm{d}t$ contributes to the right-hand side in (20) either the value 0 or the value ∞, where the latter case would violate the validity. Hence, the term only acts as a side condition asking that $-\xi(t) \in K(z(t))$ for a.a. $t \in [0,T]$. In conclusion, (20) can be rewritten as follows

$$\begin{aligned} &z(t) \in \mathcal{S}_{\text{loc}}(t) \overset{\text{def}}{=} \{\, z{\in}\boldsymbol{Z} \mid 0{\in}\partial_v\mathcal{R}(z(t),0){+}\overline{\partial}_z\mathcal{I}(t,z)\,\} \text{ a.e. on } [0,T], \\ &\mathcal{I}(T,z(T)) + \textstyle\int_0^T \mathcal{R}(z(t),\dot z(t))\,\mathrm{d}t \le \mathcal{I}(0,z(0)) + \int_0^T \partial_t\mathcal{I}(t,z(t))\,\mathrm{d}t. \end{aligned} \tag{21}$$

The first line constitutes a local stability condition that does not involve any time derivative and thus is a purely static condition, while the second condition is the usual upper energy estimate. As before, the chain rule applied to $t \mapsto \mathcal{I}(t,z(t))$ implies that the two lines provide an exact energy balance.

The problem with RIS is that in general we cannot expect solutions to be absolutely continuous with respect to time. Hence we derive notions of solutions that allow solutions with jumps. It is a common feature of all these notions that they consist of a static stability condition and an energy inequality, which we often formulate directly as an energy balance.

2.6 Solution Concepts in the One-Dimensional Case

Here we introduce the different solutions concepts for a very trivial situation, namely the case $\boldsymbol{Q} = \boldsymbol{Z} = \mathbb{R}$, i.e. we work directly with the reduced functional $\mathcal{I}$. The aim is to discuss their mutual relations already in this easy context, where functional analytical questions do not yet show up. We let

$$\mathcal{R}(z,\dot z) = \begin{cases} r_+(z)\dot z & \text{for } \dot z \geq 0, \\ r_-(z)|\dot z| & \text{for } \dot z \leq 0; \end{cases} \qquad \text{and} \quad \mathcal{I}(t,z) = \mathcal{U}(z) - \ell(t)z, \tag{22}$$

where $r_+, r_-\mathrm{BC}(\mathbb{R})$, $r_\pm(z) \geq \rho > 0$, and the function ℓ will be specified in the different examples. With the local dissipation metric $\mathcal{R}$ we associate the dissipation distance $\mathcal{D}$ defined via

$$\mathcal{D}(z_0,z_1) = \begin{cases} \int_{z_0}^{z_1} r_+(z)\,\mathrm{d}z & \text{for } z_0 \leq z_1, \\ \int_{z_1}^{z_0} r_-(z)\,\mathrm{d}z & \text{for } z_0 \geq z_1. \end{cases}$$

All definitions for solutions are for general $\mathcal{I}$, but for examples we use (22) with the nonconvex potential $\mathcal{U}$ and the initial datum z_0 given via

$$\mathcal{U}(z) = \begin{cases} \frac{1}{2}(z+4)^2 & \text{for } z \leq -2, \\ 4-\frac{1}{2}z^2 & \text{for } |z| \leq 2, \\ \frac{1}{2}(z-4)^2 & \text{for } z \geq 2; \end{cases} \qquad \text{and} \quad z_0 = -5. \tag{23}$$

If $\ell : [0,T] \to \mathbb{R}$ is specified, the RIS $(\boldsymbol{Z}, \mathcal{I}, \mathcal{R})$ is fully given.

We now introduce the main solution concepts, which are somewhat less involved in the present one-dimensional setting. Note that in all cases the solution $z : [0,T] \to \mathbb{R}$ is defined for *all* t, while some conditions need to hold only a.e. in $[0,T]$.

(1) A *differential solution* $z : [0,T] \to \mathbb{R}$ is defined via $z \in \mathrm{W}^{1,1}([0,T])$ and

$$0 \in \partial\mathcal{R}(z(t),\dot z(t)) + \mathrm{D}_z\mathcal{I}(t,z(t)) \quad \text{for a.a. } t \in [0,T].$$

(2) A *CD solution* $z : [0,T] \to \mathbb{R}$ (for 'C'ontinuous 'D'issipation) is defined via

cont. dissipation $t \mapsto \mathrm{Diss}_{\mathcal{D}}(z,[0,t])$ is continuous,
local stability $0 \in \partial\mathcal{R}(z(t),0) + \mathrm{D}_z\mathcal{I}(t,z(t))$ for a.a. $t \in [0,T]$;
energy balance $\forall t \in [0,T]:$
$\mathcal{E}(t,z(t)) + \mathrm{Diss}_{\mathcal{D}}(z,[0,t]) = \mathcal{E}(0,z(0)) - \int_0^t \dot\ell(s)z(s)\,\mathrm{d}s,$

where $\mathrm{Diss}_{\mathcal{D}}(z,[r,t]) = \sup \sum_{j=1}^N \mathcal{D}(z(t_{j-1}),z(t_j))$ with the supremum taken over all finite partitions $r \leq t_0 < t_1 < \cdots y < t_{N-1} < t_N \leq t$.

(3) A *local solution* $z : [0,T] \to \mathbb{R}$ is defined via

local stability $0 \in \partial\mathcal{R}(z(t),0) + \mathrm{D}_z\mathcal{I}(t,z(t))$ for a.a. $t \in [0,T]$;
energy inequality $\forall\, r,t$ with $0 \leq r < t \leq T:$
$\mathcal{E}(t,z(t)) + \mathrm{Diss}_{\mathcal{D}}(z,[r,t]) \leq \mathcal{E}(r,z(r)) - \int_r^t \dot\ell(s)z(s)\,\mathrm{d}s.$

(4) An *energetic solution* $z : [0,T] \to \mathbb{R}$ is defined via, for all $t \in [0,T]$,

global stability $\mathcal{I}(t,z(t)) \leq \mathcal{I}(t,\widetilde{z}) + \mathcal{D}(z(t),\widetilde{z})$ for all $\widetilde{z} \in Z$;
energy balance $\mathcal{E}(t,z(t)) + \mathrm{Diss}_{\mathcal{D}}(z,[0,t]) \leq \mathcal{E}(0,z(0)) + \int_0^t \partial_t\mathcal{I}(s,z(s))\,\mathrm{d}s.$

(5) An *approximable solution* $z : [0,T] \to \mathbb{R}$ is defined as a pointwise limit of a sequence $(z^{\varepsilon_k})_{k\in\mathbb{N}}$ with $\varepsilon_k \to 0$ of solutions z^ε of the viscous problems

$$0 \in \partial\mathcal{R}(\dot z^\varepsilon) + \varepsilon\dot z^\varepsilon + \mathrm{D}_z\mathcal{I}(t,z^\varepsilon(t)) \quad \text{for a.a. } t \in [0,T].$$

Approximable solutions are also called *vanishing-viscosity solutions.*

(6) A pair $(\widehat{t},\widehat{z}) : [0,S] \to [0,T]\times\mathbb{R}$ is defined to be a *parametrized solution*, if $(\widehat{t},\widehat{z}) \in \mathrm{W}^{1,1}([0,S],\mathbb{R}^2)$ and if for a.a. $s \in [0,S]$ we have

(i) $\widehat{t}(0) = 0,\ \widehat{t}(S) = T,\ \widehat{t}\,'(s) \geq 0,$
(ii) $\widehat{t}\,'(s) + |\widehat{z}\,'(s)| = 1,$
(iii) $0 \in \partial_{\widehat{z}'}\widehat{\mathcal{R}}(\widehat{z}(s),\widehat{z}\,'(s)) + \mathrm{D}_z\mathcal{I}(\widehat{t}(s),\widehat{z}(s)),$

where $\widehat{\mathcal{R}}(z,v) = \mathcal{R}(z,v)$ for $|v| \leq 1$ and ∞ otherwise.

(7) A *BV solution* $z : [0,T] \to \mathbb{R}$ is defined via $z \in \mathrm{BV}([0,T])$ and

(i) $0 \in \partial_{\dot z}\mathcal{R}(z(t),0) + \mathrm{D}_z\mathcal{I}(t,z(t))$ a.e. in $[0,T]$;
(ii) for all $t \in [0,T]$ we have

$$\mathcal{I}(t,z(t)) + \mathrm{Diss}_{\mathcal{D}}(z,[0,t]) + \mathrm{Jmp}_{\mathcal{I}}(z,[0,t]) = \mathcal{I}(0,z(0)) + \int_0^t \partial_s\mathcal{I}(s,z(s))ds$$

where $\mathrm{Jmp}_{\mathcal{I}}$ is defined via the jump set $J(z) \stackrel{\text{def}}{=} \{t \mid z \text{ not contin. in } t\}$ as

$$\mathrm{Jmp}_{\mathcal{I}}(z,[r,t]) \stackrel{\text{def}}{=} \Delta(r,z(r),z(r^+)) + \Delta(t,z(t^-),z(t)) + \sum_{s\in J(z)\cap]r,t[} \Delta(s,z(s^-),z(s)) + \Delta(s,z(s),z(s^+)),$$

where $\Delta(t,z_0,z_1) = |\int_{z_0}^{z_1} \mathrm{dist}(-\mathrm{D}_z\mathcal{I}(t,z), \partial_{\dot z}\mathcal{R}(z,0))\;\mathrm{d}z|$ and $z(t^{\pm})$ denotes one-sided limits, see (49).

For the general definitions for these solutions types we refer to Definition 4.5 for differentiable, CD, and local solutions, Definition 3.1 for energetic solutions, Definitions 4.11 and 5.4 for parametrized solutions, and Definitions 4.21, 4.26, and 5.7 for BV solutions.

The above definition (7) for BV solutions is very implicit, but it highlights the similarity to the other solutions concepts in relying on a (i) static stability concept and (ii) an energy balance. The discussions in Sect. 4.5 shows that (ii) asks that along jumps from $z(t^-)$ to $z(t^+)$ the driving force $\mathrm{D}_z\mathcal{I}(t,z)$ is sufficiently large, e.g. for $z \in [z(t^+), z(t^-)]$ we must have $-\mathrm{D}_z\mathcal{I}(t,z) \leq -r_-(z)$ and $\Delta(t,z(t^-),z(t^+)) = \int_{z(t^+)}^{z(t^+)} \mathrm{D}_z\mathcal{I}(t,z) - r_-(z)\,\mathrm{d}z$.

We now comment on the relation between the different solution concepts. The first fact is that the notion of local solutions includes all the others. Differential solutions may not exist, but if they do then they are also BV solutions. All approximable solutions are BV solutions, but the opposite is in general not true.

If energetic solutions are differentiable, then they coincide with differential solutions. If energetic solutions have jumps, then they jump as soon as possible, whereas BV solutions jump as late as possible. So these two solutions types should be seen as two opposite extremes in the set of all local solutions. For both these extremes we have a rather complete existence theory, see Sects. 3 and 4.5, respectively.

Parametrized solutions are special, since they are defined as curves in the extended space $\boldsymbol{Z}_T \stackrel{\mathrm{def}}{=} [0,T]\times\boldsymbol{Z}$ given in arclength parametrization. In fact, they are in correspondence to BV solutions. Under suitable technical assumptions, the latter can be turned into parametrized solutions by filling in the jumps and arclength parametrization. Vice versa, every parametrized solution generates a BV solution via $\sigma(t) = \inf\{\, s \in [0,S] \mid \widehat{t}(s) = t\,\}$ and $z(t) = \widehat{z}(\sigma(t))$.

The following examples show that these notions are genuinely different. In Examples 2.3–2.6 we have $(\boldsymbol{Z},\mathcal{I},\mathcal{R})$ as defined in (22) and (23) with $r_+ = r_- \equiv 1$, but ℓ changes from case to case. In Example 2.7 we consider varying $r_\pm$.

Example 2.3. *We consider* $(\mathbb{R},\mathcal{I},\mathcal{R})$ *according to* (22) *with* $r_+ = r_- \equiv 1$, *and* $\ell(t) = t$ *for all* $t \geq 0$. *We claim that the approximable, the parametrized, and the BV solutions on* $[0,\infty[$ *are essentially unique and coincide. However, the unique energetic solution is different. Moreover, we show that there is a uncountable family of different local solutions. With direct calculations, one sees that the energetic solutions take the form*

$$z(t) = t-5 \;\; \textit{for } t \in [0,1[, \quad z(1) \in \{-4,4\}, \quad \textit{and} \quad z(t) = t+3 \;\; \textit{for } t > 1.$$

Choose any $t_* \in [1,3]$ *and any* $z_* \in \big[3{+}t_*, 3{+}t_*{+}\min\{2, 4\sqrt{t_*{-}1}\}\big]$. *Then,*

$$z(t) = \begin{cases} t{-}5 \text{ for } t \in [0, t_*[\,, \\ z_* \ \text{ for } t \in [t_*, z_*{-}3], \\ t{+}3 \text{ for } t \geq z_*{-}3, \end{cases}$$

is a local solution. Note that the starting point of the jump at $z(t_*^-) = t_*{-}5$ *can be chosen in a full interval. Moreover, for a fixed* $t_* \in \,]1,3]$ *we still have the possibility to choose the ending point* $z_* = z(t_*^+)$ *of the jump in a full interval. All the other solution types essentially lead (up to definition in one point) to the same solution. The approximable and BV solutions read*

$$z(t) = \begin{cases} t{-}5 \text{ for } t \in [0, 3[\,, \\ z_* \ \text{ for } t = 3, \\ t{+}3 \text{ for } t > 3, \end{cases}$$

where $z_* \in [-2, 6]$ *is arbitrary. The associated arclength-parametrized solution takes the form*

$$\big(\widehat{t}(s), \widehat{z}(s)\big) = \begin{cases} \big(\frac{s}{2}, \frac{s}{2}{-}5\big) & \text{for } s \in [0,6], \\ (3, s{-}8) & \text{for } s \in [6,14], \\ \big(\frac{s}{2}{-}4, \frac{s}{2}{-}1\big) & \text{for } s \geq 14. \end{cases}$$

Example 2.4. *We take* $(\mathbb{R}, \mathcal{I}, \mathcal{R})$ *as in Example 2.3 but with* $\ell(t) = \min\{t, 4{-}t\}$ *and obtain the differential solution* z_{diff}, *which is different from the energetic solution* z_{energ}, *namely*

$$z_{\text{diff}}(t) = \begin{cases} t{-}5 \text{ for } t \in [0,2], \\ -3 \ \text{ for } t \in [2,4], \\ 1{-}t \text{ for } t \geq 4; \end{cases} \qquad z_{\text{energ}}(t) = \begin{cases} t{-}5 \text{ for } t \in [0,1], \\ t{+}3 \text{ for } t \in \,]1,2[\,, \\ 5 \ \text{ for } t \in [2,4], \\ 9{-}t \text{ for }]4,5[\,, \\ 1{-}t \text{ for } t \geq 5. \end{cases}$$

Thus, even the existence of a differentiable solution does not guarantee that this is also the energetic solution.

Example 2.5. *In this example we show that not all BV solutions are approximable solutions. Again,* $(\mathbb{R}, \mathcal{I}, \mathcal{R})$ *is as in Example 2.3 but with* $\ell(t) \overset{\text{def}}{=} \min\{t, 6{-}t\}$, *i.e., the loading reduces exactly when the solution reaches the jump point. It is easy to see that there are two different BV solutions:* z_1, *which jumps at* $t = 3$, *and* z_2, *which does not jump. We have*

$$z_1(t) = \begin{cases} t{-}5 \ \text{ for } \ t \in [0,3[\,, \\ 6 \ \text{ for } \ t \in \,]3,5]\,, \\ 11{-}t \text{ for } \ t \in [5,9], \\ 3{-}t \ \text{ for } t > 9; \end{cases} \qquad z_2(t) = \begin{cases} t{-}5 \text{ for } t \in [0,3], \\ -2 \ \text{ for } t \in [3,5], \\ 3{-}t \text{ for } t \geq 5. \end{cases}$$

For $\varepsilon > 0$ the viscous solution q^ε of the differential inclusion

$$0 \in \mathrm{Sign}(\dot z) + \varepsilon \dot z + \mathcal{U}'(z) - \ell(t), \quad z(0) = -5,$$

is unique and can be found by matching solutions of linear ODEs. We find

$$z^\varepsilon(t) = \begin{cases} t{-}5{+}\varepsilon(\mathrm{e}^{-t/\varepsilon}{-}1) & \text{for } t \in [0,3], \\ z_*^\varepsilon & \text{for } t \in [3, t_*^\varepsilon], \\ 3{-}t{+}\varepsilon(\mathrm{e}^{-(t-t_*^\varepsilon)/\varepsilon}{-}1) & \text{for } t \geq t_*^\varepsilon, \end{cases}$$

where $z_*^\varepsilon = q^\varepsilon(3-) \lesssim -2$ and $t_*^\varepsilon = 3 - z_*^\varepsilon \gtrsim 5$. Thus, we have $z^\varepsilon(t) \to z_2(t)$ for every $t \geq 0$ as $\varepsilon \downarrow 0$. Hence, z_2 is a vanishing-viscosity solution, whereas z_1 is not. As a general principle, we conjecture that viscosity slows down solutions and thus approximable solutions tend to avoid jumps if there is a choice.

Example 2.6. *Here, we study the parameter dependence of solutions under the loading*

$$\ell_\delta(t) = \min\{t, 6{+}2\delta{-}t\} \quad \textit{for } t \geq 0,$$

where δ is a small parameter. In the case $\delta = 0$ we have two BV solutions z_1 and z_2, see Example 2.5. But only z_2 is an approximable solution. For $0 < \delta < 1$ there is only one BV solution, which is then also the unique approximable solution, namely

$$z^\delta(t) = \begin{cases} t{-}5 \ \textit{for } t \in [0,3[\,, \\ t{+}3 \ \textit{for } t \in \,]3, 3{+}\delta]\,, \\ 6{+}\delta \ \textit{for } t \in [3{+}\delta, 5{-}\delta]. \end{cases}$$

Taking the limit $\delta \to 0^+$ we see that the pointwise limit of the approximable solutions z^δ is z_1, which is not an approximable solution for $\delta = 0$. Thus, the set of approximable solutions is not upper semicontinuous with respect to variations of the data.

Example 2.7. Here, $\mathcal{I}(t,\cdot)$ is uniformly convex, in fact even quadratic, but $\mathcal{R}$ depends on z such that the joint convexity condition (15) does not hold. As a consequence we obtain solutions with jumps. For $\gamma > 0$ we let

$$\mathcal{I}(t,z) = \frac{1}{2}z^2 - tz \quad \text{and} \quad \mathcal{R}(z,\dot z) = \mu(z)|\dot z| \text{ with } \mu(z) = \max\{1, \min\{2-\gamma z, 3\}\}.$$

For $z \in [-1/\gamma, 1/\gamma]$ the joint convexity condition $\alpha > \rho$ (see (15)) holds for $\rho = \gamma < 1 = \alpha$. Thus, for $\gamma > 1$ solutions have to jump across the region $[-1/\gamma, 1/\gamma]$, since there are no locally stable points. We start from the initial condition $z(0) = z_0 = -3$. Then, the energetic solution z_{energ} and the BV solution z_{BV} are different, namely

$$z_{\text{energ}}(t) = \begin{cases} t{-}3 & \text{for } t \in [0,2[\,, \\ t{-}1 & \text{for } t > 2; \end{cases} \qquad z_{\text{BV}}(t) = \begin{cases} t{-}3 & \text{for } t \in [0, 3-1/\gamma[\,, \\ t{-}1 & \text{for } t > 3-1/\gamma. \end{cases}$$

2.7 Infinite-Dimensional Examples

Here we provide the simplest and most canonical infinite-dimensional example of a RIS $(\mathcal{Z}, \mathcal{I}, \mathcal{R})$ including also a viscosity term. It is used in each of the abstract sections to discuss the different solutions concepts. We first give an abstract Banach-space setting and afterwards present a special case which connects the theory to a particular PDE.

Example 2.8 (Standard semilinear example). We consider a Banach space $\boldsymbol{X}$ and two Hilbert spaces $\boldsymbol{Z}$ and $\boldsymbol{V}$, which are densely and continuously embedded as follows:

$$\boldsymbol{Z} \Subset \boldsymbol{V} \subset \boldsymbol{X},$$

where "$\Subset$" denotes compact embedding. The different Banach spaces and their norms $\|\cdot\|_{\boldsymbol{Z}}$, $\|\cdot\|_{\boldsymbol{V}}$, and $\|\cdot\|_{\boldsymbol{X}}$ are associated with the energy functional, the viscous dissipation, and the rate-independent dissipation, respectively. We further assume that there are symmetric, bounded linear operators $\mathbb{A} \in \mathscr{L}(\boldsymbol{Z}, \boldsymbol{Z}^*)$ and $\mathbb{V} \in \mathscr{L}(\boldsymbol{V}, \boldsymbol{V}^*)$, which are invertible with bounded inverses. Without loss of generality (after choosing an equivalent Hilbert norm) one may assume that they equal the corresponding Riesz isomorphisms.

The problem under investigation is the doubly nonlinear equation

$$\begin{aligned} &0 \in \partial\Psi(\dot z(t)) + \varepsilon\mathbb{V}\dot z(t) + \mathrm{D}_z\mathcal{I}(t, z(t)) \\ &\text{with } \mathcal{I}(t,z) = \tfrac{1}{2}\langle \mathbb{A}z, z\rangle + \Phi(z) - \langle \ell(t), z\rangle. \end{aligned} \tag{24}$$

Here, $\Phi \in \mathrm{C}^2(\boldsymbol{Z};\mathbb{R})$ is a non-quadratic potential of lower order in a sense to be made precise below. The function $\ell : [0,T] \to \boldsymbol{V}^*$ is the loading. We assume that there exists $c, C > 0$, an interpolation exponent $\theta \in \,]0,1[$, and a growth exponent $q \geq 0$, such that for all $v, z, w \in \boldsymbol{Z}$ we have

$$\boldsymbol{Z} \Subset \boldsymbol{V} \subset \boldsymbol{X} \text{ with dense embeddings;} \tag{25a}$$

$$\|v\|_{\boldsymbol{V}} \leq C\|v\|_{\boldsymbol{X}}^{\theta}\|v\|_{\boldsymbol{Z}}^{1-\theta}; \tag{25b}$$

$$c\|z\|_{\boldsymbol{Z}}^2 \leq \langle \mathbb{A}z, z\rangle \leq C\|z\|_{\boldsymbol{Z}}^2, \ \|v\|_{\boldsymbol{V}}^2 = \langle \mathbb{V}v, v\rangle; \tag{25c}$$

$$\Psi : \boldsymbol{V} \to [0,\infty[\text{ convex, 1-homogeneous, } c\|v\|_{\boldsymbol{X}} \leq \Psi(v) \leq C\|v\|_{\boldsymbol{X}}; \tag{25d}$$

$$\Phi(z) \geq 0, \ \Phi : \boldsymbol{Z} \to \mathbb{R} \text{ is weakly continuous;} \tag{25e}$$

$$\mathrm{D}\Phi \in \mathrm{C}^1(\boldsymbol{Z}; \boldsymbol{V}^*), \ \|\mathrm{D}^2\Phi(z)v\|_{\boldsymbol{V}^*} \leq C(1{+}\|z\|_{\boldsymbol{Z}})^q\|v\|_{\boldsymbol{Z}}; \tag{25f}$$

$$\ell \in \mathrm{W}^{1,p}([0,T], \boldsymbol{V}^*) \text{ for some } p \geq 2. \tag{25g}$$

Condition (25f) on $\mathrm{D}\Phi$ can be weakened by replacing $\boldsymbol{V}^*$ with an interpolation space $[\boldsymbol{V}^*, \boldsymbol{Z}^*]_\eta$, $\eta \in]0,1[$, see [53]. We stay with $\boldsymbol{V}^*$ for notational simplicity. We introduce additional Hilbert spaces

$$\boldsymbol{Z}_1 \stackrel{\text{def}}{=} \{ z \in \boldsymbol{Z} \mid \mathbb{A}z \in \boldsymbol{V}^* \} \qquad \text{with } \|z\|_1 \stackrel{\text{def}}{=} \|\mathbb{A}z\|_{\boldsymbol{V}^*}, \tag{26a}$$

$$\boldsymbol{Z}_{-1} \stackrel{\text{def}}{=} \overline{\boldsymbol{V}}^{\|\cdot\|_{-1}} \qquad \text{with } \|z\|_{-1} \stackrel{\text{def}}{=} \|\mathbb{V}z\|_{\boldsymbol{Z}^*}. \tag{26b}$$

We obtain a scale of four Hilbert spaces

$$\boldsymbol{Z}_1 \Subset \boldsymbol{Z} \Subset \boldsymbol{V} \Subset \boldsymbol{Z}_{-1}, \quad \text{and } \mathbb{A}\boldsymbol{Z}_1 = \boldsymbol{V}^*, \quad \mathbb{V}\boldsymbol{Z}_{-1} = \boldsymbol{Z}^*,$$

with dense and compact embeddings. Moreover, the scale is equally spaced in the sense of interpolation, namely $[\boldsymbol{Z}_1, \boldsymbol{V}]_{1/2} = \boldsymbol{Z}$ and $[\boldsymbol{Z}, \boldsymbol{Z}_{-1}]_{1/2} = \boldsymbol{V}$. (If we compare to classical evolution triples $V \subset H \cong H^* \subset V^*$ with a linear selfadjoint, positive definite operator $A : D(A) \subset H \to H$ and $V = D(A^{1/2})$, we obtain the corresponding scale $D(A) \subset D(A^{1/2}){=}V \subset H \cong H^* \subset V^*$.)

This abstract setting can be applied to specific problems involving PDEs as they occur in modeling of hysteretic materials, like in magnetism, elastoplasticity, ferroelectricity, or shape-memory alloys. We refer to [42, 43] for surveys on these applications. For the simplest application we consider a smooth and bounded domain $\Omega \subset \mathbb{R}^d$ and let

$$\boldsymbol{Z} = \mathrm{H}^1_0(\Omega) \Subset \boldsymbol{V} = \mathrm{L}^2(\Omega) \subset \boldsymbol{X} = \mathrm{L}^1(\Omega).$$

We have $\boldsymbol{V}^* = \mathrm{L}^2(\Omega) \Subset \boldsymbol{Z}^* = \mathrm{H}^{-1}(\Omega)$, and the operators $\mathbb{A}$ and $\mathbb{V}$ are given by $-\Delta$ and id, respectively. This leads to the additional spaces $\boldsymbol{Z}_1 = \mathrm{H}^2(\Omega) \cap \mathrm{H}^1_0(\Omega)$ and $\boldsymbol{Z}_{-1} = \mathrm{H}^{-1}(\Omega)$. The functionals take the form

$$\Psi(v) = \int_\Omega |v(x)|\,\mathrm{d}x, \quad \mathcal{I}(t,z) = \int_\Omega \tfrac{1}{2}|\nabla z(x)|^2 + \phi(z(x)) + f(t,x)z(x)\,\mathrm{d}x,$$

where $f \in \mathrm{W}^{1,p}([0,T], \mathrm{L}^2(\Omega))$ defines the loading. The function $\phi \in \mathrm{C}^2(\mathbb{R};\mathbb{R})$ is assumed to satisfy $0 \le \phi(s) \le C(1{+}|s|)^q$ with $q < \infty$ for $d \le 2$ and $q < 2d/(d{-}2)$ for $d \ge 3$. Further, we assume $|\phi'(s)| \le C(1{+}|s|)^{q/2}$ and $|\phi''(s)| \le C(1{+}|s|)^{q/d}$. Then, all the conditions of the abstract theory are satisfied, cf. [53].

3 Energetic Solutions

In this section we consistently work with a state space $\mathcal{Q} = \mathcal{Y} {\times} \mathcal{Z}$ with states $q = (y, z)$, which has its reason in applications in continuum mechanics, where the z-component is dissipative while the y-component is not, cf. [43].

Whenever possible we write q instead of (y,z) to shorten notation. The state space $\mathcal{Q}$ is equipped with a Hausdorff topology, and we denote by $q_k \overset{\mathcal{Q}}{\to} q$, $y_k \overset{\mathcal{Y}}{\to} y$ and $z_k \overset{\mathcal{Z}}{\to} z$ the corresponding convergence of sequences. Throughout it is sufficient to consider sequential closedness, compactness and continuity. For notational convenience, we will not write this explicitly.

A main tool for the analysis of such systems is the interplay between the full RIS $(\mathcal{Q},\mathcal{E},\mathcal{D})$ and its reduced version $(\mathcal{Z},\mathcal{I},\mathcal{D})$, where the *reduced energy* $\mathcal{I}:[0,T]\times\mathcal{Z}\to\mathbb{R}_\infty$ is defined in (8). We define energetic solutions to $(\mathcal{Q},\mathcal{E},\mathcal{D})$ and $(\mathcal{Z},\mathcal{I},\mathcal{D})$ in such a way that each solution $q=(y,z)$ for the former system gives rise to a solution z for the latter. Vice versa each solution z for $(\mathcal{Z},\mathcal{I},\mathcal{D})$ can be made to a solution $q=(y,z)$ by a suitable choice of y. We emphasize that it is not enough to choose an arbitrary $y(t)\in\operatorname{Arg\,min}\mathcal{E}(t,\cdot,z(t))$, since further restrictions arise.

At first glance, it might seem reasonable to first consider the reduced system $(\mathcal{Z},\mathcal{I},\mathcal{D})$ and establish an existence theory there and then derive the desired existence result for the full problem $(\mathcal{Q},\mathcal{E},\mathcal{D})$. However, it turns out that in the reduction process certain natural properties (like differentiability in t) are lost. To compensate for that, stronger assumptions would be necessary, which can be avoided by working on the full system instead. Thus we present the existence theory for $(\mathcal{Q},\mathcal{E},\mathcal{D})$, which is also more natural in material modeling, cf. [43].

3.1 Abstract Setup of the Problem

The first ingredient of the energetic formulation is the *dissipation distance* $\mathcal{D}:\mathcal{Z}\times\mathcal{Z}\to[0,\infty]$, which is an extended quasi-distance. Here 'extended' means that the value ∞ is allowed and 'quasi' means that we do not ask for symmetry. The following conditions are the main assumptions on $\mathcal{D}$.

Extended quasi-distance:

$$\begin{aligned}&\text{(i)}\quad \forall\, z_1,z_2,z_3\in\mathcal{Z}:\ \mathcal{D}(z_1,z_3)\le\mathcal{D}(z_1,z_2)+\mathcal{D}(z_2,z_3),\\ &\text{(ii)}\quad \forall\, z_1,z_2\in\mathcal{Z}:\ \mathcal{D}(z_1,z_2)=0 \iff z_1=z_2;\end{aligned}\tag{D1}$$

$$\mathcal{D}:\mathcal{Z}\times\mathcal{Z}\to[0,\infty] \text{ is lower semicontinuous.}\tag{D2}$$

Here (D1) says that $\mathcal{D}$ is a distance except for the symmetry and the fact that the value ∞ is allowed. Relation (i) is the triangle inequality, and (ii) is the positivity. The unsymmetry is needed in many applications like in elastoplasticity or damage.

For curves $z:[0,T]\to\mathcal{Z}$ we define the *dissipation functional* $\mathrm{Diss}_\mathcal{D}$ via

$$\begin{aligned}\mathrm{Diss}_\mathcal{D}(z,[s,t]) \overset{\text{def}}{=} \sup\big\{ \textstyle\sum_{j=1}^N \mathcal{D}(z(t_{j-1}),z(t_j)) \ \big|\ N\in\mathbb{N},\\ s=t_0<t_1<\cdots<t_N=t\,\big\}.\end{aligned}\tag{27}$$

Further we define the following set of functions:

$$\mathrm{BV}_{\mathcal{D}}([0,T],\mathcal{Z}) \overset{\text{def}}{=} \{\, z : [0,T] \to \mathcal{Z} \mid \mathrm{Diss}_{\mathcal{D}}(z,[0,T]) < \infty \,\}. \tag{28}$$

The functions are defined everywhere and changing them at one point may increase the dissipation. Moreover, the dissipation is additive:

$$\mathrm{Diss}_{\mathcal{D}}(z,[r,t]) = \mathrm{Diss}_{\mathcal{D}}(z,[r,s]) + \mathrm{Diss}_{\mathcal{D}}(z,[s,t]) \ \text{ for all } r<s<t. \tag{29}$$

Later on, we sometimes use the notation $\mathcal{D}(q_0,q_1)$ instead of $\mathcal{D}(z_0,z_1)$ where $q_j = (y_j,z_j)$. This slight abuse of notation never leads to confusion, since $\mathcal{D}$ as a function on $\mathcal{Q} = \mathcal{Y}\times\mathcal{Z}$ still satisfies all assumptions but one has to remember that $\mathcal{D}$ satisfies the positivity (D1) only on $\mathcal{Z}$ and not on $\mathcal{Q}$.

The second ingredient is the energy-storage functional $\mathcal{E} : \mathcal{Q}_T \to \mathbb{R}_\infty$. Here $t \in [0,T]$ plays the role of a (very slow) process time which changes the loading. The following conditions form the basic assumptions on $\mathcal{E}$:

Compactness of sublevels:
$\forall\, t \in [0,T]: \ \mathcal{E}(t,\cdot) : \mathcal{Q} \to \mathbb{R}_\infty$ has compact sublevels; (E1)

Energetic control of power:
$\exists\, \lambda_{\mathcal{E}} > 0\ \forall\, (t,q)$ with $\mathcal{E}(t,q) < \infty$: (E2)
$\mathcal{E}(\cdot,q) \in \mathrm{C}^1([0,T])$ and $\big|\partial_s \mathcal{E}(s,q)\big| \le \lambda_{\mathcal{E}}\, \mathcal{E}(s,q)$ for all $s \in [0,T]$.

Condition (E2) implies $\mathrm{dom}\, \mathcal{E} = [0,T]\times \mathrm{dom}\, \mathcal{E}(0,\cdot)$, i.e. $\mathrm{dom}\, \mathcal{E}(t,\cdot)$ is independent of t. From (E2) and Gronwall's inequality we easily derive

$$\mathcal{E}(t,q) \le \mathcal{E}(s,q)\, \mathrm{e}^{\lambda_{\mathcal{E}}|t-s|} \ \text{ and } \ |\partial_t \mathcal{E}(t,q)| \le \lambda_{\mathcal{E}}\, \mathcal{E}(s,q)\, \mathrm{e}^{\lambda_{\mathcal{E}}|t-s|}. \tag{30}$$

Most typically, $\mathcal{Q}$ is a closed, convex and bounded subset of a reflexive Banach space (like $\mathrm{W}^{1,p}(\Omega,\mathbb{R}^m)$ or $\mathrm{L}^p(\Omega,\mathbb{R}^m)$ with $p \in (1,\infty)$) equipped with its weak topology $\mathcal{T}$. Then, lower semicontinuity of $\mathcal{E}$ and $\mathcal{D}$ in $(\mathcal{Q},\mathcal{T})$ is the same as the classical weak lower semicontinuity in the calculus of variations.

Definition 3.1 (Energetic solution). A function $q = (y,z) : [0,T] \to \mathcal{Q} = \mathcal{Y}\times\mathcal{Z}$ is called an *energetic solution* of the rate-independent system $(\mathcal{Q},\mathcal{E},\mathcal{D})$, if $t \mapsto \partial_t \mathcal{E}(t,q(t))$ is integrable and if the *global stability (S)* and the *energy equality (E)* hold for all $t \in [0,T]$:

(S) $q(t) \in \mathcal{S}(t)$;
(E) $\mathcal{E}(t,q(t)) + \mathrm{Diss}_{\mathcal{D}}(z,[0,t]) = \mathcal{E}(0,q(0)) + \int_0^t \partial_t \mathcal{E}(\tau,q(\tau))\,\mathrm{d}\tau$.

Condition (S) means global stability, because the *set* $\mathcal{S}(t)$ *of stable states at time* t is defined such that all $\widehat{q} \in \mathcal{Q}$ are considered as competitors:

$$\mathcal{S}(t) \stackrel{\text{def}}{=} \{q \in \mathcal{Q} \mid \mathcal{E}(t,q) < \infty,\ \mathcal{E}(t,q) \leq \mathcal{E}(t,\widehat{q}) + \mathcal{D}(q,\widehat{q}) \text{ for all } \widehat{q} \in \mathcal{Q}\}. \quad (31)$$

We shortly call $\mathcal{S}(t)$ the *stability set* at time t. The properties of the stability sets turn out to be crucial for deriving existence results.

The definition of energetic solutions is such that we obtain a rate-independent multi-valued evolutionary system in the sense of Definition 2.1, in particular we have the concatenation and the restriction property. It is clear that the stability condition (S) has the restriction and concatenation property. To see that (E) also shares these conditions we define $\mathfrak{E}_r(t) = \mathcal{E}(t,q(t)) + \text{Diss}_{\mathcal{D}}(q,[r,t]) - \int_r^t \partial_s \mathcal{E}(s,q(s))\,\text{d}s$. Then, (E) simply states that the function $\mathfrak{E}$ is equal to the constant value $\mathfrak{E}(r)$ on the whole interval. This constancy certainly remains true after restriction. When concatenating two solutions q_1 and q_2, the condition $q_1(t_2) = q_2(t_2)$ guarantees that the two constants are the same.

Rate independence manifests itself by the fact that the problem has no intrinsic time scale. It is easy to show that q is a solution to $(\mathcal{Q},\mathcal{E},\mathcal{D})$ if and only if the reparametrized curve $\widetilde{q} : t \mapsto q(\alpha(t))$, where $\dot{\alpha} > 0$, is a solution to $(\mathcal{Q},\widetilde{\mathcal{E}},\mathcal{D})$ with $\widetilde{\mathcal{E}}(t,q) = \mathcal{E}(\alpha(t),q)$. In particular, the stability (S) is a static concept, and the energy balance (E) is rate-independent, since the dissipation defined via (27) is scale-invariant.

Before discussing the question of existence of solutions we want to point out, that the concept of energetic solutions provides a priori bounds on the solutions. For the time-continuous problem these bounds are easy to derive, and the main structure becomes more transparent. Of course, similar estimates are crucial in the time-discrete setting. Using the assumption (E2) the energy balance (E) gives

$$\mathcal{E}(t,q(t)) + \text{Diss}_{\mathcal{D}}(z,[r,t]) \leq \mathcal{E}(r,q(r)) + \int_r^t \lambda_{\mathcal{E}}\mathcal{E}(s,q(s)\,\text{d}s \qquad (32)$$

for $0 \leq r < t \leq T$. Omitting the dissipation and applying Gronwall's lemma yield $\mathcal{E}(t,q(t)) \leq \mathcal{E}(0,q(0))\,\text{e}^{\lambda_{\mathcal{E}}t}$. Inserting this into (32) we estimate the dissipation via $\text{Diss}_{\mathcal{D}}(z,[0,T]) \leq \mathcal{E}(0,q(0))\text{e}^{\lambda_{\mathcal{E}}T}$, since $\mathcal{E}(t,q(t)) \geq 0$ by (E2).

3.2 The Time-Incremental Minimization Problem

The most natural approach to solve (S)&(E) is via time discretization using the fact that incremental problems exist, which are minimization problems. It is then possible to find their solutions as global minimizers of certain lower semicontinuous functionals on $\mathcal{Q}$. For this we make use of the lower semicontinuity assumptions (D2) and (E1).

For the time discretization we use the notation $\mathrm{Part}([r,s])$ for all finite partitions of the interval $[r,s] \subset \mathbb{R}$, i.e.

$$\mathrm{Part}([r,s]) \stackrel{\mathrm{def}}{=} \{\,(t_0,t_1,...,t_N) \mid r=t_0<t_1<...<t_N=r\,\}. \tag{33}$$

For a partition $\Pi \in \mathrm{Part}([r,s])$, we define N_Π as the number of subintervals and $\phi(\Pi)$ as its *fineness*, namely as the length of its largest interval:

$$\phi(\Pi) \stackrel{\mathrm{def}}{=} \max\{\,t_k - t_{k-1} \mid k=1,\dots,N_\Pi\,\}. \tag{34}$$

Note that $\phi(\Pi) = 2\max_{t\in[0,T]} \mathrm{dist}(t,\Pi)$. In particular, always $\mathrm{dist}(t,\Pi) \le \phi(\Pi)$. Having fixed a partition $\Pi = (t_0,t_1,...,t_N) \in \mathrm{Part}([0,T])$, we seek for some q_k, $k=1,...,N_\Pi$, which approximate the solution q at t_k, i.e., $q_k \approx q(t_k)$.

Our energetic approach has the major advantage that the values q_k can be found incrementally via the *incremental minimization problem*

$$(\mathrm{IMP}^\Pi) \qquad \begin{array}{l} \text{For } q_0 \in \mathcal{S}(0) \subset \mathcal{Q} \text{ find } q_1,\dots,q_N \in \mathcal{Q} \text{ such that} \\ q_k \text{ minimizes } q \mapsto \mathcal{E}(t_k,q)+\mathcal{D}(q_{k-1},q). \end{array} \tag{35}$$

We briefly write $q_k \in \mathrm{Arg\,min}\{\,\mathcal{E}(t_k,q)+\mathcal{D}(q_{k-1},q) \mid q \in \mathcal{Q}\,\}$, where "Arg min" denotes the set of all minimizers. The following result shows that (IMP^Π) is intrinsically linked to (S)&(E). Without any smallness assumptions on the time steps, the solutions of (IMP^Π) satisfy properties, which are closely related to (S)&(E).

Proposition 3.2 (Estimates for the incremental problem). *Let* (D1) *and* (E2) *hold. All solutions of* (IMP^Π) *from* (35) *satisfy the following properties:*

(i) For $k=1,\dots,N_\Pi$ *we have that* q_k *is stable at time* t_k*, i.e.,* $q_k \in \mathcal{S}(t_k)$*.*
(ii) With $e_k = \mathcal{E}(t_k,q_k)$ *and* $\delta_k = \mathcal{D}(z_{k-1},z_k)$ *we have, for* $k=1,\dots,N_\Pi$*,*

$$\int_{t_{k-1}}^{t_k} \partial_s\mathcal{E}(s,q_k)\,\mathrm{d}s \le e_k - e_{k-1} + \delta_k \le \int_{t_{k-1}}^{t_k} \partial_s\mathcal{E}(s,q_{k-1})\,\mathrm{d}s. \tag{36}$$

(iii) If (D2) *and* (E1) *hold additionally, then solutions of* (IMP^Π) *exist.*

Proof. Ad (i). The stability follows from minimization properties of the solutions and the triangle inequality. For all $\widehat{q} \in \mathcal{Q}$ we have

$$\begin{aligned} &\mathcal{E}(t_k,\widehat{q}) + \mathcal{D}(z_k,\widehat{z}) = \mathcal{E}(t_k,\widehat{q}) + \mathcal{D}(z_{k-1},\widehat{z}) + \mathcal{D}(z_k,\widehat{z}) - \mathcal{D}(z_{k-1},\widehat{z}) \\ &\ge \mathcal{E}(t_k,q_k) + \mathcal{D}(z_{k-1},z_k) + \mathcal{D}(z_k,\widehat{z}) - \mathcal{D}(z_{k-1},\widehat{z}) \ge \mathcal{E}(t_k,q_k). \end{aligned}$$

Ad (ii). The first estimate is deduced from $q_{k-1} \in \mathcal{S}(t_{k-1})$ as follows:

$$\begin{aligned}&\mathcal{E}(t_k, q_k) + \mathcal{D}(z_{k-1}, z_k) - \mathcal{E}(t_{k-1}, q_{k-1})\\&= \mathcal{E}(t_{k-1}, q_k) + \int_{t_{k-1}}^{t_k} \partial_s \mathcal{E}(s, q_k)\,\mathrm{d}s + \mathcal{D}(z_{k-1}, z_k) - \mathcal{E}(t_{k-1}, q_{k-1})\\&\geq \int_{t_{k-1}}^{t_k} \partial_s \mathcal{E}(s, q_k)\,\mathrm{d}s.\end{aligned}$$

Since $q_k \in \operatorname{Arg\,min}_{\mathcal{Q}} \mathcal{E}(t_k, q) + \mathcal{D}(z_{k-1}, z)$, the second estimate follows via

$$\begin{aligned}&\mathcal{E}(t_k, q_k) - \mathcal{E}(t_{k-1}, q_{k-1}) + \mathcal{D}(z_{k-1}, z_k)\\&\leq \mathcal{E}(t_k, q_{k-1}) - \mathcal{E}(t_{k-1}, q_{k-1}) + \mathcal{D}(z_{k-1}, z_{k-1}) \;=\; \int_{t_{k-1}}^{t_k} \partial_s \mathcal{E}(s, q_{k-1})\,\mathrm{d}s.\end{aligned}$$

Ad (iii). The minimizers are constructed inductively. In the k-th step, q_{k-1} is known and any minimizer y has to satisfy $\mathcal{J}_k(y) \overset{\text{def}}{=} \mathcal{E}(t_k, q) + \mathcal{D}(z_{k-1}, z) \leq \mathcal{E}(t_k, q_{k-1}) = \mathcal{J}_k(q_{k-1})$, since $q = q_{k-1}$ is a candidate. Using $\mathcal{D} \geq 0$ it suffices to minimize the lower semicontinuous functional $\mathcal{J}_k$ on the compact sublevel $\mathcal{E}(t_k, \cdot) \leq \mathcal{E}(t_k, q_{k-1})$. Hence, Weierstraß' extremum principle provides the existence of a minimizer q_k. □

Now we use assumption (E2) to obtain a priori bounds on the energy and the dissipation for the solution of (IMP$^{\Pi}$). Combining (E2), (30) and the upper estimate in (ii) of Proposition 3.2 give

$$e_k + \delta_k \;\leq\; e_{k-1} + e_{k-1}\big(\mathrm{e}^{\lambda_{\mathcal{E}}(t_k - t_{k-1})} - 1\big) \;=\; e_{k-1}\,\mathrm{e}^{\lambda_{\mathcal{E}}(t_k - t_{k-1})}. \tag{37}$$

Using $\delta_k \geq 0$ induction over k leads to

$$e_k \leq e_0 \prod_{j=1}^{k} \mathrm{e}^{\lambda_{\mathcal{E}}(t_j - t_{j-1})} = e_0\,\mathrm{e}^{\lambda_{\mathcal{E}} t_k} \text{ for } k = 1, \ldots, N_{\Pi}. \tag{38}$$

Summing (37) from $k = 1$ to n we find, after cancellations and using (38),

$$\begin{aligned}e_n + \textstyle\sum_{j=1}^{n} \delta_j &\leq e_0 + \textstyle\sum_{j=1}^{n} e_{j-1}\,(\mathrm{e}^{\lambda_{\mathcal{E}}(t_j - t_{j-1})} - 1)\\&\leq e_0 + e_0 \textstyle\sum_{1}^{n} (\mathrm{e}^{\lambda_{\mathcal{E}} t_j} - \mathrm{e}^{\lambda_{\mathcal{E}} t_{j-1}}) \;=\; e_0\,\mathrm{e}^{\lambda_{\mathcal{E}} t_n}.\end{aligned}$$

For each incremental solution $(q_k)_{k=1,\ldots,N}$ of (IMP$^{\Pi}$) associated with a partition $\Pi = (t_0, t_1, ..., t_N) \in \mathrm{Part}([0, T])$, we define the piecewise constant interpolant $\underline{q}^{\Pi}$ with

$$\underline{q}^{\Pi}(t) \overset{\text{def}}{=} q_{k-1} \text{ for } t \in [t_{k-1}, t_k[\text{ and } k = 1, ..., N, \text{ and } \underline{q}^{\Pi}(T) \overset{\text{def}}{=} q_N, \tag{39}$$

which is continuous from the right.

Corollary 3.3. *Assume that* (D1) *and* (E2) *hold and let* $\Pi \in \mathrm{Part}([0, T])$. *Then, for any solution* $(q_k)_{k=0,\ldots,N_{\Pi}}$ *of (IMP*$^{\Pi}$*) the interpolant* $\underline{q}^{\Pi} : [0, T] \to \mathcal{Q}$ *satisfies the following relations.*

(i) $(S)_{discr}$ *For* $t \in \Pi$ *we have* $\underline{q}^\Pi(t) \in \mathcal{S}(t)$.
(ii) $(E)_{discr}$ *For* $s, t \in \Pi$ *with* $s < t$ *we have the energy estimate*

$$\mathcal{E}(t, \underline{q}^\Pi(t)) + \mathrm{Diss}_{\mathcal{D}}(\underline{z}^\Pi, [s,t]) \leq \mathcal{E}(s, \underline{q}^\Pi(s)) + \int_s^t \partial_\tau \mathcal{E}(\tau, \underline{q}^\Pi(\tau))\,\mathrm{d}\tau.$$

(iii) *For all* $t \in [0,T]$ *we have the a priori estimate*

$$\mathcal{E}(t, \underline{q}^\Pi(t)) + \mathrm{Diss}_{\mathcal{D}}(\underline{z}^\Pi, [0,t]) \leq \mathrm{e}^{\lambda_{\mathcal{E}} t}\, \mathcal{E}(0, q_0).$$

3.3 Statement of the Main Existence Result

The existence theory developed below is based on the incremental minimization problem (IMP^Π) and the a priori estimates derived above. Choosing a sequence of partitions whose fineness tends to 0, we obtain a sequence of approximations and need to extract a suitable subsequence that converges. This can be done for the z-component only, since the dissipation provides an a priori estimate of BV-type, which allows for an application of a suitable version of Helly's selection principle as stated in Theorem 3.13.

Since the y-component allows for no control of the temporal oscillations, it has to be handled differently. We could use a technique developed in [21, 26], which chooses additional subsequences for each $t \in [0,T]$ and thus is relying on the axiom of choice. Instead, we rely on a metrizability assumption of the underlying topology, which guarantees the existence of measurable solutions. This idea uses the fact that for stable states $q = (y, z)$ the energy $\mathcal{E}(t, \cdot)$ depends only on the component z. In particular, for the reduced functional $\mathcal{I}$ defined in (8) we have

$$\mathcal{I}(t, z) = \mathcal{E}(t, y, z) \quad \text{for all } (y, z) \in \mathcal{S}(t). \tag{40}$$

We also define the *reduced power* via

$$\mathcal{P}_{\mathrm{red}}(t, z) \overset{\mathrm{def}}{=} \sup\{\, \partial_t \mathcal{E}(t, y, z) \mid y \in \operatorname*{Arg\,min}_{y \in \mathcal{Y}} \mathcal{E}(t, \cdot, z)\,\}. \tag{41}$$

The important new observation is that along all energetic solutions the reduced power $\mathcal{P}_{\mathrm{red}}(t, z)$ is realized, see (44).

After having identified a subsequence and a limit function, it is necessary to show that this limit is an energetic solution. For this we need further conditions on the functionals $\mathcal{E}$ and $\mathcal{D}$ expressing a certain compatibility between these two functionals. To define these conditions, we introduce the notion of a *stable sequence* $(t_k, q_k)_{k \in \mathbb{N}}$ via

$$\sup_{k \in \mathbb{N}} \mathcal{E}(t_k, q_k) < \infty \quad \text{and} \quad \forall\, k \in \mathbb{N}: \; q_k \in \mathcal{S}(t_k). \tag{42}$$

The *compatibility conditions* between $\mathcal{E}$ and $\mathcal{D}$ rely on convergent stable sequences and read as follows:

$$\forall \text{ stable sequences } (t_k, q_k)_{k\in\mathbb{N}} \text{ with } (t_k, q_k) \overset{\mathcal{Q}_T}{\to} (t,q) \text{ it holds:}$$
$$\partial_t \mathcal{E}(t,q) = \lim_{k\to\infty} \partial_t \mathcal{E}(t,q_k), \tag{C1}$$
$$q \in \mathcal{S}(t). \tag{C2}$$

Condition (C2) is called the *closedness of the stability set.* Condition (C1) is called *conditioned continuity of the power of the external forces.* Note that in the limit in (C1) the time is fixed to t, although $q_k \in \mathcal{S}(t_k)$. These central conditions are discussed more in detail in Sect. 3.5.

We now state our existence result for energetic solutions. After preparing a few intermediate results, the proof is completed on pp. 122–124 below.

Theorem 3.4 (Existence of energetic solutions). *Assume that* $\mathcal{E}$ *and* $\mathcal{D}$ *satisfy the assumptions* (D1)–(D2), (E1)–(E2), *and the compatibility conditions* (C2) *and* (C1). *Further assume that*

$$\text{the topology of } \mathcal{Q} \text{ restricted to compact sets is separable metrizable.} \tag{43}$$

(i) Then, for each $q_0 \in \mathcal{S}(0)$ *there exists an energetic solution* $q = (y,z) : [0,T] \to \mathcal{Q}$ *to the RIS* $(\mathcal{Q}, \mathcal{E}, \mathcal{D})$ *with* $q(0) = q_0$. *Moreover,* $q : [0,T] \to \mathcal{Q}$ *is measurable and*

$$\partial_t \mathcal{E}(t, y(t), z(t)) = \mathcal{P}_{\text{red}}(t, z(t)) \text{ for a.a. } t \in [0,T]. \tag{44}$$

(ii) If $\Pi^l \in \text{Part}([0,T])$ *is a sequence of partitions with fineness* $\phi(\Pi^l) \to 0$ *for* $l \to \infty$, *and* $\underline{q}^{\Pi_l}$ *is the interpolant of any solution of the associated* (IMP$^{\Pi^l}$), *then there exist a subsequence* $q_k = q^{\Pi_{l_k}}$ *and a solution* $\widetilde{q} = (\widetilde{y}, \widetilde{z})$ *to the initial-value problem* $(\mathcal{Q}, \mathcal{E}, \mathcal{D}, q_0)$ *such that the following holds:*

$$\forall t \in [0,T] : \ z_k(t) \overset{\mathcal{Z}}{\to} \widetilde{z}(t); \tag{45a}$$
$$\forall t \in [0,T] : \ \text{Diss}_{\mathcal{D}}(z_k, [0,t]) \to \text{Diss}_{\mathcal{D}}(\widetilde{z}, [0,t]); \tag{45b}$$
$$\forall t \in [0,T] : \ \mathcal{E}(t, q_k(t)) \to \mathcal{E}(t, \widetilde{q}(t)); \tag{45c}$$
$$\partial_t \mathcal{E}(\cdot, q_k(\cdot)) \to \partial_t \mathcal{E}(\cdot, \widetilde{q}(\cdot)) \ \text{ in } \ \mathrm{L}^1((0,T)). \tag{45d}$$

(iii) If additionally the functional $\mathcal{E}$ *is such that for each stable point* $q = (y,z) \in \mathcal{S}(t)$ *the functional* $\mathcal{E}(t,\cdot,z)$ *has the unique minimizer* y, *then the convergence in* (45a) *can be improved to* $q_k(t) \overset{\mathcal{Q}}{\to} \widetilde{q}(t)$.

We also provide an easy applicable version of the existence result, where we strengthen the assumptions considerably, but still allow for a big variety of applications. By making $\mathcal{D}$ continuous on $\mathcal{Z}$, it is possible to decouple the assumptions on $\mathcal{E}$ and $\mathcal{D}$ completely, and the compatibility conditions (C2) and (C1) can be established easily.

Theorem 3.5 (Simplified existence result for energetic solutions). *Assume that* $(\mathcal{Q}, \mathcal{E}, \mathcal{D})$ *satisfy* (D1), (E1), (E2), (43) *as well as the following conditions:*

$$\mathcal{D} : \mathcal{Z}\times\mathcal{Z} \to [0, \infty[\text{ is continuous;} \tag{46}$$

$$\begin{aligned} &\exists\, C_E^* > 0\ \forall q \in \mathcal{Q} \text{ with } \mathcal{E}(0, q) < \infty : \\ &\mathcal{E}(\cdot, q) \in \mathrm{C}^1([0, T]),\ |\partial_t\mathcal{E}(t, q)-\partial_t\mathcal{E}(s, q)| \le C_E^*|t-s|\mathcal{E}(0, q). \end{aligned} \tag{47}$$

Then, all assumptions of the main existence result Theorem 3.4 are fulfilled and, hence, existence of energetic solutions to $(\mathcal{Q}, \mathcal{E}, \mathcal{D})$ *is guaranteed for all initial conditions* $q(0) = q_0 \in \mathcal{S}(0)$.

Proof. Clearly, (46) implies (D2). To establish the compatibility conditions we use Corollary 3.9, since our present assumptions (46) and (47) are exactly the conditions (54) and (56) imposed there. □

3.4 Jump Conditions for Energetic Solutions

Here we discuss some basic properties of solutions.

Let $q : [0, T] \to \mathcal{Q}$ be an energetic solution to $(\mathcal{Q}, \mathcal{E}, \mathcal{D})$. First, we exploit the energy balance to show that q satisfies simple a priori estimates for the energy and the dissipation. For this, we use that (E) holds for all intervals $[s, t]$. Omitting the nonnegative dissipation in (E) and employing (30) in the power term give $\mathcal{E}(t, q(t)) \le \mathcal{E}(s, q(s))\, \mathrm{e}^{\lambda_{\mathcal{E}}(t-s)}$ for all $0 \le s < t \le T$. Inserting this into the right-hand side of the energy balance yields

$$\mathcal{E}(t, q(t)) + \mathrm{Diss}_{\mathcal{D}}(q, [s, t]) \le \mathcal{E}(s, q(s))\, \mathrm{e}^{\lambda_{\mathcal{E}}(t-s)} \quad \text{for } 0 \le s < t \le T. \tag{48}$$

Second, we derive a simple lemma, which implies continuity a.e. in $[0, T]$ of the z component. For $f : [a, b] \to Y$ we use the following definition of *one-sided limits*:

$$\begin{aligned} f(t^+) &\overset{\text{def}}{=} \lim_{h\to 0^+} f(t+h), \quad f(b^+) \overset{\text{def}}{=} f(b), \\ f(t^-) &\overset{\text{def}}{=} \lim_{h\to 0^+} f(t-h), \quad f(a^-) \overset{\text{def}}{=} f(a). \end{aligned} \tag{49}$$

To analyze the behavior at jump points, first note that $\mathrm{Diss}_{\mathcal{D}}(z, [0, T]) < \infty$ implies that $\delta : t \mapsto \mathrm{Diss}_{\mathcal{D}}(z, [0, t])$ has at most a countable number of jump points. At a continuity point of δ we have

$$\mathcal{D}(z(t-\varepsilon), z(t)) + \mathcal{D}(z(t), z(t+\varepsilon)) \leq \mathrm{Diss}_{\mathcal{D}}(z, [t-\varepsilon, t+\varepsilon]) \to 0 \text{ for } \varepsilon \to 0.$$

Because $\{ z(t-\varepsilon) \mid 0 < \varepsilon < \varepsilon_0 \}$ lies in a compact sublevel, we may assume $z(t-\varepsilon_j) \overset{\mathcal{Z}}{\to} z_*$. By the lower semicontinuity (D2) we find $\mathcal{D}(z_*, z(t)) \leq \liminf_{j\to\infty} \mathcal{D}(z(t-\varepsilon_j), z(t)) = 0$. Using (D1) we conclude $z_* = z(t)$. By uniqueness of the limit we find $z(t^-) = \lim z(t-\varepsilon) = z(t)$. Similarly, we have $z(t^+) = z(t)$. Hence, we conclude that $z : [0,T] \to \mathcal{Z}$ is continuous at every continuity point of δ. Moreover, at every jump point of δ the left-hand and right-hand limits $z(t^-)$ and $z(t^+)$ exist. In general, the three values $z(t^-)$, $z(t)$, and $z(t^+)$ may be different.

Since the jump conditions are most easily formulated in terms of the reduced system $(\mathcal{Z}, \mathcal{I}, \mathcal{D})$ we define the *reduced stability sets*

$$\widehat{\mathcal{S}}(t) \overset{\text{def}}{=} \{ z \in \mathcal{Z} \mid \mathcal{I}(t,z) < \infty,\ \forall \widetilde{z} \in \mathcal{Z} :\ \mathcal{I}(t,z) \leq \mathcal{I}(t,\widetilde{z}) + \mathcal{D}(z,\widetilde{z}) \}. \tag{50}$$

Lemma 3.6 (Jump conditions for energetic solutions). *Assume that* (D1), (D2), (E1), (E2) *and* (C2) *hold. Let* $q = (y,z) : [0,T] \to \mathcal{Q}$ *be an energetic solution to* $(\mathcal{Q}, \mathcal{E}, \mathcal{D})$. *Then, for all* $t \in [0,T]$ *we have the relations*

$$\mathcal{I}(t, z(t)) + \mathcal{D}(z(t^-), z(t)) = \mathcal{I}(t, z(t^-)), \tag{51a}$$

$$\mathcal{I}(t, z(t^+)) + \mathcal{D}(z(t), z(t^+)) = \mathcal{I}(t, z(t)), \tag{51b}$$

$$\mathcal{I}(t, z(t^-)) = \lim_{\tau \to t^-} \mathcal{I}(\tau, z(\tau)), \tag{51c}$$

$$\mathcal{I}(t, z(t^+)) = \lim_{\tau \to t^+} \mathcal{I}(\tau, z(\tau)), \tag{51d}$$

$$\mathcal{D}(z(t^-), z(t)) + \mathcal{D}(z(t), z(t^+)) = \mathcal{D}(z(t^-), z(t^+)). \tag{51e}$$

Moreover, we have $z(t^-), z(t), z(t^+) \in \widehat{\mathcal{S}}(t) \subset \mathcal{Z}$ *for all* $t \in [0,T]$.

Proof. We consider only the first statement for $t > 0$, since the second works analogously for $t < T$. We subtract the energy balance for $\tau < t$ from that of t and use (29) to obtain $\mathcal{I}(t, z(t)) + \mathrm{Diss}_{\mathcal{D}}(z, [\tau, t]) = \mathcal{E}(\tau, z(\tau)) + \int_\tau^t \mathcal{P}_{\mathrm{red}}(s, z(s))\,\mathrm{d}s$. Passing to the limit $\tau \to t^-$, the last term disappears, and we find $\mathcal{I}(t, z(t)) + \mathcal{D}(z(t^-), z(t)) = \lim_{\tau\to t^-} \mathcal{I}(\tau, z(\tau))$.

We claim $\mathcal{I}(t, z(t^-)) = \lim_{\tau\to t^-} \mathcal{I}(\tau, z(\tau))$. By the lower semicontinuity (E1) we know $\mathcal{I}(t, z(t^-)) \leq \liminf_{\tau\to t^-} \mathcal{I}(\tau, z(\tau))$ as $z(\tau) \overset{\mathcal{Z}}{\to} z(t^-)$. The opposite inequality follows from stability of $z(\tau)$ with respect to $z(t^-)$, namely $\mathcal{I}(\tau, z(\tau)) \leq \mathcal{I}(\tau, z(t^-)) + \mathcal{D}(z(\tau), z(t^-))$. Using $\mathcal{D}(z(\tau), z(t^-)) \leq \lim_{s\to t^-} \mathrm{Diss}_{\mathcal{D}}(z, [\tau, s])$ we obtain $\mathcal{D}(z(\tau), z(t^-)) \to 0$ for $\tau \to t^-$. This implies $\limsup_{\tau\to t^-} \mathcal{I}(\tau, z(\tau)) \leq \mathcal{I}(t, z(t^-)) + 0$, and (51a) and (51c) are established. Assertions (51b) and (51d) are obtained analogously.

To establish the last statement fix $t \in]0,T]$ and consider $q_n = q(t-\frac{1}{n}) \in \mathcal{S}(t-\frac{1}{n})$. Using (E1) and (C2) there exists a convergent subsequence such that $q_{n_m} \overset{\mathcal{Q}}{\to} (\widetilde{y}, z(t^-)) \in \mathcal{S}(t)$, i.e., $z(t^-) \in \widehat{\mathcal{S}}(t)$. Analogously, we show $z(t^+) \in \widehat{\mathcal{S}}(t)$.

To establish (51e) it suffices to show "$\leq$", since the triangle inequality (D1) implies "$\geq$". For this we use $z(t^-) \in \widehat{\mathcal{S}}(t)$ and test with $z(t^+)$ to obtain $\mathcal{I}(t, z(t^-)) \leq \mathcal{I}(t, z(t^+)) + \mathcal{D}(z(t^-), z(t^+))$. Inserting (51a) and (51b) the desired estimate follows. □

3.5 On Compatibility Conditions (C1) and (C2)

The central condition that makes the whole theory working is the conditioned closedness of the stability set (C2). For this, the interplay of the chosen topology and the properties of $\mathcal{E}$ and $\mathcal{D}$ are essential. The main philosophy of this condition is that stable sequences behave better than usual sequences.

In many applications the power continuity (C1) is really a condition on $\mathcal{E}$ alone, namely if $\partial_t\mathcal{E} : \{ (t,q) \mid \mathcal{E}(t,q) \leq E \} \to \mathbb{R}$ is continuous for all $E > 0$. Such cases typically occur if the space $\mathcal{Q}$ is a reflexive Banach space equipped with the weak topology and if the loading of the problem is lower order or even linear. However, there are also important applications where the full generality of (C1) is needed, in particular, if $\mathcal{D}$ is not continuous.

A fairly general way of establishing the crucial closedness condition (C2) is given in terms of finding a *joint recovery sequence* $(\widetilde{q}_l)_{l\in\mathbb{N}}$:

$$\begin{aligned}&\forall \text{ stab.seq. } (t_l, q_l) \overset{\mathcal{Q}_T}{\to} (t,q)\ \forall \widetilde{q} \in \mathcal{Q}\ \exists \widetilde{q}_l \overset{\mathcal{Q}}{\to} \widetilde{q} :\\ &\limsup_{l\to\infty} \big(\mathcal{E}(t_l,\widetilde{q}_l)+\mathcal{D}(q_l,\widetilde{q}_l)-\mathcal{E}(t_l,q_l)\big) \leq \mathcal{E}(t,\widetilde{q})+\mathcal{D}(q,\widetilde{q})-\mathcal{E}(t,q).\end{aligned} \tag{52}$$

We also provide two stronger conditions, namely

$$\begin{aligned}&\forall \text{ stab.seq. } (t_l, q_l) \overset{\mathcal{Q}_T}{\to} (t,q)\ \forall \widetilde{q} \in \mathcal{Q}\ \exists \widetilde{q}_l \overset{\mathcal{Q}}{\to} \widetilde{q} :\\ &\limsup_{l\to\infty} \big(\mathcal{E}(t_l,\widetilde{q}_l)+\mathcal{D}(q_l,\widetilde{q}_l)\big) \leq \mathcal{E}(t,\widetilde{q})+\mathcal{D}(q,\widetilde{q});\end{aligned} \tag{53}$$

$$\left.\begin{aligned}&q_k \overset{\mathcal{Q}}{\to} q,\ \widetilde{q}_k \overset{\mathcal{Q}}{\to} \widetilde{q} \text{ and}\\ &\sup_{k\in\mathbb{N}} \big(\mathcal{E}(t,q_k)+\mathcal{E}(t,\widetilde{q}_k)\big) < \infty\end{aligned}\right\} \implies \mathcal{D}(q_k,\widetilde{q}_k) \to \mathcal{D}(q,\widetilde{q}). \tag{54}$$

Since the following results are straightforward, we refer to [48] for a full proof.

Proposition 3.7 (Sufficient conditions for (C2)**).** *Assume* (E1).

(i) If for each stable sequence $(t_l, q_l) \overset{\mathcal{Q}_T}{\to} (t,q)$ *there exists a sequence* $(\widetilde{q}_l)_{l\in\mathbb{N}}$ *such that* $\limsup_{l\to\infty} \mathcal{E}(t_l,\widetilde{q}_l)+\mathcal{D}(q_l,\widetilde{q}_l) \leq \mathcal{E}(t,q)$, *then the energy converges along stable sequences, i.e.*

$$\forall \text{ stab.seq. } (t_l, q_l) \overset{\mathcal{Q}_T}{\to} (t,q): \quad \mathcal{E}(t_l, q_l) \to \mathcal{E}(t,q). \tag{55}$$

In particular, (53) *implies* (55).

(ii) *We have the implications* (54) $\Longrightarrow$ (53) $\Longrightarrow$ (52) $\Longrightarrow$ (C2).

(iii) *If* (E2) *holds additionally, then the conditions* (52) *and* (53) *remain the same if* $\mathcal{E}(t_l,\cdot)$ *is replaced by* $\mathcal{E}(t,\cdot)$.

Concerning the conditioned continuity of the power, we mention that often the case is considered that $\mathcal{Y}$ is a weakly closed subset of a reflexive Banach space $\boldsymbol{Y}$ equipped with the weak topology. Moreover the energy takes the form $\mathcal{E}(t,q) = \Phi(q) - \langle \ell(t), y\rangle$, where $\ell \in \mathrm{W}^{1,1}([0,T],\boldsymbol{Y}^*)$; then it is easy to establish (C1) even without using the stability.

The following abstract result establishes the continuity of the power (C1) under more general conditions. It purely relies on semicontinuity properties, is independent of a linear structure, and goes back to an idea in [21] for showing that the stresses in nonlinear elasticity converge weakly, if the functions y^n as well as the energy converge. The following result is an abstract and much simpler version of this fact, see [42, Prop. 5.6] for the proof.

Proposition 3.8 (Sufficient conditions for (C1)**).** *If* $\mathcal{E}$ *satisfies* (E1)–(E2), *then* (56) *implies* (57)*:*

$$\begin{aligned} &\forall E>0\ \forall \varepsilon>0\ \exists \delta>0\ \forall t_1,t_2\in[0,T]\ \forall q\in\mathcal{Q}: \\ &\mathcal{E}(0,q)\le E,\ |t_2-t_1|\le\delta \implies \big|\partial_t\mathcal{E}(t_1,q)-\partial_t\mathcal{E}(t_2,q)\big|\le\varepsilon, \end{aligned} \tag{56}$$

$$\left.\begin{aligned} &(t_m,q_m)\overset{\mathcal{Q}_T}{\to}(t,q) \text{ and} \\ &\mathcal{E}(t_m,q_m)\to\mathcal{E}(t,q)<\infty \end{aligned}\right\} \implies \partial_t\mathcal{E}(t,q_m)\to\partial_t\mathcal{E}(t,q). \tag{57}$$

Together with Proposition 3.7 we obtain the following result.

Corollary 3.9. *Assume that* (D1), (D2), (E1), (E2), (53), *and* (56) *hold. Then both compatibility conditions* (C1) *and* (C2) *are satisfied.*

3.6 Proof of Theorem 3.4

The proof follows the main steps in [26], however it includes a new argument, namely a precise characterization of the power, see (44) and Proposition 3.11. This approach allows us to simplify the assumptions considerably in the case that $\mathcal{Q}$ satisfies the metrizability condition (43).

The proof of the main existence result makes extensive use of the reduced energy functional $\mathcal{I}: \mathcal{Z}_T \to \mathbb{R}_\infty$, since the stability condition (S) as well as the energy balance (E) can be formulated easily for the reduced RIS $(\mathcal{Z},\mathcal{I},\mathcal{D})$. For this recall the reduced stability set $\widehat{\mathcal{S}}(t)$ defined in (50). Clearly, we have $q=(y,z)\in\mathcal{S}(t)$ if and only if $z\in\widehat{\mathcal{S}}(t)$ and $y\in\operatorname{Argmin}\mathcal{E}(t,\cdot,z)$. The only difficulty in reducing from $\mathcal{Q}$ to $\mathcal{Z}$ is that in general $\mathcal{I}(\cdot,z)$ is no longer differentiable. Thus, we define energetic solutions $z : [0,T] \to \mathcal{Z}$ of the reduced RIS $(\mathcal{Z}, \mathcal{I}, \mathcal{D})$ via the reduced power $\mathcal{P}_{\mathrm{red}}$ defined in (41) as follows:

$$\begin{aligned}
&(\mathrm{S})_{\mathrm{red}} \quad z(t) \in \widehat{\mathcal{S}}(t),\\
&(\mathrm{E})_{\mathrm{red}} \quad \mathcal{I}(t, z(t)) + \mathrm{Diss}_{\mathcal{D}}(z, [0,t]) = \mathcal{I}(0, z(0)) + \textstyle\int_0^t \mathcal{P}_{\mathrm{red}}(s, z(s))\,\mathrm{d}s.
\end{aligned} \tag{58}$$

Because of (44) each energetic solution $q = (y,z) : [0,T] \to \mathcal{Q}$ gives rise to a reduced energetic solution $z : [0,T] \to \mathcal{Z}$ for $(\mathcal{Z}, \mathcal{I}, \mathcal{D})$. The next lemma shows that the opposite is also true. Each solution z for $(\mathcal{Z}, \mathcal{I}, \mathcal{D})$ can be made into a full solution $q = (y,z)$ for the RIS $(\mathcal{Q}, \mathcal{E}, \mathcal{D})$ by selecting a suitable measurable y-component.

Lemma 3.10 (Selection of the y-component). *Assume that* (D1), (D2), (E1), (E2), (C2), (C1), *and* (43) *are satisfied. Let $z : [0,T] \to \mathcal{Z}$ be measurable with* $\mathrm{Diss}_{\mathcal{D}}(z, [0,T]) + \sup_{t\in[0,T]} \mathcal{I}(t, z(t)) < \infty$ *and $z(t) \in \widehat{\mathcal{S}}(t)$ for each $t \in [0,T]$. Then there exists a measurable function $y : [0,T] \to \mathcal{Y}$ such that for all $t \in [0,T]$ we have*

$$(y(t), z(t)) \in \mathcal{S}(t) \quad \textit{and} \quad \mathcal{P}_{\mathrm{red}}(t, z(t)) = \partial_t \mathcal{E}(t, y(t), z(t)).$$

Proof. Our proof is based on a variant of Filippov's selection theorem, which we use here with the complete measure space $([0,T], \mathfrak{S}, \mu)$ with $\mathfrak{S}$ = the σ-algebra of the Lebesgue measurable subsets and $\mu = \mathfrak{L}^1(\cdot)$ the one-dimensional Lebesgue measure. For given $(t,z) \in \mathcal{Z}_T$ we define $M(t,z) \stackrel{\mathrm{def}}{=} \mathrm{Arg\,min}\{\, \mathcal{E}(t, \widetilde{y}, z) \mid \widetilde{y} \in \mathcal{Y}\,\}$. For the given measurable $z : [0,T] \to \mathcal{Z}$ we compose a set-valued mapping $G : [0,T] \rightrightarrows \mathcal{Y}$ via

$$G(t) \stackrel{\mathrm{def}}{=} M(t, z(t^-)) \cup M(t, z(t)) \cup M(t, z(t^+)) \subset Y \subset \mathcal{Y},$$

where Y is a compact subset of $\mathcal{Y}$, which exists due to (E1). Using assumption (43), we know that the topology on Y is complete, separable and metrizable.

Using (E1) each $M(t,z)$ is nonempty, and hence each $G(t)$ is nonempty. Employing (C2) we show that the graph $\mathrm{Gr}(G) = \{\, (t,y) \mid y \in G(t)\,\}$ is closed in $[0,T]\times Y$ and hence is measurable. Indeed, consider $(t_k, y_k) \in \mathrm{Gr}(G)$ with $t_k \to t_*$ and $y_k \to y_*$, then there exists $z_k = z(t_k^\nu)$ with $t_k^\nu \in \{t_k, t_k^-, t_k^+\}$ such that $y_k \in M(t_k, z_k)$. Using the last statement in Lemma 3.6 (which is valid for every measurable $z : [0,T] \to \mathcal{Z}$ with $z(t) \in \widehat{\mathcal{S}}(t)$ and not only for energetic solutions), we conclude $(y_k, z_k) \in \mathcal{S}(t_k)$. After taking a subsequence (not relabeled) we may assume $z_k \stackrel{\mathcal{Z}}{\to} z_*$ and (C2) provides $(y_*, z_*) \in \mathcal{S}(t_*)$, which implies $y_* \in M(t_*, z_*)$. Moreover, (t_*, z_*) lies in the closure of $\mathrm{Gr}(z) \subset \mathcal{Z}_T$, which means $z_* = z(t_*^\nu)$ with $t_k^\nu \in \{t_k, t_k^-, t_k^+\}$. Thus, we have established $y_* \in G(t_*)$ as desired.

The set-valued mapping $F : [0,T] \rightrightarrows Y$ is defined via $F(t) = M(t, z(t))$. Clearly, $F(t)$ is nonempty and closed for each t. Since z is continuous outside an at most countable set $J(z) \subset [0,T]$, we have $F(t) = G(t)$ for $t \in [0,T]\backslash J(z)$. Thus, F is a measurable set-valued mapping as well.

We now define the function $g : \mathrm{Gr}(F) \to \mathbb{R}$ via

$$g(t,y) \stackrel{\text{def}}{=} \partial_t \mathcal{E}(t,y,z(t)) - \mathcal{P}_{\text{red}}(t,z(t)) \text{ for } t \in [0,T] \text{ and } y \in F(t).$$

For fixed $t \in [0,T]$ the function $g(t,\cdot) : F(t) \to \mathbb{R}$ is continuous because of (C1). Since $\mathcal{E}$ has compact sublevels by (E1), g is Borel-measurable. Moreover, $z : [0,T] \to \mathcal{Z}$ is Borel-measurable, since it is continuous except for a countable number of points. Thus, for each $h \geq 0$ the functions $\gamma_h : [0,T-h]\times Y \to \mathbb{R}; (t,y) \mapsto \mathcal{E}(t+h,y,z(t))$ are $(\mathfrak{L} \otimes \mathfrak{B}(Y), \mathfrak{B}(\mathbb{R}))$-measurable. Since g is the difference of the pointwise limit of the measurable difference quotients $\frac{1}{h}(\gamma_h - \gamma_0)$ and $t \mapsto \mathcal{P}_{\text{red}}(t,z(t))$ (which is measurable), it is measurable as well. Hence, the restriction of g to $\mathrm{Gr}(F)$ is measurable.

Next we show that for each t there exists $y \in F(t)$ with $g(t,y) = 0$. Indeed, by (E1) the set $M(t,z(t)) = \mathrm{Arg\,min}\, \mathcal{E}(t,\cdot,z(t))$ is a nonempty compact set. Choose a sequence $(y_m)_m$ approaching the supremum in the definition (41) of the reduced power, viz., $\mathcal{P}_{\text{red}}(t,z(t)) = \sup\{\, \partial_t \mathcal{E}(t,\widetilde{y},z(t)) \mid \widetilde{y} \in M(t,z(t))\,\}$. Taking a subsequence, we may assume $y_{m_n} \stackrel{\mathcal{Y}}{\to} y_* \in M(t,z(t))$. Since $(y_{m_n}, z(t)) \in \mathcal{S}(t)$ we have a stable sequence, and (C1) gives $\mathcal{P}_{\text{red}}(t,z(t)) = \lim_{n\to\infty} \partial_t \mathcal{E}(t,y_{m_n},z(t)) = \partial_t \mathcal{E}(t,y_*,z(t))$, as desired.

We are now able to apply Filippov's theorem (cf. [5, Thm. 8.2.9+10] and obtain the desired measurable selection $y : [0,T] \to Y$ with $y(t) \in F(t)$ and $g(t,y(t)) = 0$. □

We next present a lower energy estimate that is valid for all stable processes. The fact that stability implies such a lower energy estimate was first observed in [51]. Here we use a stronger version that replaces the work of the external forces on the right-hand side by the integral of the reduced power $\mathcal{P}_{\text{red}}$. Since the left-hand side in (59) does not depend on the y-component of the stable process (see (40)), it is clear that the lower bound on the right-hand side should also be expressible in terms of z alone. It is this seemingly simple observation that allowed us to simplify the assumptions of the main existence result.

Proposition 3.11 (Lower energy estimate). *Assume that* (D1), (D2), (E1), (E2), (C2), (C1), *and* (43) *hold. Let* $q = (y,z) : [0,T] \to \mathcal{Q}$ *be measurable with* $\sup_{t\in[0,T]} \mathcal{E}(t,q(t)) + \mathrm{Diss}_{\mathcal{D}}(z,[0,T]) < \infty$, *and* $q(t) \in \mathcal{S}(t)$ *for all* $t \in [0,T]$. *Then, for all* $0 \leq r < s \leq T$ *we have the lower energy inequality*

$$\begin{aligned}&\mathcal{E}(s,q(s)) + \mathrm{Diss}_{\mathcal{D}}(z,[r,s]) - \mathcal{E}(r,q(r))\\ &\geq \int_r^s \mathcal{P}_{\text{red}}(t,z(t))\,\mathrm{d}t \geq \int_r^s \partial_t \mathcal{E}(t,q(t))\,\mathrm{d}t.\end{aligned} \tag{59}$$

Proof. We use the fact that the left-hand side is independent of the y-component, viz. (40). By the stability of $q(r)$ and $q(s)$ it can be written as $\mathcal{I}(s, z(s)) + \mathrm{Diss}_{\mathcal{D}}(z, [r, s]) - \mathcal{I}(r, z(r))$. Thus, it suffices to show the first inequality in (59), since the second one follows directly from the definition (41) of the reduced power, namely $\partial_t \mathcal{E}(t, q(t)) \leq \mathcal{P}_{\mathrm{red}}(t, z(t))$ for all $t \in [0, T]$. By Lemma 3.10 we may choose y such that we have equality.

Hence we assume now that q satisfies this equality, i.e. $\partial_t \mathcal{E}(t, q(t)) = \mathcal{P}_{\mathrm{red}}(t, z(t))$. Take any partition $\Pi = (t_0, t_1, \ldots, t_N) \in \mathrm{Part}([r, s])$. For each $t_{j-1} \in \Pi$ we use $q(t_{j-1}) \in \mathcal{S}(t_{j-1})$ to obtain $\mathcal{E}(t_{j-1}, q(t_{j-1})) \leq \mathcal{E}(t_{j-1}, q(t_j)) + \mathcal{D}(q(t_{j-1}), q(t_j))$, which is the same as

$$\mathcal{E}(t_j, q(t_j)) + \mathcal{D}(z(t_{j-1}), z(t_j)) - \mathcal{E}(t_{j-1}, q(t_{j-1})) \geq \mathcal{E}(t_j, q(t_j)) - \mathcal{E}(t_{j-1}, q(t_j)).$$

Summing over $j \in \{1, \ldots, N_\Pi\}$ we find

$$\begin{aligned}
&\mathcal{E}(s, q(s)) + \mathrm{Diss}_{\mathcal{D}}(z, [r, s]) - \mathcal{E}(r, q(r)) \\
&\geq \mathcal{E}(s, q(s)) + \textstyle\sum_{j=1}^{N_\Pi} \mathcal{D}(z(t_{j-1}), z(t_j)) - \mathcal{E}(r, q(r)) \\
&\geq \textstyle\sum_{j=1}^{N_\Pi} \big(\mathcal{E}(t_j, q(t_j)) - \mathcal{E}(t_{j-1}, q(t_j))\big) = \sum_{j=1}^{N_\Pi} \int_{t_{j-1}}^{t_j} \partial_t \mathcal{E}(\tau, q(t_j))\, \mathrm{d}\tau \\
&= \textstyle\int_r^s \partial_t \mathcal{E}(\tau, \overline{q}^\Pi(\tau))\, \mathrm{d}\tau,
\end{aligned}$$

where $\overline{q}^\Pi$ is the left-continuous interpolant with $\overline{q}^\Pi(t) = q(t_j)$ for $t \in]t_{j-1}, t_j]$. Since the partition Π was arbitrary, we can apply Lemma 3.12 to obtain the desired result. □

Lemma 3.12. *Let the conditions* (E2), (C1) *and the metrizability condition* (43) *hold. Moreover, assume that* $q : [0, T] \to \mathcal{Q}$ *is measurable, and there is a* $C > 0$ *such that for all* $t \in [0, T]$ *we have* $\mathcal{E}(t, q(t)) \leq C$ *and* $q(t) \in \mathcal{S}(t)$. *Then, for all* $r, s \in [0, T]$ *with* $r < s$ *we have*

$$\sup_{\Pi \in \mathrm{Part}([r,s])} \int_r^s \partial_t \mathcal{E}(\tau, \overline{q}^\Pi(\tau))\, \mathrm{d}\tau \geq \int_s^t \partial_\tau \mathcal{E}(\tau, q(\tau))\, \mathrm{d}t.$$

Proof. Since each function $t \mapsto \mathcal{E}(t+h, q(t))$ is measurable, the power $\tau \mapsto \partial_\tau \mathcal{E}(\tau, q(\tau))$ is measurable as well, because it is a pointwise limit of measurable difference quotients. Moreover, there is a constant $c_0 > 0$ such that $|\partial_t \mathcal{E}(t, \overline{q}^\Pi(t))| \leq c_0 \lambda_{\mathcal{E}}$.

Using (43) we may apply Lusin's theorem to q, which takes values in a compact set since the energy is bounded. For $\varepsilon > 0$ we find a compact set $K \subset [r, s]$ with

$$c_0 \mathfrak{L}^1([r, s] \setminus K) \lambda_{\mathcal{E}} < \varepsilon, \quad \text{and} \quad q|_K : K \to \mathcal{Q} \text{ is continuous.} \qquad (60)$$

This implies $\int_r^s \partial_t \mathcal{E}(t, \overline{q}^\Pi(t))\, \mathrm{d}t \geq \int_K \partial_t \mathcal{E}(t, \overline{q}^\Pi(t))\, \mathrm{d}t - \varepsilon$ for all partitions Π.

We now construct a sequence of partitions $(\Pi_n)_n$ that allows us to prove the assertion. Let $t_0^n = r$ and define the other points inductively, namely as long as $t_j^n < s$ we set

$$t_{j+1}^n = \begin{cases} \max\{t \in K \mid t_j^n < t \le t_j^n + \frac{1}{n}\} & \text{if } K \cap \,]t_j^n, t_j^n+\frac{1}{n}] \neq \emptyset, \\ \min\{t_j^n+\frac{1}{n}, s\} & \text{else.} \end{cases}$$

On the one hand, there cannot be two adjacent intervals that are small: if $t_{j+1}^n < t_j^n + \frac{1}{n}$, then $K \cap \,]t_{j+1}^n, t_j^n+\frac{1}{n}]$ is empty. Now, if $t_{j+1}^n < s$, then t_{j+2}^n exceeds $\min\{t_j^n+\frac{1}{n}, s\}$. Hence, Π_n has at most $2(s-r)n+1$ intervals, and by construction the fineness satisfies $\phi(\Pi_n) \le \frac{1}{n}$. On the other hand, the choice of the nodes in Π_n is such that for $t \in K$ we always have $\overline{\tau}^\Pi(t) \in K$ as well. Indeed, $t_{j+1}^n \in \Pi \backslash K$ occurs only, if $]t_j^n, t_{j+1}^n]$ has empty intersection with K. Thus, we have shown $\overline{\tau}^{\Pi_n}(t) \in K$ and $\overline{\tau}^{\Pi_n}(t) \to t^+$ for $n \to \infty$ for all $t \in K$.

Recall $\overline{q}^{\Pi_n}(t) = q(\overline{\tau}^{\Pi_n}(t))$ and use the stability of q to conclude that $(\overline{\tau}^{\Pi_n}(t), \overline{q}^{\Pi_n}(t))_{n\in\mathbb{N}}$ is a stable sequence converging to $(t, q(t))$ because of (60). Exploiting (C1) we find $\partial_t \mathcal{E}(t, \overline{q}^{\Pi_n}(t)) \to \partial_t \mathcal{E}(t, q(t))$ and $\lim_{n\to\infty} \int_K \partial_t \mathcal{E}(t, \overline{q}^{\Pi_n}(t))\, dt = \int_K \partial_t \mathcal{E}(t, q(t))\, dt$ by Lebesgue's dominated convergence theorem. In summary we have

$$\begin{aligned} &\sup_\Pi \textstyle\int_r^s \partial_t \mathcal{E}(t, \overline{q}^\Pi(t))\,dt \ge \limsup_{n\to\infty} \int_r^s \partial_t \mathcal{E}(t, \overline{q}^{\Pi_n}(t))\,dt \\ &\ge -\varepsilon + \limsup_{n\to\infty} \textstyle\int_K \partial_t \mathcal{E}(t, \overline{q}^{\Pi_n}(t))\,dt = -\varepsilon + \int_K \partial_t \mathcal{E}(t, q(t))\,dt \\ &\ge -2\varepsilon + \textstyle\int_r^s \partial_t \mathcal{E}(t, q(t))\,dt. \end{aligned}$$

Because $\varepsilon > 0$ was arbitrary, this is the desired result. □

The existence theory developed below builds on the (IMP$^\Pi$) and a priori estimates. The general strategy for constructing solutions to (S)&(E) is to choose a sequence of partitions Π^m with $\phi(\Pi^m) \to 0$, to extract a convergent subsequence of $(z^l)_{l\in\mathbb{N}}$ of $(z^{\Pi^m})_{m\in\mathbb{N}}$, and then to show that the limit $z : [0,T] \to \mathcal{Z}$ solves (S)&(E). The existence of a convergent subsequence is guaranteed by the following version of Helly's selections principle, see [38,58] for a full proof.

Theorem 3.13 (Generalized version of Helly's selection principle). *Let $\mathcal{D} : \mathcal{Z}\times\mathcal{Z} \to [0,\infty]$ satisfy* (D1) *and* (D2) *and let $\mathcal{K}$ be a (sequentially) compact subset of $\mathcal{Z}$. Then, for every sequence $(z^l)_{l\in\mathbb{N}}$ with $z^l : [0,T] \to \mathcal{K}$ and $\sup_{l\in\mathbb{N}} \mathrm{Diss}_{\mathcal{D}}(z^l, [0,T]) \le \infty$, there exist a subsequence $(z^{l_n})_{n\in\mathbb{N}}$ and functions $z_\infty : [0,T] \to \mathcal{K}$ and $\delta_\infty : [0,T] \to [0,C]$ such that the following holds:*

(i) $\delta_{l_n}(t) \overset{\text{def}}{=} \mathrm{Diss}_{\mathcal{D}}(z^{l_n}, [0,t]) \to \delta_\infty(t)$ *for all* $t \in [0,T]$*;*
(ii) $z_{l_n}(t) \overset{\mathcal{Z}}{\to} z_\infty(t)$ *for all* $t \in [0,T]$*;*
(iii) $\mathrm{Diss}_{\mathcal{D}}(z_\infty, [t_0,t_1]) \le \delta_\infty(t_1) - \delta_\infty(t_0)$ *for all* $0 \le t_0 < t_1 \le T$.

We are now ready to prove the main existence result stated in Theorem 3.4.

Proof (of Theorem 3.4). We divide the proof into 6 steps.

Step 1: A priori estimates. We choose an arbitrary sequence of partitions Π^m whose fineness $f_m = \phi(\Pi^m)$ tends to 0. The time-incremental minimization problems (IMP$^\Pi$) are solvable and the piecewise constant interpolants $\underline{q}^m = (\underline{y}^m, \underline{z}^m) : [0,T] \to \mathcal{Q}$ defined in (39) satisfy the a priori estimates

$$\mathrm{Diss}_{\mathcal{D}}(\underline{z}^m, [0,T]) \leq C_{\mathcal{D}} \quad \text{and} \quad \forall t \in [0,T]: \ \mathcal{E}(t, \underline{q}^m(t)) \leq C_{\mathcal{E}},$$

where $C_{\mathcal{D}}$ and $C_{\mathcal{E}}$ are given explicitly in Corollary 3.3.

Step 2: Selection of subsequences. Our version of Helly's selection principle in Theorem 3.13 allows us to select a subsequence of $(\underline{z}^m)_{m\in\mathbb{N}}$ that converges pointwise and that makes the dissipation converge as well. Moreover, the functions $P^m : t \mapsto \partial_t \mathcal{E}(t, \underline{q}^m(t))$ form an equibounded sequence in $\mathrm{L}^1((0,T))$. Thus, by choosing a further subsequence $(\underline{q}^{m_k})_{k\in\mathbb{N}}$ we may assume the following convergence properties for $k \to \infty$, where we write q_k as shorthand for $\underline{q}^{m_k}$ and p_k for P^{m_k}:

$$p_k \rightharpoonup p_{\mathrm{weak}} \text{ in } \mathrm{L}^1((0,T)),$$

$$\forall t \in [0,T]: \ \delta_k(t) \overset{\mathrm{def}}{=} \mathrm{Diss}_{\mathcal{D}}(z_k, [0,t]) \to \delta(t) \text{ and } z_k(t) \overset{\mathcal{Z}}{\rightharpoonup} z(t).$$

Since the limit $z : [0,T] \to \mathcal{Z}$ satisfies $\mathrm{Diss}_{\mathcal{D}}(z, [0,T]) \leq \delta(T) \leq C_{\mathcal{D}} < \infty$, we know that z is measurable and that it satisfies the energetic bound $\mathcal{I}(t, z(t)) \leq C_{\mathcal{E}}$. Thus, Lemma 3.10 provides a measurable $y : [0,T] \to \mathcal{Y}$ such that

$$y(t) \in \operatorname{Arg\,min} \mathcal{E}(t, \cdot, z(t)) \quad \text{and} \quad \partial_t \mathcal{E}(y(t), z(t)) = \mathcal{P}_{\mathrm{red}}(t, z(t)), \tag{61}$$

and $y(0) = y_0$, where $q_0 = (y_0, z_0)$ is the given initial value with $q_0 \in \mathcal{S}(0)$. By construction $z(0) = z^m(0) = z_0$ such that $y(0) = y_0$ is an admissible choice satisfying the first relation in (61) but not necessarily the second.

Step 3: Stability of the limit function. We use the compatibility condition (C2). For fixed $t \in [0,T]$ we define τ_k to be the largest value in $\Pi^{m_k} \cap [0,t]$ such that $q_k(t) = q_k(\tau_k)$. Then, $q_k(t) \in \mathcal{S}(\tau_k)$, $\tau_k \leq t$, and $\tau_k \to t$. By choosing a further subsequence, if necessary, we obtain $q_{k_l}(t) \overset{\mathcal{Q}}{\rightharpoonup} \widetilde{q} = (\widetilde{y}, z(t))$. In particular, $(\tau_{k_l}, q_{k_l}(t))_{l\in\mathbb{N}}$ forms a convergent stable sequence. Now, (C2) yields $\widetilde{q} \in \mathcal{S}(t)$, whence $\widetilde{y} \in \operatorname{Arg\,min} \mathcal{E}(t, \cdot, \widetilde{z}(t))$. However, this also implies $q(t) = (y(t), z(t)) \in \mathcal{S}(t)$, since for all $\widehat{q} = (\widehat{y}, \widehat{z}) \in \mathcal{Q}$ we have $\mathcal{E}(t, q(t)) = \mathcal{I}(t, z(t)) = \mathcal{E}(t, \widetilde{q}) \leq \mathcal{E}(t, \widehat{q}) + \mathcal{D}(z(t), \widehat{z})$.

Step 4: Upper energy estimate. We define the functions

$$
\begin{aligned}
&e_k(t) \overset{\text{def}}{=} \mathcal{E}(t,q_k(t)), \quad \delta_k(t) \overset{\text{def}}{=} \mathrm{Diss}_{\mathcal{D}}(z_k,[0,t]), \quad e_\infty(t) \overset{\text{def}}{=} \liminf_{k\to\infty} e_k(t),\\
&E(t) \overset{\text{def}}{=} \mathcal{E}(t,q(t)), \quad \varDelta(t) \overset{\text{def}}{=} \mathrm{Diss}_{\mathcal{D}}(z,[0,t]), \quad \delta_\infty(t) \overset{\text{def}}{=} \lim_{k\to\infty} \delta_k(t),\\
&w_k(t) \overset{\text{def}}{=} \int_0^t \partial_s\mathcal{E}(s,q_k(s))\,\mathrm{d}s = \int_0^t p_k(s)\,\mathrm{d}s,\\
&W(t) \overset{\text{def}}{=} \int_0^t \partial_s\mathcal{E}(s,q(s))\,\mathrm{d}s = \int_0^t \mathcal{P}_{\mathrm{red}}(s,z(s))\,\mathrm{d}s,
\end{aligned}
$$

where by construction $e_k(0) = E(0) = e_\infty(0)$. Employing Corollary 3.3(ii) and the boundedness of $\partial_t\mathcal{E}$ by $C_1\lambda_{\mathcal{E}}$ (use (E2) and Step 1) give $e_k(t)+\delta_k(t) \leq E(0) + w_k(t) + C_1\lambda_{\mathcal{E}}\phi(\varPi^{m_k})$. Since $\mathcal{E}$ and $\mathrm{Diss}_{\mathcal{D}}$ are lower semicontinuous (cf. Theorem 3.13(iii)) and since by weak convergence we have $w_\infty(t) = \lim_{k\to\infty} w_k(t) = \int_0^t p_{\mathrm{weak}}(s)\,\mathrm{d}s$, the limit $k \to \infty$ leads to

$$E(t) + \varDelta(t) \leq e_\infty(t) + \delta_\infty(t) \leq E(0) + w_\infty(t) = E(0) + \int_0^t p_{\mathrm{weak}}(s)\,\mathrm{d}s. \quad (62)$$

The next step is now to relate p_{weak} and $\mathcal{P}_{\mathrm{red}}(\cdot,z(\cdot))$ using the compatibility condition for the power (C1). As in Step 3 we choose a subsequence of $(q_k(t))_k$ such that $\mathcal{S}(\tau_{k_l}) \ni q_{k_l}(t) \overset{\mathcal{Q}}{\to} \widetilde{q}$ and $p_{k_l}(t) \to p_{\mathrm{sup}}(t) \overset{\text{def}}{=} \limsup_{k\to\infty} p_k(t)$. Thus, (C1) is applicable and we find

$$p_{k_l}(t) = \partial_t\mathcal{E}(t,q_{k_l}(t)) \to \partial_t\mathcal{E}(t,\widetilde{q}) = p_{\mathrm{sup}}(t) \leq \mathcal{P}_{\mathrm{red}}(t,z(t)),$$

where the latter estimate follows from $\widetilde{q} = (\widetilde{y},z(t)) \in \mathcal{S}(t)$ (use (C2)) and the definition of $\mathcal{P}_{\mathrm{red}}$. Fatou's lemma gives $w_\infty(t) \leq \int_0^t p_{\mathrm{sup}}(s)\,\mathrm{d}s$, and we conclude the upper energy estimate

$$E(t) + \varDelta(t) \leq e_\infty(t) + \delta_\infty(t) \leq E(0) + w_\infty(t) \leq E(0) + W(t).$$

Step 5: Lower energy estimate. Because of our construction of the function $q : [0,T] \to \mathcal{Q}$, we are able to apply Proposition 3.11 and obtain the lower energy estimate $E(t) + \varDelta(t) = \mathcal{E}(t,q(t)) + \mathrm{Diss}_{\mathcal{D}}(z,[0,t]) \geq \mathcal{E}(0,z(0)) + \int_0^t \partial_s\mathcal{E}(s,q(s))\,\mathrm{d}s = E(0)+W(t)$. Thus, we have shown that the limit function $q : [0,T] \to \mathcal{Q}$ satisfies stability and energy balance for all times, whence it is an energetic solution.

Step 6: Improved convergence. Finally we show that the convergences (45) stated at the end of the theorem hold. The convergence (45a) is already shown. The lower and upper energy estimate imply $E(0)+W(t) \leq E(t) + \varDelta(t) \leq e_\infty(t)+\delta_\infty(t) \leq E(0)+ \int_0^t p_{\mathrm{weak}}\,\mathrm{d}s \leq E(0)+ \int_0^t p_{\mathrm{sup}}\,\mathrm{d}s \leq E(0)+W(t)$. Hence, all inequalities are in fact equalities. Using $E(t) \leq e_\infty(t)$, $\varDelta(t) \leq$

$\delta_\infty(t)$, and $p_{\rm weak} \leq p_{\rm sup} \leq \mathcal{P}_{\rm red}$ we conclude $\Delta(t) = \delta_\infty(t)$ and $E(t) = e_\infty(t)$, which proves the convergence statements (45b) and (45c). Moreover, we also find $p_{\rm weak}(t) = p_{\rm sup}(t) = \mathcal{P}_{\rm red}(t, z(t))$ a.e. in $[0, T]$. Since the weak limit and the pointwise limsup of the sequence p_k coincide, the strong convergence (45d) holds, cf. [26, Prop. A2]. Thus, Theorem 3.4 is proved. □

3.7 Γ-Convergence for Sequences of Rate-Independent Systems

We now consider sequences of rate-independent systems $((\mathcal{Q}, \mathcal{E}_k, \mathcal{D}_k))_{k\in\mathbb{N}}$ and study the question under what assumptions energetic solutions $q_k : [0, T] \to \mathcal{Q}$ converge to a limit, which is an energetic solution to a limit system $(\mathcal{Q}, \mathcal{E}_\infty, \mathcal{D}_\infty)$. As already revealed in [58], this theory is still very close to the existence theory for energetic solutions above, so the proof of Γ-convergence follows essentially the same six steps of the proof of Theorem 3.4.

The notion of *Γ-convergence*, introduced by De Giorgi [20], exclusively applies to functionals. It is sometimes also called *variational convergence* or *epigraph convergence*, cf. [4, 5, 10, 16]. Here we just give a brief outline that is sufficient for our purposes. We consider a metrizable topological space $\mathcal{Q}$, which means for our application that we restrict to a compact sublevel and use the metrizability assumption (43). For a sequence $(\mathcal{J}_k)_{k\in\mathbb{N}}$ of functionals $\mathcal{J}_k : \mathcal{Q} \to \mathbb{R}_\infty$ we are interested in the behavior for $k \to \infty$, which reflects the behavior of minimizers. In particular, the Γ-limit $\mathcal{J}$ is defined in such a way that if q_k minimizes $\mathcal{J}_k$ and $q_k \overset{\mathcal{Q}}{\to} q_\infty$, then q_∞ minimizes $\mathcal{J}$.

Definition 3.14 (Γ-convergence). A sequence $(\mathcal{J}_k)_{k\in\mathbb{N}}$ of functionals on a metrizable topological space $\mathcal{Q}$ *Γ-converges to* $\mathcal{J} : \mathcal{Q} \to \mathbb{R}_\infty$, written $\mathcal{J} = \Gamma\text{-}\lim_{k\to\infty} \mathcal{J}_k$ or $\mathcal{J}_k \overset{\Gamma}{\longrightarrow} \mathcal{J}$, if

($\Gamma_{\rm inf}$) *Γ-liminf estimate:*
$$q_k \overset{\mathcal{Q}}{\to} q \quad \Longrightarrow \quad \mathcal{J}(q) \leq \liminf_{k\to\infty} \mathcal{J}_k(q_k),$$
($\Gamma_{\rm sup}$) *Γ-limsup estimate or "existence of recovery sequences":*
$$\forall \widehat{q} \in \mathcal{Q}\ \exists\, (\widehat{q}_k)_{k\in\mathbb{N}} \text{ with } \widehat{q}_k \overset{\mathcal{Q}}{\to} \widehat{q}: \quad \mathcal{J}(\widehat{q}) \geq \limsup_{k\to\infty} \mathcal{J}_k(\widehat{q}_k).$$

The sequence $(\widehat{q}_k)_{k\in\mathbb{N}}$ is called a *recovery sequence* for $\widehat{q}$ since ($\Gamma_{\rm inf}$) and ($\Gamma_{\rm sup}$) imply $\mathcal{J}_k(\widehat{q}_k) \to \mathcal{J}(\widehat{q})$, i.e., $\widehat{q}_k$ recovers the correct energy level. The following results are fundamental in the theory of Γ-convergence.

Proposition 3.15. *Under the above assumptions we have the following:*

(i) $\mathcal{J} = \Gamma\text{-}\liminf_{k\to\infty} \mathcal{J}_k$ *is always lower semicontinuous.*
(ii) For $\mathcal{J}, \mathcal{J}_k : \mathcal{Q} \to \mathbb{R}_\infty$ *with* $\mathcal{J} = \Gamma\text{-}\lim_{k\to\infty} \mathcal{J}_k$*, set* $\alpha = \inf_{\mathcal{Q}} \mathcal{J}$ *and* $\alpha_k = \inf_{\mathcal{Q}} \mathcal{J}_k$*. Assume* $\alpha \in \mathbb{R}$ *and that there exist* $\delta > 0$ *and a compact set*

$C \subset \mathcal{Q}$ such that all sublevels $\{q \mid \mathcal{J}_k(q) \le \alpha+\delta\}$ are contained in C. Then, $\alpha_k \to \alpha$ and for each sequence q_k with $q_k \to \widetilde{q}$ and $\limsup_{k\to\infty} \mathcal{J}_k(\widetilde{q}_k) = \alpha$ we have $\mathcal{J}(\widetilde{q}) = \alpha$, i.e., $\widetilde{q}$ is a minimizer of $\mathcal{J}$. In particular, if q_k are minimizers of $\mathcal{J}_k$, we conclude that all accumulation points of $(q_k)_k$ are minimizers of $\mathcal{J}$.

It is surprising that Γ-convergence can be used in a rather easy way for energetic solutions of RIS. This is certainly due to the fact that the evolution is strongly governed by the static stability condition. Applications of Γ-convergence occur naturally in space-time discretizations [33, 47, 57] or homogenization [52] of rate-independent material models. See also [6, 28] for applications in fracture or [8, 44] for damage.

We now list the assumptions on the rate-independent systems $(\mathcal{Q}, \mathcal{E}_k, \mathcal{D}_k)$, $k \in \mathbb{N}_\infty \stackrel{\text{def}}{=} \mathbb{N} \cup \{\infty\}$, which are sufficient for our convergence theory. They are in complete analogy to the assumptions in the existence theory above; however, certain assumptions need to be uniform in k, while other assumptions are only needed for the limiting system with $k = \infty$. Since we are already dealing with a sequence of problems and we have to choose subsequences several times, we need to adjust the notion of stable sequences. The stability sets $\mathcal{S}_k(t)$ are defined for $(\mathcal{Q}, \mathcal{E}_k, \mathcal{D}_k)$ as in (31). A sequence $((t_l, q_{k_l}))_{l\in\mathbb{N}}$ is a *stable sequence* (abbreviated as "stab.seq." further on), if

$$q_{k_l} \in \mathcal{S}_{k_l}(t_l) \text{ for all } l \in \mathbb{N} \quad \text{and} \quad \sup_{l\in\mathbb{N}} \mathcal{E}_{k_l}(t_l, q_{k_l}) < \infty. \tag{63}$$

Note that $(q_{k_l})_{l\in\mathbb{N}}$ denotes a subsequence to indicate the index k_l for which we have stability. As in the previous sections, we say that $((t_l, \widetilde{q}_l))_{l\in\mathbb{N}}$ is a stable sequence for $(\mathcal{Q}, \mathcal{E}, \mathcal{D}_\infty)$ if $\widetilde{q}_l \in \mathcal{S}_\infty(t_l)$, and we shortly write "stab.seq.$^\infty$" in that case.

We collect all our assumptions and comment on them afterwards.

Quasi-distance: $\forall\, k \in \mathbb{N}_\infty\ \forall\, z, \widetilde{z}, \widehat{z} \in \mathcal{Z}$:
$$\mathcal{D}_k(z, \widetilde{z}) = 0 \Leftrightarrow z = \widetilde{z} \text{ and } \mathcal{D}_k(z, \widehat{z}) \le \mathcal{D}_k(z, \widetilde{z}) + \mathcal{D}_k(\widetilde{z}, \widehat{z}). \tag{64a}$$

Lower semicontinuity of $\mathcal{D}_k$:
$$\forall\, k \in \mathbb{N}_\infty: \quad \mathcal{D}_k : \mathcal{Z}\times\mathcal{Z} \to [0, \infty] \text{ is lower semicontinuous.} \tag{64b}$$

Lower Γ-limit for $\mathcal{D}_k$:
$$\forall \text{ stab.seq. } (t_l, q_{k_l}) \stackrel{\mathcal{Q}_T}{\rightharpoondown} (t, q) \text{ and } (\widetilde{t}_l, \widetilde{q}_{k_l}) \stackrel{\mathcal{Q}_T}{\rightharpoondown} (\widetilde{t}, \widetilde{q}): \quad \mathcal{D}_\infty(q, \widetilde{q}) \le \liminf_{l\to\infty} \mathcal{D}_{k_l}(q_{k_l}, \widetilde{q}_{k_l}). \tag{64c}$$

Compactness of energy sublevels:

For all $t \in [0, T]$ and all $E \in \mathbb{R}$ we have (64d)

(i) $\forall\, k \in \mathbb{N}_\infty: \quad \{q \in \mathcal{Q} \mid \mathcal{E}_k(t, q) \le E\}$ is compact;

(ii) $\bigcup_{k=1}^{\infty} \{q \in \mathcal{Q} \mid \mathcal{E}_k(t, q) \le E\}$ is relatively compact.

Separability and metrizability: The topology restricted to sublevels of $\mathcal{E}(t,\cdot)$ is compact, separable and metrizable. (64e)

Uniform control of the power $\partial_t\mathcal{E}_k$:

$$\begin{aligned}&\exists\,\lambda_{\mathcal{E}}>0\;\forall\,k\in\mathbb{N}_\infty\;\forall\,(t,q)\text{ with }\mathcal{E}_k(t,q)<\infty:\\&\mathcal{E}_k(\cdot,q)\in\mathrm{C}^1([0,T]),\;|\partial_t\mathcal{E}_k(t,q)|\le\lambda_{\mathcal{E}}\,\mathcal{E}_k(s,q)\text{ for }s\in[0,T].\end{aligned}\tag{64f}$$

Lower Γ-limit for $\mathcal{E}_k$:

$$\forall\text{ stab.seq. }(t_l,q_{k_l})\overset{\mathcal{Q}_T}{\rightharpoonup}(t,q):\;\mathcal{E}_\infty(t,q)\le\liminf_{l\to\infty}\mathcal{E}_{k_l}(t_l,q_{k_l}).\tag{64g}$$

Conditioned semicontinuity of the power: $\forall\,t\in[0,T]:$

$$\forall\text{ stab.seq. }(t_l,q_{k_l})\overset{\mathcal{Q}_T}{\rightharpoonup}(t,q):\;\limsup_{l\to\infty}\partial_t\mathcal{E}_{k_l}(t,q_{k_l})\le\partial_t\mathcal{E}_\infty(t,q),\tag{64h}$$

$$\forall\text{ stab.seq.}^\infty\;(t_l,\widetilde{q}_l)\overset{\mathcal{Q}_T}{\rightharpoonup}(t,q):\;\liminf_{l\to\infty}\partial_t\mathcal{E}_\infty(t,\widetilde{q}_l)\ge\partial_t\mathcal{E}_\infty(t,q).\tag{64i}$$

Conditioned upper semicontinuity of stability sets:

$$\forall\text{ stab.seq. }(t_l,q_{k_l})\overset{\mathcal{Q}_T}{\rightharpoonup}(t,q):\quad q\in\mathcal{S}_\infty(t).\tag{64j}$$

Assumptions (64a)–(64c) mainly concern the dissipation distances $\mathcal{D}_k$: the first two correspond to the earlier conditions (D1) and (D2), whereas (64c) is the new Γ-liminf condition. Assumptions (64d)–(64g) are mainly on the stored-energy functionals $\mathcal{E}_k$: the first two correspond to the earlier (E1) and (E2), whereas (64g) is the new Γ-liminf condition. Conditions (64j) and (64h)–(64i) correspond to the compatibility conditions (C1) and (C2), respectively.

It may seem strange that we do not ask the Γ-convergences of $\mathcal{D}_k\xrightarrow{\Gamma}\mathcal{D}_\infty$ and $\mathcal{E}_k(t,\cdot)\xrightarrow{\Gamma}\mathcal{E}_\infty(t,\cdot)$ for $k\to\infty$. In fact, we do not need this in general, because the compatibility conditions (64h), (64i), and (64j) implicitly provide the Γ-limsup estimates when restricted to the stability sets $\mathcal{S}_\infty(t)$. In fact, condition (64j) is almost identical to the compatibility condition (C2). Hence, the construction of joint-recovery sequences in Sect. 3.5 applies equally here. Whereas in many practical applications $\mathcal{D}_k\xrightarrow{\Gamma}\mathcal{D}_\infty$ and $\mathcal{E}_k(t,\cdot)\xrightarrow{\Gamma}\mathcal{E}_\infty$ holds, the importance of the interlinked assumptions is that we are automatically forced to consider Γ-convergence in the intrinsic topology, namely the one induced by convergence of stable sequences.

We present two convergence results and refer to [58] for the proofs. The first result concerns exact solutions q_k of the RIS $(\mathcal{Q},\mathcal{E}_k,\mathcal{D}_k)$, and we already assume that these solutions converge. This is not a restrictive assumption, since from the proof it becomes clear that any sequence of solutions has a subsequence for which the z-component pointwise converges, and that is the only important assumption.

Theorem 3.16 (($\mathcal{Q}, \mathcal{E}_k, \mathcal{D}_k$) **converges to** ($\mathcal{Q}, \mathcal{E}_\infty, \mathcal{D}_\infty$)). *Assume that (64) holds and that $q_k : [0,T] \to \mathcal{Q}$ are energetic solutions of $(\mathcal{Q}, \mathcal{E}_k, \mathcal{D}_k)$. Let us further assume that, for all $t \in [0,T]$, we have $q_k(t) \overset{\mathcal{Q}}{\to} q(t)$ and $\mathcal{E}_k(0, q_k(0)) \to \mathcal{E}_\infty(0, q(0))$ for $k \to \infty$. Then $q : [0,T] \to \mathcal{Q}$ is an energetic solution of $(\mathcal{Q}, \mathcal{E}_\infty, \mathcal{D}_\infty)$, and for all $t \in [0,T]$ we have*

$$\mathcal{E}_k(t, q_k(t)) \to \mathcal{E}_\infty(t, q(t)), \quad \mathrm{Diss}_k(q_k, [0,t]) \to \mathrm{Diss}_\infty(q, [0,t]),$$
$$\partial_t \mathcal{E}_k(\cdot, q_k(\cdot)) \to \partial_t \mathcal{E}_\infty(\cdot, q(\cdot)) \text{ in } \mathrm{L}^1([0,T]).$$

Next we show that even incremental solutions of $(\mathcal{Q}, \mathcal{E}_k, \mathcal{D}_k)$ for a given sequence $(\Pi^k)_{k\in\mathbb{N}}$ of partitions with fineness $\phi(\Pi^k) \to 0$ have subsequences converging to solutions of $(\mathcal{Q}, \mathcal{E}_\infty, \mathcal{D}_\infty)$. Thus, we do not need exact solutions of each $(\mathcal{Q}, \mathcal{E}_k, \mathcal{D}_k)$ to guarantee that the limiting functions are solutions. For the partitions $\Pi^k = (0 = t_0^k, t_1^k, \ldots, t_{N_k-1}^k, t_{N_k}^k = T)$, we use the fully implicit incremental minimization problem $(\mathrm{IMP})_k^{\Pi^k}$

$$\text{Given } q_0^k \in \mathcal{Q}, \text{ for } j = 1, ..., N_k \text{ find } \; q_j^k \in \operatorname*{Arg\,min}_{q\in\mathcal{Q}} \Big(\mathcal{E}_k(t_j^k, q) + \mathcal{D}_k(q_{j-1}^k, q) \Big).$$

For each solution $((t_j^k, q_j^k))_{j=0,1,\ldots,N_k}$ we define the piecewise constant interpolants $\underline{q}^k : [0,T] \to \mathcal{Q}$ as in (39). The following result states the convergence of subsequences of the solutions $\underline{q}^k$ to energetic solutions of the limit system $(\mathcal{Q}, \mathcal{E}, \mathcal{D})$.

Theorem 3.17 (($\mathbf{IMP}_k^{\Pi^k}$) **converges to** ($\mathcal{Q}, \mathcal{E}_\infty, \mathcal{D}_\infty$)). *Let conditions (64a)–(64j) hold. Let the sequence of partitions Π^k satisfy $\phi(\Pi^k) \to 0$, and let the sequence of initial conditions q_0^k satisfy*

$$q_0^k \overset{\mathcal{Q}}{\to} q_0 \quad \text{and} \quad \mathcal{E}_k(0, q_0^k) \to \mathcal{E}_\infty(0, q_0) \in \mathbb{R}. \tag{65}$$

Then, each (IMP)^k has at least one solution $\underline{q}^k : [0,T] \to \mathcal{Q}$ and there exist a subsequence $(\underline{q}^{k_l})_{l\in\mathbb{N}}$ and a measurable, energetic solution $q : [0,T] \to \mathcal{Q}$ for the RIS $(\mathcal{Q}, \mathcal{E}_\infty, \mathcal{D}_\infty)$ with $q(0) = q_0$, such that (i)–(iv) hold:

$$\begin{aligned}
&\text{(i) } \forall t \in [0,T] : \mathcal{E}_{k_l}(t, \underline{q}^{k_l}(t)) \to \mathcal{E}_\infty(t, q(t)), \\
&\text{(ii) } \forall t \in [0,T] : \mathrm{Diss}_{k_l}(\underline{q}^{k_l}, [0,t]) \to \mathrm{Diss}_\infty(q, [0,t]), \\
&\text{(iii) } \forall t \in [0,T] : \underline{z}^{k_l}(t) \overset{\mathcal{Z}}{\to} z(t), \\
&\text{(iv) } \partial_t \mathcal{E}_{k_l}(\cdot, \underline{q}^{k_l}(\cdot)) \to \partial_t \mathcal{E}_\infty(\cdot, q(\cdot)) \text{ in } \mathrm{L}^1([0,T]).
\end{aligned} \tag{66}$$

Moreover, any $\widetilde{q} : [0,T] \to \mathcal{Q}$ obtained as such a limit is an energetic solution of $(\mathcal{Q}, \mathcal{E}_\infty, \mathcal{D}_\infty)$, if additionally $y(t) \in \operatorname{Arg\,min} \mathcal{E}(t, \cdot, z(t))$ for all t and $\partial_t \mathcal{E}_\infty(t, y(t), z(t)) = \mathcal{P}_{\mathrm{red}}^\infty(t, z(t))$ a.e. in $[0,T]$.

4 Rate-Independent Systems in Banach Spaces

In Banach spaces we have two important additional tools deriving from the linear structure. First, the functionals at hand may have differentials or subdifferentials such that it is possible to formulate force balances and rate equations rather than comparing energies, as in the energetic formulation. Second, we can employ convexity and duality methods like the Legendre-Fenchel transform as indicated in Sect. 2.5.

In Sect. 4.2 we discuss several weakened versions of the subdifferential problem

$$0 \in \partial_{\dot{q}}\mathcal{R}(q(t),\dot{q}(t)) + \varepsilon\mathbb{V}\dot{z} + \overline{\partial}_q\mathcal{E}(t,q(t)), \tag{67}$$

with $\varepsilon = 0$ and provide some comparison to energetic solutions. Using convexity arguments we derive temporal continuity properties in Sect. 4.3. Finally the vanishing-viscosity approach considers first $\varepsilon > 0$ and then the limit $\varepsilon \to 0$ is used to derive new types of solutions, namely the notion of parametrized and BV solutions.

4.1 The Basic Banach-Space Setup

Before starting with details we explain the usage of the different Banach spaces $\boldsymbol{X}$, $\boldsymbol{V}$, $\boldsymbol{C}$, and $\boldsymbol{Q} = \boldsymbol{Y}\times\boldsymbol{Z}$. The smallest space $\boldsymbol{Q}$ is a reflexive Banach space, such that $\mathcal{E} : \boldsymbol{Q}_T \overset{\text{def}}{=} [0,T]\times\boldsymbol{Q} \to \mathbb{R}_\infty$ has bounded and weakly closed sublevels. For $\mathcal{E} : \boldsymbol{Q}_T \to \mathbb{R}_\infty$ (similarly for the reduced functional $\mathcal{I} : \boldsymbol{Z}_T \to \mathbb{R}_\infty$) we make the following assumptions, which are used without further mentioning in the sequel:

$$\exists\, c, C > 0\ \forall\, (t,q) \in \boldsymbol{Q}_T: \quad \mathcal{E}(t,q) \geq c\|q\| - C; \tag{68a}$$

$$\begin{aligned}&\exists\, \lambda_{\mathcal{E}} > 0\ \forall\, (s,q) \in \mathcal{Q}_T \text{ with } \mathcal{E}(s,q) < \infty :\\ &\mathcal{E}(\cdot,q) \in \mathrm{C}^1([0,T]) \text{ and } |\partial_t\mathcal{E}(t,q)| \leq \lambda_{\mathcal{E}}\,\mathcal{E}(t,q) \text{ for all } t.\end{aligned} \tag{68b}$$

Then, the basic energy estimate, cf. Sect. 2.4, shows that all solutions and approximations of interest lie in a bounded set in $\boldsymbol{Q}$, which is useful for extracting subsequences that weakly converge in $\boldsymbol{Q}$.

The function space $\boldsymbol{X}$ is chosen to make the dissipation coercive, i.e. $\mathcal{R}(z,v) \geq c\|v\|_{\boldsymbol{X}}$. The space $\boldsymbol{C}$ is used to provide uniform convexity properties like $\langle \mathrm{D}_q^2\mathcal{E}(t,q)w, w\rangle \geq c\|w\|_{\boldsymbol{C}}^2$. Finally, the Hilbert space $\boldsymbol{V}$ is used for measuring viscosity, e.g. in the small viscosity approximation we use the dissipation potential $\mathcal{R}_\varepsilon(z,v) = \Psi(v) + \frac{\varepsilon}{2}\|v\|_{\boldsymbol{V}}^2$. Throughout we assume that the embeddings $\boldsymbol{Q} \subset \boldsymbol{C}$, $\boldsymbol{Z} \subset \boldsymbol{X}$, and $\boldsymbol{Z} \subset \boldsymbol{V}$ hold. Having in mind the PDE version of our standard Example 2.8, namely

$$0 \in \mathrm{Sign}(\dot{z}) + \varepsilon\mathbb{V}\dot{z} - \Delta z + \Phi'(z) - \ell(t,x), \quad (t,x) \in [0,T]\times\Omega, \quad z(t,\cdot)|_{\partial\Omega} = 0,$$

we have $\mathcal{R}_\varepsilon(z,v) = \|v\|_{\mathrm{L}^1} + \frac{\varepsilon}{2}\|v\|^2_{\mathrm{L}^2}$ and $\mathcal{I}(t,z) = \int_\Omega \frac{1}{2}|\nabla z|^2 + \Phi(z) - \ell(t)z\,\mathrm{d}x$. This leads to the typical choice $\boldsymbol{Z} = \mathrm{H}^1_0(\Omega) \Subset \boldsymbol{V} = \mathrm{L}^2(\Omega) \subset \boldsymbol{X} = \mathrm{L}^1(\Omega)$.

A major difficulty for rate-independent systems arising in applications in continuum mechanics is that the rate-independent norm of $\boldsymbol{X}$ is usually given in terms of a (weighted) L^1 norm. Thus, in general $\boldsymbol{X}$ is not reflexive and does not enjoy the Radon-Nikodym property, which would provide differentiability of Lipschitz functions. In principle, it would be possible to use the weak* derivative $\frac{1}{h}(z(t+h) - z(t)) \overset{*}{\rightharpoonup} \dot{z}$ in a bigger space $\boldsymbol{X}_0$ which contains $\boldsymbol{X}$ as a closed subspace and is the dual of a separable space, e.g. $\mathrm{M}(\Omega) \supset \mathrm{L}^1(\Omega)$. However, we are able to avoid this concept by using the dissipation Diss in derivative-free form. Another way to handle the missing weak closedness of $\boldsymbol{X}$ is discussed in Sect. 5, where $\boldsymbol{X}$ is treated as a complete metric space.

Example 4.1. A typical example for $\boldsymbol{X} = \mathrm{L}^1(\mathbb{R})$ is obtained by the functions $z(t) = \chi_{[\alpha(t),\beta(t)]}$, where $\alpha, \beta \in \mathrm{W}^{1,1}([0,T])$ with $\alpha \le \beta$ a.e. Letting $f(t) = |\dot{\alpha}(t)| + |\dot{\beta}(t)|$ we obtain $\|z(t_2)-z(t_1)\|_{\mathrm{L}^1} \le |\alpha(t_2)-\alpha(t_1)| + |\beta(t_2)-\beta(t_1)| \le \int_{t_1}^{t_2} f(t)\,\mathrm{d}t$. Hence z lies in $\mathrm{AC}([0,T],\mathrm{L}^1(\mathbb{R}))$ but not in $\mathrm{W}^{1,1}([0,T],\mathrm{L}^1(\mathbb{R}))$, since $\dot{z}(t) = \dot{\beta}(t)\delta_{\beta(t)} - \dot{\alpha}(t)\delta_{\alpha(t)}$ is a Radon measure but not in $\mathrm{L}^1(\mathbb{R})$.

The linear Banach space structure allows for the usage of subdifferentials. For the dissipation potential $\mathcal{R}$ we always use the convex subdifferential $\partial_v\mathcal{R}(z,\cdot)$ for the convex function $\mathcal{R}(z,\cdot) : \boldsymbol{X} \to \mathbb{R}_\infty$. For the energy functional $\mathcal{E}(t,\cdot) : \boldsymbol{Q} \to \mathbb{R}_\infty$ there are several possible choices of subdifferentials. For simplicity we restrict to the limiting subdifferential used in [64], also called Mordukhovich differential. It is a suitable closure of the Fréchet subdifferential

$$\partial_q^{\mathrm{Fr}}\mathcal{E}(t,q) \overset{\mathrm{def}}{=} \{\, \eta \in \boldsymbol{Q}^* \mid \mathcal{E}(t,q+w) \ge \mathcal{E}(t,q) + \langle \eta, w\rangle + o(\|w\|_{\boldsymbol{Q}})_{z\to 0} \,\}$$

and is given in the form

$$\overline{\partial}_q\mathcal{E}(t,q) \overset{\mathrm{def}}{=} \{\, \eta \in \boldsymbol{Q}^* \mid \exists\, (q_n,\eta_n)_{n\in\mathbb{N}} : \ \eta_n \in \partial_q^{\mathrm{Fr}}\mathcal{E}(t,q_n),\ q_n \rightharpoonup q \text{ in } \boldsymbol{Q}, \\ \eta_n \rightharpoonup \eta \text{ in } \boldsymbol{Q}^*,\ \sup_{n\in\mathbb{N}} \mathcal{E}(t,q_n) < \infty \,\}.$$

For the sum of a convex function $\mathcal{J}_1$ and a C^1 function $\mathcal{J}_2$ we have the sum rule

$$\overline{\partial}(\mathcal{J}_1+\mathcal{J}_2)(q) = \overline{\partial}\mathcal{J}_1(q) + \mathrm{D}\mathcal{J}_2(q). \tag{69}$$

Of course, most definitions and some of the results can be transferred to other subdifferentials, but this would complicate the presentation unnecessarily.

The difficulty in finding a suitable notion of subdifferential lies in the two opposite requirements. First, we want the subdifferential to be sufficiently large, such that it has good closure properties. If approximations satisfy $\eta_n(t) \in \overline{\partial}_q\mathcal{E}(t,q_n(t))$ a.e. in $[0,T]$, $q_n \rightsquigarrow q$, and $\eta_n \rightsquigarrow \eta$, we want to be able to conclude $\eta(t) \in \overline{\partial}_q \mathcal{E}(t, q(t))$ a.e. in $[0, T]$. Second, we want the subdifferential to be not too big, such that we still can show a counterpart to the classical chain rule $\frac{\mathrm{d}}{\mathrm{d}t}\mathcal{E}(t, q(t)) = \langle \mathrm{D}_q\mathcal{E}(t, q(t)), \dot{q}(t)\rangle + \partial_t \mathcal{E}(t, q(t))$.

As a start, we define a suitable generalized chain rule, which will be useful in the following sections. Here $\boldsymbol{Y}$ is a general Banach space, which can play the role of $\mathbb{R}\times\boldsymbol{Z}$ for time-dependent functionals.

Definition 4.2 (Chain rules). Let $\boldsymbol{Y}$ be a Banach space and $\mathcal{J} : \boldsymbol{Y} \to \mathbb{R}_\infty$ a functional with (sub)differential $\overline{\partial}\mathcal{J} : \boldsymbol{Y} \rightrightarrows \boldsymbol{Y}^*$. We say that the triple $(\boldsymbol{Y}, \mathcal{J}, \overline{\partial}\mathcal{J})$ satisfies the *chain-rule equality*, if for all $y \in \mathrm{W}^{1,1}([0, T]; \boldsymbol{Y})$ and all measurable $\eta : [0, T] \to \boldsymbol{Y}^*$ the following holds:

$$\begin{aligned}
&\text{if } \sup_{t\in[0,T]} \mathcal{J}(y(t)) < \infty,\ \textstyle\int_0^T \|\dot{y}(t)\|_{\boldsymbol{Y}} \|\eta(t)\|_{\boldsymbol{Y}^*}\,\mathrm{d}t < \infty,\\
&\qquad \text{and } \eta(t) \in \overline{\partial}\mathcal{J}(y(t)) \text{ a.e. in } [0, T],\\
&\text{then } t \mapsto \mathcal{J}(y(t)) \text{ is absolutely continuous and}\\
&\qquad \frac{\mathrm{d}}{\mathrm{d}t}\mathcal{J}(y(t)) = \langle \eta(t), \dot{y}(t)\rangle \text{ a.e. in } [0, T].
\end{aligned} \tag{70}$$

We say that $(\boldsymbol{Y}, \mathcal{J}, \overline{\partial}\mathcal{J})$ satisfies the *chain-rule inequality*, if for all $y \in \mathrm{W}^{1,1}([0, T]; \boldsymbol{Y})$ and all measurable $\eta : [0, T] \to \boldsymbol{Y}^*$ we have

$$\begin{aligned}
&\text{if } \sup_{t\in[0,T]} \mathcal{J}(y(t)) < \infty,\ \textstyle\int_0^T \|\dot{y}(t)\|_{\boldsymbol{Y}} \|\eta(t)\|_{\boldsymbol{Y}^*}\,\mathrm{d}t < \infty,\\
&\qquad \text{and } \eta(t) \in \overline{\partial}\mathcal{J}(y(t)) \text{ a.e. in } [0, T],\\
&\text{then } \textstyle\int_0^T \langle \dot{y}(t), \eta(t)\rangle\,\mathrm{d}t \ \geq\ \mathcal{J}(y(0)) - \mathcal{J}(y(T)).
\end{aligned} \tag{71}$$

The chain rule holds for functionals $\mathcal{E} \in \mathrm{C}^1(\mathcal{Q}_T)$, but also in much more general situations, see e.g. [64]. We apply the chain rules to the space $\boldsymbol{Y} = \mathbb{R}\times\boldsymbol{V}$, and we always assume classical differentiability of $\mathcal{I}$ with respect $t \in [0, T]$. Hence we have to be careful only with respect to $\overline{\partial}_z\mathcal{I}$ and usually split off the work of the external forces $\int_0^T \partial_t\mathcal{I}(t, z(t))\,\mathrm{d}t$.

We now provide a few more examples for the Banach-space setting. They are rather degenerate and will serve as counterexamples.

Example 4.3. We let $\Omega = \,]0, 1[\, \subset \mathbb{R}^1$, $\boldsymbol{X} = \mathrm{L}^1(\Omega)$, and $\boldsymbol{Z} = \mathrm{L}^2(\Omega)$. Hence, $\boldsymbol{Z}$ is not compactly embedded in $\boldsymbol{X}$. The functionals are

$$\begin{aligned}
&\mathcal{I}_\alpha(t, z) = \textstyle\int_\Omega \Phi_\alpha(z(x)) - (t+x)z(x)\,\mathrm{d}x,\\
&\mathcal{R}_\varepsilon(z, \dot{z}) = \textstyle\int_\Omega |\dot{z}(x)| + \frac{\varepsilon}{2}|\dot{z}(t)|^2\,\mathrm{d}x, \quad \text{and } z_0(x) = 0.
\end{aligned}$$

We consider the case $\alpha \in [2, \infty]$ and

$$\Phi_\alpha(z) = \tfrac{1}{\alpha}|z|^\alpha \text{ for } |z| \leq 1 \ \text{ and } \Phi_\alpha(z) = \tfrac{1}{2}|z|^2 + \tfrac{1}{\alpha} - \tfrac{1}{2} \text{ for } |z| \geq 1.$$

The point of this example is that we are able to calculate the solution by solving uncoupled scalar differential inclusions for each $x \in \Omega$, namely

$$0 \in \mathrm{Sign}(\dot z(t,x)) + \varepsilon \dot z(t,x) + \Phi'_\alpha(z(t,x)) - t - x, \quad z(0,x) = z_0(x) = 0. \quad (72)$$

Because of $\Phi'_\alpha(0) = 0$ we see that z_0 is locally stable at $t = 0$.

For $\varepsilon = 0$ we obtain the solution $z_\alpha(t,x) = M_\alpha(\max\{0, t+x-1\})$ where $M_\alpha(\sigma) = \sigma^{1/(\alpha-1)}$ for $\sigma \in [0,1]$ and $M_\alpha(\sigma) = \sigma$ for $\sigma \geq 1$. For $\alpha = 2$ the abstract theory of uniformly convex problems gives $z \in \mathrm{C}^{\mathrm{Lip}}([0,T], \mathrm{L}^2(\Omega))$. However, using the explicit formula

$$\dot z_\alpha(t,x) = \begin{cases} 0 & \text{for } t+x<1, \\ \frac{1}{\alpha-1}(t+x-1)^{-\frac{\alpha-2}{\alpha-1}} & \text{for } 1<t+x<2, \\ 1 & \text{for } t+x>2, \end{cases}$$

we find $\dot z_\alpha \in \mathrm{L}^\infty([0,T], \mathrm{L}^p(\Omega))$ whenever $p > \frac{\alpha-1}{\alpha-2}$. In particular, $p = 1$ is always possible, and we are able to write down the abstract differential inclusion

$$0 \in \partial \mathcal{R}_0(\dot z_\alpha(t)) + \mathrm{D}_z \mathcal{I}_\alpha(t, z_\alpha(t)) \quad \subset \mathrm{L}^2(\Omega) = \boldsymbol{Z}^*, \quad (73)$$

since $\partial \mathcal{R}_0(\dot z_\alpha(t)) \in \mathrm{L}^\infty(\Omega) = \boldsymbol{X}^* \subset \boldsymbol{Z}^*$.

The limit case $\alpha \to \infty$ is more interesting, because it leads to the degenerate convex limit potential $\Phi_\infty(z) = \frac{1}{2}\max\{0, |z|^2-1\}$ and the limit solution z with

$$z(t,x) = \begin{cases} 0 & \text{for } t+x<1, \\ 1 & \text{for } 1<t+x\leq 2, \\ t+x-1 & \text{for } t+x \geq 2; \end{cases} \qquad \dot z(t) = \begin{cases} \delta_{1-t} & \text{for } t \in \,]0,1[\,, \\ \chi_{]2-t,1[} & \text{for } t \in \,]1,2[\,, \\ 1 & \text{for } t>2 \end{cases}$$

Thus, we have $z \in \mathrm{C}^{\mathrm{Lip}}([0,T], \boldsymbol{X})$ with $\boldsymbol{X} = \mathrm{L}^1(\Omega)$, but the derivative only exists in the weak* sense, namely $\dot z \in \mathrm{L}^\infty_{\mathrm{w}*}([0,T], \mathrm{M}(\Omega))$ with $\mathrm{M}(\Omega) = \mathrm{C}_0(\Omega)^*$. We are no longer able to give any sense to the subdifferential inclusion (73), since after extending $\mathcal{R} : \boldsymbol{X} \to \mathbb{R}_\infty$ to $\mathcal{R}_\circ : \mathrm{M}(\Omega) \to \mathbb{R}_\infty$ via the weak* lower semicontinuous hull, the subdifferential $\partial \mathcal{R}_\circ(\dot z_\alpha)$ needs to be treated in $\mathrm{M}(\Omega)^*$ which is no longer comparable to $\boldsymbol{Z}^*$. Thus, we do not have a *differential solution*, but z is still a *CD solution* in the sense of Definition 4.5.

For completeness and later reference, we also give the viscous approximations for the case $\alpha = \infty$. The solution z^ε reads

$$z^\varepsilon(t,x) = U^\varepsilon(t+x) \quad \text{with } U^\varepsilon(\tau) = \begin{cases} 0 & \text{for } \tau \leq 1, \\ (\tau-1)/\varepsilon & \text{for } 1 \leq \tau \leq 1+\varepsilon, \\ 1 & \text{for } 1+\varepsilon \leq \tau \leq 2, \\ \tau-1+\varepsilon \mathrm{e}^{-(\tau-2)/\varepsilon} - \varepsilon & \text{for } \tau \geq 2. \end{cases}$$

Obviously, we have $z^\varepsilon(t,\cdot) \to z(t,\cdot)$ in $\boldsymbol{Z}$. For each $\varepsilon > 0$ we have $z^\varepsilon \in \mathrm{C}^{\mathrm{Lip}}([0,T],\mathrm{L}^p(\Omega))$. However, the Lipschitz bound blows up for $\varepsilon > 0$ if $p > 1$. Only in the case $p = 1$ we obtain a uniform bound.

Example 4.4. This example uses exactly the same function spaces as the previous Example 4.3, but the potential defining $\mathcal{I}$ is replaced by the nonconvex function $\Phi_{\mathrm{nc}} : z \mapsto \mathcal{U}(z)$ with $\mathcal{U}$ from (23). For the viscous problem (72) with the initial condition $z_0(x) = -4$ we obtain the unique solution $z^\varepsilon(t,x) = V^\varepsilon(t{+}x)$ with

$$V^\varepsilon(\tau) = \begin{cases} -4 & \text{for } \tau \in [0,1{+}\varepsilon], \\ \tau{-}5{-}\varepsilon & \text{for } \tau \in [1{+}\varepsilon, 3{+}\varepsilon], \\ 2\varepsilon \mathrm{e}^{(\tau-3-\varepsilon)/\varepsilon} {-}\tau{+}1{-}\varepsilon & \text{for } \tau \in [3{+}\varepsilon, 3{+}\varepsilon{+}\delta_\varepsilon], \\ \tau{+}3{-}\varepsilon{-}(4{+}\delta_\varepsilon)\mathrm{e}^{-(\tau-3-\varepsilon-\delta_\varepsilon)/\varepsilon} & \text{for } \tau \geq 3{+}\varepsilon{+}\delta_\varepsilon, \end{cases}$$

where $\delta_\varepsilon = \varepsilon \log(2/\varepsilon) + O(\varepsilon)$ for $\varepsilon \to 0$. For $\varepsilon \to 0$, the solutions $z^\varepsilon(t,\cdot)$ converge to z strongly in $\boldsymbol{Z} = \mathrm{L}^2(\Omega)$, where $z(t,x) = V^0(t{+}x)$ a.e. in Ω, with $V^0(\tau) = \max\{-4, \tau{-}5\}$ for $\tau < 3$ and $V^0(\tau) = \tau{+}3$ for $\tau > 3$.

As in the previous example, the functions z^ε have a uniform Lipschitz bound (namely 8), if the values are considered in $\boldsymbol{X} = \mathrm{L}^1(\Omega)$. However, the total length in $\boldsymbol{Z} = \mathrm{L}^2(\Omega)$ tends to ∞ like $1/\sqrt{\varepsilon}$. More important is the fact, that the limit function $z \in \mathrm{C}^{\mathrm{Lip}}([0,5];\boldsymbol{X}) \cap \mathrm{C}^0([0,5];\boldsymbol{Z})$ does not satisfy the simple energy balance. All quantities can be calculated explicitly, and we find

$$\mathcal{I}(t,z(t)) + \mathrm{Diss}_\Psi(z,[0,t]) + \varrho(t) = \mathcal{I}(0,z(0)) + \textstyle\int_0^t \partial_s \mathcal{I}(s,z(s))\,\mathrm{d}s$$

with $\varrho(t) = \max\{0, 16\min\{t{-}2, 1\}\}$. This means that in the time interval $t \in [2,3]$, which is exactly where z jumps from -2 to $+6$, there is an additional limit dissipation from the "infinitesimal viscous jumps". However, these jumps are distributed continuously in time such that z remains continuous in time as a function with values in $\boldsymbol{Z} = \mathrm{L}^2(\Omega)$.

This example will be reconsidered in Example 4.27 to highlight the difference between strong and weak BV solutions.

4.2 Differential, CD, and Local Solutions

To define solutions concepts that avoid derivatives we use the special feature of the subdifferential for 1-homogeneous dissipation potentials given in Lemma 2.2, i.e. the rate-independent differential inclusion (67) is equivalent to

$$\begin{aligned} &q(t) \in \mathcal{S}_{\mathrm{loc}}(t) = \{\, q = (y,z) \in \mathcal{Q} \mid 0 \in \partial_v \mathcal{R}(z,\dot z) + \overline{\partial}_q \mathcal{E}(t,q) \,\}, \\ &\mathcal{E}(t,q(t)) + \int_0^t \mathcal{R}(z,\dot z)\,\mathrm{d}s = \mathcal{E}(0,z(0)) + \int_0^t \partial_s \mathcal{E}(s,z(s))\,\mathrm{d}s. \end{aligned}$$

However, the point is that the derivative of z in the energy balance can be replaced by using the dissipation functional $\mathrm{Diss}_{\mathcal{D}}$. This leads to a derivative-free formulation. Thus, we consider compatible pairs $(\mathcal{R}, \mathcal{D})$ which satisfy

$$\exists\, c_R, C_R > 0\ \forall\, z \in \boldsymbol{Z},\ v \in \boldsymbol{X} : c_R \|v\|_{\boldsymbol{X}} \leq \mathcal{R}(z, v) \leq C_R \|v\|_{\boldsymbol{X}}, \tag{74a}$$

$$\forall\, z, v \in \mathcal{Z}: \ \lim_{\varepsilon \to 0^+} \tfrac{1}{\varepsilon}\mathcal{D}(z, z+\varepsilon v) = \mathcal{R}(z, v), \tag{74b}$$

$$\begin{aligned}&\forall\, z \in \mathrm{W}^{1,1}([r,t], \boldsymbol{X}) \cap \mathrm{C}_{\mathrm{w}}([r,t], \boldsymbol{Z}) : \\ &\qquad\qquad \int_r^t \mathcal{R}(z(s), \dot{z}(s))\,\mathrm{d}s = \mathrm{Diss}_{\mathcal{D}}(z, [r,t]).\end{aligned} \tag{74c}$$

As usual we assume the Banach-space structure $\boldsymbol{Q} = \boldsymbol{Y} \times \boldsymbol{Z}$ with continuous and dense embedding $\boldsymbol{Z} \subset \boldsymbol{X}$. Then, $\partial_{\dot{z}} \mathcal{R}(z, v) \subset \boldsymbol{X}^* \subset \boldsymbol{Z}^*$ can also be embedded into $\boldsymbol{Q}^*$ by putting 0 in the component $\boldsymbol{Y}^*$. The main problem in the notions of solutions to RIS arises from the fact, that we want to allow for solutions $q = (y, z) : [0, T] \to \boldsymbol{Y} \times \boldsymbol{Z}$ which are not necessarily differentiable, maybe not even continuous. In this section, discontinuity is only allowed for local solutions, while the other solution concepts ask for weak continuity.

Definition 4.5 (Differentiable, CD, and local solutions). Consider $(\boldsymbol{Q}, \mathcal{E}, \mathcal{R})$ with compatible $\mathcal{D}$. A function $q = (y, z) : [0, T] \to \boldsymbol{Q} = \boldsymbol{Y} \times \boldsymbol{Z}$ is called

(i) a *differential solution* to $(\boldsymbol{Q}, \mathcal{E}, \mathcal{R})$, if $q \in \mathrm{W}^{1,1}([0,T], \boldsymbol{Q})$ and

$$0 \in \begin{pmatrix} 0 \\ \partial_{\dot{z}} \mathcal{R}(z(t), \dot{z}(t)) \end{pmatrix} + \overline{\partial}_q \mathcal{E}(t, q(t)) \quad \text{for a.a. } t \in [0,T]; \tag{75}$$

(ii) a *semi-differential solution* to $(\boldsymbol{Q}, \mathcal{E}, \mathcal{R})$, if (75) holds, $q \in \mathrm{C}_{\mathrm{w}}([0,T], \boldsymbol{Q})$, and $z \in \mathrm{W}^{1,1}([0,T], \boldsymbol{X})$;

(iii) a *CD solution* (for 'C'ontinuous 'D'issipation) to $(\boldsymbol{Q}, \mathcal{E}, \mathcal{R})$, if $t \mapsto \mathrm{Diss}_{\mathcal{D}}(q, [0,t])$ is continuous and if for all $t \in [0,T]$ we have

$$0 \in \begin{pmatrix} 0 \\ \partial_{\dot{z}} \mathcal{R}(z(t), 0) \end{pmatrix} + \overline{\partial}_q \mathcal{E}(t, q(t)) \quad \text{and} \tag{76a}$$

$$\mathcal{E}(t, q(t)) + \mathrm{Diss}_{\mathcal{D}}(q, [0,t]) = \mathcal{E}(0, q(0)) + \int_0^t \partial_s \mathcal{E}(s, q(s))\,\mathrm{d}s; \tag{76b}$$

(iv) a *local solution* to $(\boldsymbol{Q}, \mathcal{E}, \mathcal{R})$, if

$$0 \in \begin{pmatrix} 0 \\ \partial_{\dot{z}} \mathcal{R}(z(t), 0) \end{pmatrix} + \overline{\partial}_q \mathcal{E}(t, q(t)) \quad \text{for a.a. } t \in [0,T] \ \text{ and} \tag{77a}$$

$$\mathcal{E}(t_2,q(t_2)) + \mathrm{Diss}_{\mathcal{D}}(q,[t_1,t_2]) \leq \mathcal{E}(t_1,q(t_1)) + \int_{t_1}^{t_2} \partial_s\mathcal{E}(s,q(s))\,\mathrm{d}s$$
$$\text{for all } 0 \leq t_1 < t_2 \leq T. \tag{77b}$$

Referring to Definition 2.1 we note that the solution types in (i)–(iv) define possibly multi-valued evolutionary systems in the sense that the concatenation and restriction properties hold. The restriction property will fail, if we replace (77b) by the weaker global energy inequality with $t_1 = 0$ and $t_2 = T$. The point is that the local stability condition is not strong enough to provide a lower energy estimate unless additional continuity properties are assumed.

The notion of differentiable solutions can be rewritten as the following evolutionary quasi-variational inequality, which is often used in the literature, see e.g. [7, 35]:

$$\begin{aligned} &\forall_{\text{a.a.}} t \in [0,T]\ \exists \eta(t) \in \boldsymbol{X}^* \cap \overline{\partial}_q\mathcal{E}(t,q(t))\ \forall v \in \boldsymbol{X}: \\ &\qquad \langle \eta(t), v - \dot{z}(t)\rangle_{\boldsymbol{X}^*\times\boldsymbol{X}} + \mathcal{R}(z(t),v) - \mathcal{R}(z(t),\dot{z}(t)) \geq 0. \end{aligned} \tag{78}$$

The important fact about the definitions of CD and local solutions is that they do not assume any differentiability of the solution. To see that the notions are genuinely different, we refer to Example 4.3, where for $\mathcal{I}_\infty$ we have a solution $z \in \mathrm{C}([0,T],\boldsymbol{Z}) \cap \mathrm{C}^{\mathrm{Lip}}([0,T],\boldsymbol{X})$ which does not lie in $\mathrm{W}^{1,1}([0,T],\boldsymbol{X})$. Thus, we have a CD solution which is not a differential solution.

If a suitable chain-rule condition holds for $\mathcal{E}$, we see that the above notions of solutions are ordered from strong to weak and that we are able to go backward, if the solutions have the appropriate temporal behavior.

Proposition 4.6. *Assume that* $(\boldsymbol{Q},\mathcal{E},\mathcal{R})$ *and* $\mathcal{D}$ *satisfy* (74) *and the chain rule* (70)*. Then,* $q \in \mathrm{W}^{1,1}([0,T],\boldsymbol{Q})$ *is a differential solution if and only if it is a CD solution.*

Proof. Assume that q is a differential solution. Obviously, (76a) holds by Lemma 2.2, which characterizes the subdifferentials of 1-homogeneous functionals. To establish the energy balance, take a measurable selection η with $\eta(t) \in \overline{\partial}_q\mathcal{E}(t,q(t))$ a.e. in $[0,T]$. Then, $\eta = (0,\zeta)$ with $\zeta \in \mathrm{L}^\infty([0,T],\boldsymbol{Z}^*)$, because (74a) implies $\|\zeta(t)\|_{\boldsymbol{X}^*} \leq C_R$ and $\boldsymbol{X}^* \subset \boldsymbol{Z}^*$ continuously. The same lemma also gives $\mathcal{R}(z(t),\dot{z}(t)) = -\langle\zeta(t),\dot{z}(t)\rangle$. Using the chain rule (70), integration, and (74c) give (76b).

If q is a CD solution, we choose $\eta = (0,\zeta) \in \overline{\partial}_q\mathcal{E}(t,q(t))$ according to (76a), which implies $-\zeta(t) \in \mathcal{R}(z(t),0)$. Using (76b), (74c), and (70) gives $\mathcal{R}(z(t),\dot{z}(t)) = -\langle\zeta(t),\dot{z}(t)\rangle$. With Lemma 2.2 we obtain the subdifferential formulation (75). □

Under reasonable assumptions it is possible to show that all local solutions that are also continuous are in fact CD solutions. For this one needs to show that local stability together with continuity provides a lower energy estimate.

The next result shows that continuous energetic solutions are in fact CD solutions.

Proposition 4.7. *Assume that* $(\boldsymbol{Q}, \mathcal{E}, \mathcal{R})$ *satisfies* (74)*. Then, an energetic solution* q *with* $q \in \mathrm{C_w}([0,T], \boldsymbol{Q})$ *is a CD solution.*

Proof. Since the energy balance is common in both formulations it suffices to show the local stability. For $t \in [0,T]$ let $\mathcal{J}(\widetilde{q}) = \mathcal{E}(t, \widetilde{q}) + \mathcal{D}(z(t), \widetilde{z})$. From the global stability (S) (see Definition 3.1) we see that $(y, w) = q(t)$ is the global minimizer, which implies $0 \in \overline{\partial}\mathcal{J}(q(t))$. Moreover, with (74a) and (74b) we find $\overline{\partial}\mathcal{J}(q(t)) = \overline{\partial}\mathcal{E}(t, q(t)) + (0, \partial_{\dot{z}}\mathcal{R}(z(t), 0))^{\mathsf{T}}$, which provides the desired result. □

As a corollary we obtain a first existence result for CD solutions.

Theorem 4.8. *Assume that* $\boldsymbol{Z}$ *is compactly embedded into* $\boldsymbol{X}$*, and that* $\mathcal{R}(z, v) = \Psi(v)$*, with* $\Psi : \boldsymbol{X} \to [0, \infty[$ *being coercive and strongly continuous. Moreover, let* $\mathcal{E} : [0,T] \to \boldsymbol{Q} \to \mathbb{R}_\infty$ *satisfy* (E1) *and* (E2) *with respect to the weak topology on* $\boldsymbol{Q}$*. Moreover, assume*

$$\forall\, t \in [0,T] : \ \mathcal{E}(t, \cdot) : \mathcal{Q} \to \mathbb{R}_\infty \ \textit{is strictly convex};$$

$$\exists\, C_E^* \ \forall\, q \ \textit{with}\ \mathcal{E}(0, q) < \infty : \ \mathcal{E}(\cdot, q) \in \mathrm{C}^1([0,T]),$$

$$|\partial_t \mathcal{E}(t, q) - \partial_t \mathcal{E}(s, q)| \le C_E^* |t-s| \mathcal{E}(0, q).$$

Then, for each $q_0 \in \mathcal{S}(0)$ *there exists an energetic solution* q *that is also a CD solution to* $(\boldsymbol{Q}, \mathcal{E}, \Psi)$ *with* $q(0) = q_0$.

Proof. We first employ Theorem 3.5, which provides the existence of an energetic solution. Using the strict convexity of $\mathcal{E}(t, \cdot)$ we conclude that $y(t)$ is uniquely defined from $z(t)$ as minimizer of $\mathcal{E}(t, \cdot, z(t))$. Moreover, the reduced functional $\mathcal{I}$ with $\mathcal{I}(t, z) = \min_{y \in \boldsymbol{Y}} \mathcal{E}(t, y, z)$ is still strictly convex with respect to z. Thus, the mappings $\mathcal{J}_{t,\widehat{z}} : z \mapsto \mathcal{I}(t, z) + \Psi(z - \widehat{z})$ are strictly convex as well. Lemma 3.6 states that the left limit $z(t^-)$ (in the strong $\boldsymbol{X}$-topology) is globally stable as well. Hence, it is a minimizer of $\mathcal{J}_{t, z(t^-)}$. By the jump relations (51), $z(t)$ is a minimizer as well, which now must coincide with $z(t^-)$ by continuity. Similarly, $z(t^+)$ must coincide with $z(t)$. Hence we have shown $z(t^-) = z(t) = z(t^+)$, which implies strong continuity in $\boldsymbol{X}$ and weak continuity in $\boldsymbol{Z}$.

By uniqueness of the minimizers of $\mathcal{E}(t, \cdot, z(t))$ we also obtain $y(t^-) = y(t) = y(t^+)$, where limits are taken weakly in $\boldsymbol{Y}$. We use here that weak limits of minimizers are minimizers because of the weak lower semicontinuity. Thus, $q \in \mathrm{C_w}([0,T], \boldsymbol{Q})$ is established, and Proposition 4.7 provides the CD solution. □

4.3 Systems with Convexity Properties

The temporal continuity results from the previous section can be improved under stronger convexity assumptions. For this we introduce a possibly larger function space $\boldsymbol{C}$ such that $\boldsymbol{Q}$ is continuously embedded into $\boldsymbol{C}$. We consider general energetic solutions to $(\boldsymbol{Q}, \mathcal{E}, \mathcal{D})$ under the following uniform α-convexity condition:

$$\begin{aligned} &\exists\, \alpha \geq 2,\, E>0\; \exists\, c_*>0\; \forall \theta \in [0,1]\; \forall t, q_0, q_1 \text{ with } \mathcal{E}(t,q_0), \mathcal{E}(t,q_1) \leq E: \\ &\quad \mathcal{E}(t,q_\theta) + \mathcal{D}(z_0,z_\theta) + c_*\theta(1{-}\theta)\|q_1{-}q_0\|_{\boldsymbol{C}}^{\alpha} \\ &\qquad \leq (1{-}\theta)\big(\mathcal{E}(t,q_0){+}\mathcal{D}(z_0,z_0)\big) + \theta\big(\mathcal{E}(t,q_1){+}\mathcal{D}(z_0,z_1)\big), \end{aligned} \tag{79}$$

where $q_\theta = (1{-}\theta)q_0 + \theta q_1$. Here $\alpha = 2$ is the case of classical uniform convexity. Clearly, this property is satisfied, if $\mathcal{E}(t,\cdot)$ is uniformly α-convex and $\mathcal{D} : (z_0,z_1) \mapsto \Psi(z_1{-}z_0)$ is merely convex. As a second assumption we need that the power $\partial_t\mathcal{E}$ is Lipschitz or Hölder continuous with respect to the same norm $\|\cdot\|_{\boldsymbol{C}}$, namely,

$$\begin{aligned} &\exists\, \beta \in \,]0,1]\; \forall E>0\; \exists\, C_*>0\; \forall t, q_0, q_1 \text{ with } \mathcal{E}(t,q_0), \mathcal{E}(t,q_1) \leq E: \\ &\quad |\partial_t\mathcal{E}(t,q_1) - \partial_t\mathcal{E}(t,q_0)| \leq C_*\|q_1{-}q_0\|_{\boldsymbol{C}}^{\beta}. \end{aligned} \tag{80}$$

Before stating the main time-regularity result, we emphasize that in smooth cases the convexity condition (79) with $\alpha = 2$ and $\boldsymbol{C} = \boldsymbol{Q}$ is essentially the same as the joint convexity condition (15). In fact, for $\mathcal{E} \in \mathrm{C}^2$ and $\mathcal{R}(\cdot,\widehat{z}) \in \mathrm{C}^1$ condition (79) implies

$$\langle \mathrm{D}^2\mathcal{E}(t,q)\widehat{q},\widehat{q}\rangle + \mathrm{D}_z\mathcal{R}(z,\widehat{z})[\widehat{z}] \geq 2c_*\|\widehat{q}\|_{\boldsymbol{C}}^2 \text{ for all } \widehat{q} \in \boldsymbol{Q}.$$

The following result is taken from [67, 68], where more details and some applications are given.

Theorem 4.9 (Lipschitz and Hölder continuity). *Let* $(\boldsymbol{Q}, \mathcal{E}, \mathcal{D})$ *be a RIS satisfying the power control* (E2), (79) *and* (80). *Then, for every energetic solution* $q : [0,T] \to \boldsymbol{Q}$ *there exists* $C>0$ *such that*

$$\|q(t){-}q(s)\|_{\boldsymbol{C}} \leq C|t{-}s|^{1/(\alpha-\beta)} \quad \textit{for all } s,t \in [0,T].$$

Proof. We choose $E = \sup\{\, \mathcal{E}(t,q(t)) \mid t \in [0,T]\,\}$ and obtain $c_*, C_* > 0$ from (79) and (80). Exploiting the uniform α-convexity we derive an *improved stability estimate.* Indeed, using $q(s) \in \mathcal{S}(s)$ and (79) with $q_0 = q(s)$ and $q_1 = q(t)$ where $s<t$ we obtain

$$\begin{aligned}\mathcal{E}(s,q(s)) &\le \mathcal{E}(s,q_\theta)+\mathcal{D}(z_0,z_\theta) \qquad \text{(where } q_\theta=(1-\theta)q(s)+\theta q(t)\,\text{)}\\ &\le (1-\theta)\mathcal{E}(s,q(s))+\theta\big(\mathcal{E}(s,q(t))+\mathcal{D}(z(s),z(t))\big)-c_*\theta(1-\theta)\|q(t)-q(s)\|_{\boldsymbol{C}}^\alpha.\end{aligned}$$

Subtracting $\mathcal{E}(s,q(s))$, dividing by θ, and taking the limit $\theta\to 0^+$ lead to

$$\mathcal{E}(s,q(s))+c_*\|q(t)-q(s)\|_{\boldsymbol{C}}^\alpha \le \mathcal{E}(s,q(t))+\mathcal{D}(z(s),z(t)), \tag{81}$$

which is the desired improved stability estimate. Employing the dissipation estimate $\mathcal{D}(z(s),z(t))\le \mathrm{Diss}_{\mathcal{D}}(z,[s,t])$ and the energy balance we obtain

$$\begin{aligned} c_*\|q(t)-q(s)\|_{\boldsymbol{C}}^\alpha &\le \mathcal{E}(s,q(t))+\mathcal{D}(z(s),z(t))-\mathcal{E}(s,q(s))\\ &\le \mathcal{E}(s,q(t))-\mathcal{E}(t,q(t))+\mathcal{E}(t,q(t))+\mathrm{Diss}_{\mathcal{D}}(z,[s,t])-\mathcal{E}(s,q(s))\\ &= \int_s^t \partial_r\mathcal{E}(r,q(r))-\partial_r\mathcal{E}(r,q(t))\,\mathrm{d}r \le \int_s^t C_*\|q(t)-q(r)\|_{\boldsymbol{C}}^\beta\,\mathrm{d}r.\end{aligned}$$

Letting $\eta(\tau)=\int_{t-\tau}^t \|q(t)-q(r)\|_{\boldsymbol{C}}^\beta\,\mathrm{d}r$ for $\tau\in[0,t-s]$ leads to $\eta'(\tau)\le (C_*\eta(\tau)/c_*)^{\beta/\alpha}$. Since $\eta(0)=0$ we find $\eta(\tau)\le C_1\tau^{\alpha/(\alpha-\beta)}$ and, thus,

$$\|q(t)-q(s)\|_{\boldsymbol{C}}=\eta'(t-s)^{1/\beta}\le \big(\tfrac{C_*}{c_*}\eta(t-s)\big)^{1/\alpha}\le \big(\tfrac{C_*C_1}{c_*}\big)^{1/\alpha}(t-s)^{1/(\alpha-\beta)},$$

where C_1 depends only on C_*, c_*, α, and β. This is the desired result. □

Since $\boldsymbol{Q}\subset\boldsymbol{C}$ and $\boldsymbol{Q}$ is reflexive, the solutions studied in Theorem 4.9 lie in $\mathrm{C_w}([0,T];\boldsymbol{Q})$ and thus are CD solutions, cf. Proposition 4.7. If $\alpha=2$, $\beta=1$, and $\boldsymbol{C}=\boldsymbol{Q}$ we even obtain differential solutions.

Corollary 4.10. *Let the energetic system $(\boldsymbol{Q},\mathcal{E},\mathcal{D})$ satisfy (79) and (80) with $\alpha=2$ and $\beta=1$ and some reflexive space $\boldsymbol{C}$ such that $\boldsymbol{Q}$ embeds into $\boldsymbol{C}$ continuously.*

(A) If the space $\boldsymbol{C}$ equals $\boldsymbol{Q}$, then every energetic solution is a differential solution.
(B) If there exists $C>0$ such that $\|v\|_{\boldsymbol{X}}\le C\|(0,v)\|_{\boldsymbol{C}}$ for all $v\in\boldsymbol{Z}$, then every energetic solution is a semi-differential solution.

Proof. Theorem 4.9 gives $q\in\mathrm{C^{Lip}}([0,T],\boldsymbol{C})$, which implies $\dot q\in\mathrm{L}^\infty([0,T],\boldsymbol{C})$ by reflexivity of $\boldsymbol{C}$. Now, (A) follows by employing Propositions 4.6 and 4.7.

Part (B) follows similarly using that $q\in\mathrm{L}^\infty([0,T],\boldsymbol{Q})$ and that $q\in\mathrm{C^{Lip}}([0,T],\boldsymbol{C})$ implies $q\in\mathrm{C_w}([0,T],\boldsymbol{Q})$ and $z\in\mathrm{W}^{1,\infty}([0,T],\boldsymbol{X})$. □

We finally mention some results where *existence of solutions* is established in cases of uniform convexity and smoothness properties, and thus are independent of any compactness arguments. In [51, Thm. 7.1] the case $\mathcal{E}\in\mathrm{C}^3(\boldsymbol{Q}_T)$ satisfying (E2) and the uniform convexity $\langle \mathrm{D}_z^2\mathcal{E}(t,q)w,w\rangle\ge\alpha\|w\|_{\boldsymbol{Q}}^2$ and $\Psi\in\mathrm{C}^0(\boldsymbol{Q})$ being 1-homogeneous and convex (no coercivity needed) is studied. It is shown that the RIS $(\boldsymbol{Q},\mathcal{E},\Psi)$ has for each stable initial condition a unique differential solution q, i.e., we have

$$0 \in \partial\Psi(\dot q(t)) + \mathrm{D}_q\mathcal{E}(t,q(t)) \ \text{ for a.a. } t \in [0,T]. \tag{82}$$

Clearly the joint convexity condition (15) holds, since $\rho = 0$. In [36] existence was derived for the case of quadratic energies in a Hilbert space $\boldsymbol{H}$, namely $\mathcal{I}(t,z,) = \frac{1}{2}(z|z) - (\ell(t)|z)$, where $(\cdot|\cdot)$ denotes the scalar product in $\boldsymbol{H}$. The dissipation distance is formulated in terms of the sets $K(z) = \partial_{\dot z}\mathcal{R}(z,0)$. The joint convexity now reads

$$\exists \rho \in\,]0,1[\ \forall z_1, z_2 \in \boldsymbol{H}:\ d_{\mathrm{Hausdorff}}(K(z_1), K(z_2)) \le \rho\|z_1{-}z_2\|_{\boldsymbol{H}}.$$

The definition of $\mathcal{I}$ gives $\kappa = 1$, and it is easy to see that the last condition implies $|\mathcal{R}(z_1,v){-}\mathcal{R}(z_2,v)| \le \rho\|z_1{-}z_2\|_{\boldsymbol{H}}\|v\|_{\boldsymbol{H}}$, but it is unclear whether the existence result of [36] holds under this weaker condition.

Finally we want to comment on the possibility to use convexity and smoothness to obtain *uniqueness results.* We follow the simpler result in [51] and refer to [7] and [46] for generalizations. We again consider (82) with the same specifications for $(\boldsymbol{Q}, \mathcal{E}, \Psi)$ as there. Comparing two differential solutions q_1 and q_1, we can use the monotonicity of $\partial\Psi$ and obtain, for a.a. $t \in [0,T]$, the estimate

$$\mu(t) \stackrel{\text{def}}{=} \langle \mathrm{D}_q\mathcal{E}(t,q_1(t)){-}\mathrm{D}_q\mathcal{E}(t,q_2(t)), \dot q_1(t){-}\dot q_2(t)\rangle \le 0. \tag{83}$$

For $\gamma(t) \stackrel{\text{def}}{=} \langle \mathrm{D}_q\mathcal{E}(t,q_1(t)){-}\mathrm{D}_q\mathcal{E}(t,q_2(t)), q_1(t){-}q_2(t)\rangle$ uniform 2-convexity gives $\gamma(t) \ge \kappa\|q_1(t){-}q_2(t)\|^2$. Moreover,

$$\dot\gamma(t) = \mu(t) + \langle \mathrm{D}_q^2\mathcal{E}(t,q_1)\dot q_1{-}\mathrm{D}_q^2\mathcal{E}(t,q_2)\dot q_2, q_1(t){-}q_2(t)\rangle + \tau(t),$$

where $\tau(t) = \langle \partial_t\mathrm{D}_q\mathcal{E}(t,q_1){-}\partial_t\mathrm{D}_q\mathcal{E}(t,q_2), q_1{-}q_2\rangle$. The smoothness $\mathcal{E} \in \mathrm{C}^3$ gives $|\tau(t)| \le C\|q_1{-}q_2\|^2 \le C\gamma/\alpha$. Subtracting $2\mu(t) \le 0$ and rearranging the terms we find

$$\dot\gamma \le \langle \xi_1, \dot q_1\rangle + \langle \xi_2, \dot q_2\rangle + C\gamma/\alpha$$

where $\xi_j = \mathrm{D}_q\mathcal{E}(t,q_{3-j}) - \mathrm{D}_q\mathcal{E}(t,q_j) - \mathrm{D}_q^2\mathcal{E}(t,q_j)(q_{3-j}{-}q_j)$. From $q_j \in \mathrm{C}^{\mathrm{Lip}}([0,T],\boldsymbol{Q})$ we know that $\|\dot q_j\|$ is bounded, and differentiability of $\mathcal{E}$ implies $\|\xi_j\| \le C\|q_1{-}q_2\|^2 \le C\gamma/\kappa$. Thus, $\dot\gamma \le C_*\gamma$ and Gronwall's lemma provide

$$\|q_1(t) - q_2(t)\|_{\boldsymbol{Q}} \le C\|q_1(0) - q_2(0)\|_{\boldsymbol{Q}}\mathrm{e}^{C_* t},$$

which gives the desired uniqueness result.

In [7, 46] the uniqueness results are generalized to RIS in a Hilbert space $\boldsymbol{H}$ with a dissipation potential $\mathcal{R}$ depending on $z \in \boldsymbol{H}$. The key is to use the auxiliary function $\mathcal{B}(z,\xi) = \sup\{\, \langle \xi, v\rangle - \frac{1}{2}\mathcal{R}(q,v) \mid v \in \boldsymbol{H}\,\}$, where $\frac{1}{2}{-}\mathcal{B}\big(z(t), -\mathrm{D}_z\mathcal{I}(t,z)\big)$ measures the distance of $\mathrm{D}_z\mathcal{I}(t,z)$ from the boundary of $\partial\mathcal{R}(z,0)$. Under restrictive assumption it is then possible to derive estimates of the type $\Gamma(t) \leq C\mathrm{e}^{C_* t}\Gamma(0)$ for the combined quantity

$$\Gamma(t) \stackrel{\text{def}}{=} \sqrt{\gamma(t)} + \Big|\mathcal{B}\big(z_1(t), -\mathrm{D}_z\mathcal{I}(t, z_1(t))\big) - \mathcal{B}\big(z_2(t), -\mathrm{D}_z\mathcal{I}(t, z_2(t))\big)\Big|.$$

4.4 Parametrized Solutions Via the Vanishing-Viscosity Approach

A main challenge in modeling rate-independent processes is the appearance of jumps. Since rate independence is a limit for systems under vanishing loading rates, we expect solutions of rate-independent systems to occur as limits of systems with relaxation times that are very small compared to the changes in the loading. Thus, we expect the solutions to occur as pointwise limits of time-continuous solutions. In particular, in a nonconvex situation solutions change slowly by following the loading for most of the time, but in-between there are sudden transitions from one stable regime to another one.

Here we define notions of solutions that are associated with the so-called *vanishing-viscosity approach.* For rate-independent processes this was proposed in [22] and further analyzed in [37, 53, 59, 60]. In particular, we use the idea of arclength parametrization of solutions developing jumps, which was established earlier for systems with dry friction and small viscosity, see [9, 29, 54, 61].

For simplicity we restrict the presentation in this and the following sections to the reduced case, since the vanishing-viscosity approach does not help to control the non-dissipative component $y \in \boldsymbol{Y}$ in the limit. Reasonable theories could be obtained in the case that $\mathcal{E}(t,\cdot,z)$ has a unique minimizer $y = Y(t,z)$ such that for the reduced energy functional $\mathcal{I}(t,z) = \mathcal{E}(t, Y(t,z), z)$ the power $\partial_t\mathcal{I}(t,z) = \partial_t\mathcal{E}(t, Y(t,z), z)$ is well defined.

For superlinear dissipation potentials $\mathcal{R}_{\mathrm{sl}}$ we define the rescaled potential $\mathcal{R}_\varepsilon(z,v) = \frac{1}{\varepsilon}\mathcal{R}_{\mathrm{sl}}(z,\varepsilon v)$, where the small parameter $\varepsilon > 0$ is the quotient obtained from dividing the time scale induced by the loading by the relaxation time due to viscosity (which is inverse to the viscosity). For parametrized solutions, the dissipation potential $\mathcal{R}$ must have the special form

$$\mathcal{R}_\varepsilon(z,v) = \mathcal{R}(z,v) + \tfrac{\varepsilon}{2}\langle \mathbb{V}(z)v, v\rangle,$$

where $\mathcal{R}$ is the rate-independent part and $\frac{\varepsilon}{2}\langle \mathbb{V}(z)v, v\rangle$ the small viscous part. Thus, we are led to study the differential inclusion

$$0 \in \partial_{\dot z}\mathcal{R}(z(t), \dot z(t)) + \varepsilon\mathbb{V}(z(t))\dot z(t) + \overline{\partial}_z\mathcal{I}(t, z(t)), \quad z(0) = z_0. \tag{84}$$

We simplify further by assuming that $\mathbb{V}$ is independent of z and denote by $\boldsymbol{V}$ the Hilbert space with norm $\|w\|_{\boldsymbol{V}} = \big(\langle \mathbb{V}w, w\rangle\big)^{1/2}$, i.e., $\mathbb{V}: \boldsymbol{V} \to \boldsymbol{V}^*$ is a norm-preserving bijection.

Under reasonable assumptions (see e.g., [13, 15, 63, 65]) one obtains solutions z^ε of (84) satisfying the energy identity

$$\mathcal{I}(t, z^\varepsilon(t)) + \int_0^t \mathcal{R}(z^\varepsilon(s), \dot z^\varepsilon(s)) + \varepsilon\|\dot z^\varepsilon(s)\|_{\boldsymbol{V}}^2 \,\mathrm{d}s = \mathcal{I}(0, z_0) + \int_0^t \partial_s\mathcal{I}(s, z^\varepsilon(s))\,\mathrm{d}s.$$

Using the coercivity $\mathcal{R}(z, w) \geq c_0\|w\|_{\boldsymbol{X}}$ and the coercivity of $\mathcal{I}$ provides the a priori estimates

$$\mathrm{Var}_{\boldsymbol{X}}(z^\varepsilon, [0, T]) + \varepsilon\|\dot z^\varepsilon\|^2_{\mathrm{L}^2([0,T];\boldsymbol{V})} + \|z^\varepsilon\|_{\mathrm{L}^\infty([0,T];\boldsymbol{Z})} \leq C. \tag{85}$$

Clearly for $\varepsilon > 0$ the solutions satisfy $z^\varepsilon \in \mathrm{H}^1([0, T]; \boldsymbol{V}) \subset \mathrm{C}^0([0, T]; \boldsymbol{V})$. Thus, solutions are not able to jump over potential barriers as it is the case for energetic solutions. Even in the limit $\varepsilon \to 0$ the potential barriers remain active and delay possible jumps.

These a priori bounds allow us to take the limit $\varepsilon \to 0$, if we assume that the embedding $\boldsymbol{Z} \subset \boldsymbol{X}$ is compact. Using the bound on the variation in $\boldsymbol{X}$, we are then able to apply Helly selection principle (cf. Thm. 3.13)) to extract a subsequence $(z^{\varepsilon_n})_{n\in\mathbb{N}}$ with $\varepsilon_n \to 0$ such that $z^{\varepsilon_n}(t) \rightharpoonup z(t)$ in $\boldsymbol{Z}$ for some limit $z : [0, T] \to \boldsymbol{Z}$. This limit is then called a $\mathbb{V}$-*approximable solution* of the RIS $(\boldsymbol{Z}, \mathcal{I}, \mathcal{R})$, cf. [12,34,69].

We proceed further by deriving equations that characterize such limit solutions. For this we introduce the concept of parametrized solutions that should be seen as a helpful, intermediate tool in understanding the limit procedure. The idea for resolving jumps in rate-independent systems is to consider the graph of the viscous solutions in the extended phase space $\boldsymbol{Z}_T$ and to study the limit of the whole graph. The advantage is that jumps do not shrink to a single point at jump time t, but provide a jump curve lying in the plane $\{t\}\times\boldsymbol{Z}$. In [22] it was observed that the scaling invariance of RIS can be effectively used for parametrizing these graphs.

For a viscous solution $z^\varepsilon : [0, T] \to \boldsymbol{Z}$ we consider the graph

$$\mathrm{Graph}(z^\varepsilon) \overset{\mathrm{def}}{=} \{\,(t, z^\varepsilon(t)) \mid t \in [0, T]\,\} \subset \boldsymbol{Z}_T.$$

We use an arclength parametrization that is based on the viscous norm, namely $s = \sigma^\varepsilon(t) = t + \int_0^t \|\dot z^\varepsilon(r)\|_{\boldsymbol{V}}\,\mathrm{d}r$, which has the inverse $t = \tau^\varepsilon(s)$. The choice of the $\boldsymbol{V}$-norm is crucial to maintain the structure of a generalized gradient flow. Introducing the rescaled function $Z^\varepsilon(s) = z^\varepsilon(\tau^\varepsilon(s))$ we observe that it is a solution of the transformed problem

$$\left.\begin{aligned} &\tau(0) = 0,\ Z(0) = z_0, \qquad \dot\tau(s) + \|\dot Z(s)\|_{\boldsymbol{V}} = 1, \text{ and} \\ &0 \in \partial_{\dot z}\mathcal{R}(Z(s), \dot Z(s)) + \tfrac{\varepsilon}{1-\|\dot Z(s)\|_{\boldsymbol{V}}}\mathbb{V}\dot Z(s) + \overline{\partial}_z\mathcal{I}(\tau(s), Z(s)) \end{aligned}\right\} \tag{86}$$

for a.a. $s \in [0, S^\varepsilon]$, where $S^\varepsilon = \sigma^\varepsilon(T)$. For this, it was essential that $\partial\mathcal{R}(z,\cdot)$ is 0-homogeneous.

The main observation is that the viscous term with ε as a prefactor is again a subdifferential, namely of the potential

$$\mathcal{V}_\varepsilon(w) \stackrel{\text{def}}{=} \varepsilon g(\|w\|_{\boldsymbol{V}}) \ \text{ with } g(\nu) = \begin{cases} -\log(1-\nu) - \nu & \text{for } \nu < 1 \\ \infty & \text{otherwise.} \end{cases}$$

Such a potential only exists because we used the norm $\|\cdot\|_{\boldsymbol{V}}$ for parametrizing the graph. Thus, defining the potential $\widetilde{\mathcal{R}}_\varepsilon(z,w) \stackrel{\text{def}}{=} \mathcal{R}(z,w)+\mathcal{V}_\varepsilon(w)$ we can rewrite the second equation in (86) in the form $0 \in \partial_{\dot{Z}}\widetilde{\mathcal{R}}_\varepsilon(Z,\dot{Z}) + \overline{\partial}\mathcal{I}(\tau, Z)$. Moreover, $\widetilde{\mathcal{R}}_\varepsilon$ converges monotonously to the limit functional

$$\widetilde{\mathcal{R}}_0(z,w) \stackrel{\text{def}}{=} \begin{cases} \mathcal{R}(z,w) & \text{for } \|w\|_{\boldsymbol{V}} \le 1, \\ \infty & \text{otherwise.} \end{cases}$$

In the case $\boldsymbol{V} \subset \boldsymbol{X}$ the convergence is even a Mosco convergence, and $\widetilde{\mathcal{R}}_0(z,\cdot) : \boldsymbol{V} \to [0,\infty]$ is weakly lower semicontinuous on $\boldsymbol{V}$. We let $\mathcal{R}^{\boldsymbol{V}}(z,w) = \widetilde{\mathcal{R}}_0(z,w)$ and observe that we have the sum rule for the subdifferentials

$$\partial_{\dot{z}}\mathcal{R}^{\boldsymbol{V}}(z,w) = \partial_{\dot{z}}\mathcal{R}(z,w) + \partial\mathcal{V}_0(w), \ \text{ where } \mathcal{V}_0(w) = \begin{cases} 0 & \text{for } \|w\|_{\boldsymbol{V}} \le 1, \\ \infty & \text{otherwise.} \end{cases}$$

In the case $\boldsymbol{X} \subset \boldsymbol{V}$, the functional $\widetilde{\mathcal{R}}_0(z,\cdot)$ must be extended to $\boldsymbol{V}$ via ∞ outside of $\boldsymbol{X}$. In general, $\widetilde{\mathcal{R}}_0(z,\cdot)$ is not weakly lower semicontinuous and we set

$$\mathcal{R}^{\boldsymbol{V}}(z,\cdot) = \text{wlsc}\widetilde{\mathcal{R}}_0(z,\cdot) : w \mapsto \inf\{\liminf_{k\to\infty} \widetilde{\mathcal{R}}_0(z,w_k) \mid w_k \rightharpoonup w \text{ in } \boldsymbol{V}\},$$

where "wlsc" denotes the weak lower semicontinuous hull. Taking the formal limit $\varepsilon \to 0$ in (86) we are led to the following definition.

Definition 4.11 (Parametrized solutions). Let the RIS $(\boldsymbol{Z},\mathcal{I},\mathcal{R})$ and $\boldsymbol{X}$ be given such that (74a) holds. Moreover, let $\mathbb{V}$ and $\boldsymbol{V}$ be given as above.

Then, a pair $\zeta = (\tau, Z) : [0,S] \to \boldsymbol{Z}_T$ is called a $\mathbb{V}$-*parametrized solution,* if $(\tau, Z) \in \mathrm{W}^{1,1}([0,T], \mathbb{R}\times\boldsymbol{V})$ and the following equations hold:

$$\left.\begin{aligned} &\tau(0) = 0, \ \tau(S) = T, \ \dot{\tau}(s) \ge 0, \ \dot{\tau}(s) + \|\dot{Z}(s)\|_{\boldsymbol{V}} = 1 \\ &0 \in \partial_{\dot{z}}\mathcal{R}^{\boldsymbol{V}}(Z(s),\dot{Z}(s)) + \overline{\partial}_z\mathcal{I}(\tau(s), Z(s)) \end{aligned}\right\} \text{ a.e. on } [0,S]. \tag{87}$$

We also say that ζ is a parametrized solution to the RIS $(\boldsymbol{Z},\mathcal{I},\mathcal{R},\mathbb{V})$.

The present definition follows [22] and hides the rate-independent nature by asking for a strict arclength parametrization in $\boldsymbol{V}_T \stackrel{\text{def}}{=} [0,T]\times\boldsymbol{V} \subset \mathbb{R}\times\boldsymbol{V}$, where we use the extended norm $\|(t,v)\|_{\boldsymbol{V}_T} = |t| + \|v\|_{\boldsymbol{V}}$. Following [59, 60], the parametrization may be kept free by replacing the last two relations in (87) by

$$\dot\tau(s) + \|\dot Z(s)\|_{\boldsymbol{V}} = \alpha(s), \quad 0 \in \partial_{\dot z}\mathcal{R}^{\boldsymbol{V}}(Z(s), \tfrac{1}{\alpha(s)}\dot Z(s)) + \overline{\partial}_z\mathcal{I}(\tau(s), Z(s)),$$

where the *parametrization function* $\alpha \in \mathrm{L}^1([0,S])$ satisfies $\alpha(s) > 0$ a.e. in $[0,S]$. Clearly, a rescaling of the graph does not change the problem. Moreover, a rescaling of the time dependence of $\mathcal{I}$ can be compensated by a rescaling as follows. If $\widetilde{\mathcal{I}}(\widetilde t, z) = \mathcal{I}(\phi(\widetilde t), z)$ and $\zeta : [0,S] \to \boldsymbol{V}_T$ is a parametrized solution to $(\boldsymbol{Z}, \mathcal{I}, \mathcal{R}, \mathbb{V})$ with parametrization function α, then $\widetilde\zeta : \widetilde t \mapsto \zeta(\phi(\widetilde t))$ is a parametrized solution to $(\boldsymbol{Z}, \widetilde{\mathcal{I}}, \mathcal{R}, \mathbb{V})$ with parametrization function $\widetilde\alpha : \widetilde t \mapsto \alpha(\phi(\widetilde t))\phi'(\widetilde t)$.

The main feature of parametrized solutions can be seen by discussing the subdifferential of $\mathcal{R}^{\boldsymbol{V}}$, which is done using the polar $\mathcal{R}^\circ(z,\cdot)$ of $\mathcal{R}(z,\cdot)$, where the polar Ψ° of a convex potential Ψ is defined via $\Psi^\circ(\xi) \stackrel{\text{def}}{=} \sup\{\,\langle \xi, v\rangle \mid \Psi(v) \leq 1\,\}$. Along the arclength parametrized solutions we can distinguish three different dynamical regimes:

Sticking: We have $\dot Z(s) = 0$ and $\dot\tau(s) = 1$, i.e. the potential forces $\xi(s) \in \overline{\partial}_z\mathcal{I}(\zeta(s))$ are so small that the state does not change, namely $\mathcal{R}^\circ(z(s), -\xi(s)) \leq 1$ or equivalently $0 \in \partial\mathcal{R}(Z(s), 0) + \xi(s)$.
Rate-independent slip: We have $0 < \|\dot Z\|_{\boldsymbol{V}} < 1$ and $0 < \dot\tau(s) < 1$, i.e. the state changes so slowly that the rate-independent friction is strong enough to compensate the driving force, namely $0 \in \partial\mathcal{R}(Z(s), \dot Z(s)) + \xi(s)$, which implies $\mathcal{R}^\circ(z(s), -\xi(s)) = 1$.
Viscous jump: We have $\|\dot Z\|_{\boldsymbol{V}} = 1$ and $\dot\tau(s) = 0$, i.e., the motion is faster than the loading scale, and the system moves in a jump-like fashion. During this jump phase the driving force $\xi(s)$ satisfies $0 \in \partial\mathcal{R}(Z(s), \dot Z(s)) + \lambda(s)\mathbb{V}\dot Z(s) + \xi(s)$ for some $\lambda(s) \geq 0$, which implies $\mathcal{R}^\circ(z(s), -\xi(s)) \geq 1$.

From this we also find another equivalent formulation of (87), namely

$$\left.\begin{array}{l} \tau(0) = 0,\ \tau(S) = T,\ \dot\tau(s) \geq 0,\ \dot\tau(s) + \|\dot Z(s)\|_{\boldsymbol{V}} = 1 \\ 0 \in \partial_{\dot z}\mathcal{R}(Z(s), \dot Z(s)) + \lambda(s)\mathbb{V}\dot Z(s) + \overline{\partial}_z\mathcal{I}(\tau(s), Z(s)) \\ \lambda(s) \geq 0, \qquad \lambda(s)(1 - \|\dot Z(s)\|_{\boldsymbol{V}}) = 0 \end{array}\right\} \text{ a.e. on } [0,S]. \tag{88}$$

Using the chain rule for $\mathcal{I}$ we obtain an energy balance in the form

$$\begin{aligned} &\mathcal{I}(\zeta(s_2)) + \textstyle\int_{s_1}^{s_2} \mathcal{R}(Z(s), \dot Z(s)) + \lambda(s)\|\dot Z(s)\|_{\boldsymbol{V}}^2\,\mathrm{d}s \\ &= \mathcal{I}(\zeta(s_1)) + \textstyle\int_{s_1}^{s_2} \partial_t\mathcal{I}(\zeta(s))\dot\tau(s)\,\mathrm{d}s, \end{aligned} \tag{89}$$

which shows the viscous contribution $\lambda\|\dot{Z}\|_{\boldsymbol{V}}^2$ to the total dissipation that remains in the vanishing-viscosity limit.

Passing from arclength parametrized solutions to the limit $\varepsilon \to 0$ we may arrive at limits $\zeta : [0, S] \to \boldsymbol{Z}_T$ that are not properly parametrized but satisfy

$$\left.\begin{array}{l}\tau(0)=0,\ \dot{\tau}(s)\geq 0,\ \dot{\tau}(s){+}\|\dot{Z}(s)\|_{\boldsymbol{V}}=\alpha(s)\in[0,1]\\ 0\in\partial_{\dot{z}}\mathcal{R}^{\boldsymbol{V}}(Z(s),\dot{Z}(s))+\overline{\partial}_z\mathcal{I}(\tau(s),Z(s))\end{array}\right\}\ \text{a.e. on } [0,S]. \qquad (90)$$

The following lemma states that reparametrization of the graph then leads to an arclength parametrized solution again.

Lemma 4.12 (Reparametrization). *Assume that* $\zeta \in \mathrm{C_w}([0, S]; \boldsymbol{Z}_T)$ *satisfies* (90). *Let* $\widehat{\sigma}(s) = \int_0^s \dot{\tau}(r){+}\|\dot{Z}(r)\|_{\boldsymbol{V}}\,\mathrm{d}r$, $\widehat{S} = \widehat{\sigma}(S)$. *If* $\widehat{S} > 0$, *then the* reparametrization

$$\widehat{\zeta} : \begin{cases} [0,\widehat{S}] \to \quad \boldsymbol{Z}_T; \\ \quad \widehat{s} \quad \mapsto \zeta(\sigma_{\min}(\widehat{s})) \end{cases} \quad \textit{with } \sigma_{\min}(\widehat{s}) = \inf\{\, s \in [0, S] \mid \widehat{\sigma}(s) = \widehat{s}\,\} \qquad (91)$$

satisfies $\widehat{\zeta} \in \mathrm{C^{Lip}}([0, \widehat{S}], \boldsymbol{V}_T) \cap \mathrm{C_w}([0, \widehat{S}], \boldsymbol{Z}_T)$ *and is a* $\mathbb{V}$*-parametrized solution.*

Proof. Since the result is more or less standard, we sketch the arguments and refer to [60, Lem. 4.1] and [59, Prop 6.10] for more details. Clearly, $\widetilde{\zeta}$ is well-defined, and for $0 \leq \widehat{s}_1 < \widehat{s}_2 \leq \widehat{S}$ we have

$$\begin{aligned}\|\widehat{\zeta}(\widehat{s}_1){-}\widehat{\zeta}(\widehat{s}_2)\|_{\boldsymbol{V}_T} &\overset{\text{def}}{=} |\widehat{\tau}(\widehat{s}_1) - \widehat{\tau}(\widehat{s}_2)| + \|\widehat{Z}(\widehat{s}_1){-}\widehat{Z}(\widehat{s}_2)\|_{\boldsymbol{V}}\\ &\leq \|\zeta(\sigma_{\min}(\widehat{s}_1)) - \zeta(\sigma_{\min}(\widehat{s}_2))\|_{\boldsymbol{V}_T} \leq \widehat{\sigma}(\sigma_{\min}(\widehat{s}_2)) - \widehat{\sigma}(\sigma_{\min}(\widehat{s}_1)) = \widehat{s}_2 - \widehat{s}_1.\end{aligned}$$

Thus, $\dot{\widehat{\zeta}}$ exists almost everywhere, and we obtain $\|\dot{\widehat{\zeta}}(\widehat{s})\|_{\boldsymbol{V}_T} = 1$ a.e. on $[0, \widehat{S}]$ by the standard chain rule. □

The existence theory for parametrized solutions is a delicate matter depending on the choice of the space $\boldsymbol{V}$. If we choose $\mathbb{V}$ such that $\boldsymbol{X} \subset \boldsymbol{V}$, then we have $\|w\|_{\boldsymbol{V}} \leq C\mathcal{R}(z, w)$, and the a priori estimates (85) imply that $S^\varepsilon = T{+}\int_0^T \|\dot{Z}^\varepsilon(t)\|_{\boldsymbol{V}}\,\mathrm{d}t \leq C$. Thus, it is easy to extract a converging subsequence. However, it is difficult to control the convergence of $\partial\widetilde{\mathcal{R}}_\varepsilon(Z^\varepsilon, \dot{Z}^\varepsilon)$ towards $\partial\mathcal{R}^{\boldsymbol{V}}$. In the opposite case $\boldsymbol{V} \subset \boldsymbol{X}$ the convergence of the subgradients of $\partial\widetilde{\mathcal{R}}_\varepsilon$ to $\partial\mathcal{R}^{\boldsymbol{V}}$ follows easily from the Mosco convergence, but it is unclear whether the arclength S^ε of the curves stays bounded.

Example 4.13. In fact, Example 4.3 with $\alpha = \infty$ provides a case with $\boldsymbol{V} = \mathrm{L}^2(\Omega) \subset \boldsymbol{X} = \mathrm{L}^1(\Omega)$ for which $S^\varepsilon \to \infty$ and hence $\widehat{S} = 0$. For $t \in [\varepsilon, 1]$ we have

$$\|\dot{z}^\varepsilon(t)\|_{\mathrm{L}^1} = 1 \ \text{ and } \ \|\dot{z}^\varepsilon(t)\|_{\mathrm{L}^2} = 1/\sqrt{\varepsilon}.$$

Hence, the L^2 parametrized viscous solutions $(\tau^\varepsilon, Z^\varepsilon)$ satisfy $\tau^\varepsilon(s) \to \infty$ and $Z^\varepsilon(s) \to 0$ for all $s > 0$.

In the rest of this section, we reduce the discussion to our standard Example 2.8. The main reason for this restriction is that we need to show that S^ε remains bounded. This is trivial in the finite-dimensional cases treated in [22, 59], where $\boldsymbol{X} = \boldsymbol{V}$. For infinite-dimensional cases with $\boldsymbol{V} \subsetneqq \boldsymbol{X}$ this problem is solved only in the semilinear setting discussed below.

The first step towards the existence theory is a general convergence result for the vanishing-viscosity limit in the case $\boldsymbol{V} \subset \boldsymbol{X}$. The proof is based on the energetic formulation of generalized gradient systems as introduced in Sect. 2.5. We employ the Mosco convergence of $\widetilde{\mathcal{R}}_\varepsilon$ to $\widetilde{\mathcal{R}}_0$ and the chain-rule inequality (71). Since $\widetilde{\mathcal{R}}_\varepsilon(v) = \Psi(v) + \varepsilon g(\|v\|_{\boldsymbol{V}})$ and $\Psi^*(\xi) = 0$ for $\xi \in \partial\Psi(0)$ and ∞ otherwise, the Fenchel transform reads

$$\widetilde{\mathcal{R}}^*_\varepsilon(\xi) = \min\{\, \varepsilon g^*(\tfrac{1}{\varepsilon}\|\xi{-}\kappa\|_{\boldsymbol{V}^*}) \mid \kappa \in \partial\Psi(0)\,\}$$

where $g^*(\varrho) = \sup_{r\geq 0} \varrho r{-}g(r)$. In the limit $\varepsilon \to 0$ we find the Γ-limits

$$\begin{aligned}\widetilde{\mathcal{R}}_0(v) &= \Psi(v) + \mathcal{V}_0(v), \quad \text{and}\\ \widetilde{\mathcal{R}}^*_0(\xi) &= M^{\boldsymbol{V}}_\Psi(\xi) \overset{\text{def}}{=} \min\{\, \|\xi{-}\kappa\|_{\boldsymbol{V}^*} \mid \kappa \in \partial\Psi(0)\,\},\end{aligned} \tag{92}$$

where $\mathcal{V}_0(v) = 0$ for $\|v\|_{\boldsymbol{V}} \leq 1$ and ∞ otherwise.

Proposition 4.14 (Vanishing-viscosity limit). *Let $(\boldsymbol{Z}, \mathcal{J}, \Psi)$, $\boldsymbol{X}$, and $\boldsymbol{V}$ be given as in Example 2.8. Assume that $\zeta^\varepsilon = (\tau^\varepsilon, Z^\varepsilon) : [0,S] \to \boldsymbol{Z}_T$ satisfy (86) on $[0,S]$ with $Z^\varepsilon(0) = z_0 \in \boldsymbol{Z}$. Then there exist a subsequence $(\varepsilon_n)_{n\in\mathbb{N}}$ and functions $\zeta : [0,S] \in \mathrm{C}^{\mathrm{Lip}}([0,S], \boldsymbol{V})$ and $\alpha \in \mathrm{L}^\infty([0,S])$ such that (90) and the convergences $\zeta^{\varepsilon_n}(s) \rightharpoonup \zeta(s)$ in $\boldsymbol{Z}$ for all $s \in [0,S]$ and $\dot\zeta^{\varepsilon_n} \overset{*}{\rightharpoonup} \dot\zeta$ in $\mathrm{L}^\infty([0,S]; \boldsymbol{V}_T)$ hold.*

If $\widehat{S} = \int_0^S \dot\tau(r){+}\|\dot Z(r)\|_{\boldsymbol{V}}\,\mathrm{d}r > 0$, then the reparametrization $\widehat{\zeta} : [0,\widehat{S}]{\times}\boldsymbol{Z}_T$ defined in (91) is a $\mathbb{V}$-parametrized solution.

Proof. We have uniform a priori bounds in $\mathrm{C}^{\mathrm{Lip}}([0,S]; \boldsymbol{V})$ and $\mathrm{L}^\infty([0,T]; \boldsymbol{Z}_T)$ for ζ^ε. Thus, we find a $\zeta \in \mathrm{C}^{\mathrm{Lip}}([0,T]; \boldsymbol{V}) \cap \mathrm{L}^\infty([0,T]; \boldsymbol{Z}_T)$ such that

(i) $\zeta^\varepsilon \to \zeta$ in $\mathrm{C}([0,T]; \boldsymbol{V}_T)$,
(ii) $\dot\zeta^\varepsilon \overset{*}{\rightharpoonup} \dot\zeta$ in $\mathrm{L}^\infty([0,T]; \boldsymbol{V}_T)$,
(iii) $Z^\varepsilon(t) \overset{*}{\rightharpoonup} Z(t)$ in $\boldsymbol{Z}$ for all $t \in [0,T]$.

along a suitable subsequence (not relabeled). For (iii) note that $\zeta^\varepsilon([0,S]; \boldsymbol{V}_T) \cap \mathrm{L}^\infty([0,S]; \boldsymbol{Z}_T)$ implies $\zeta^\varepsilon \in \mathrm{C}_{\mathrm{w}}([0,S]; \boldsymbol{Z}_T)$. The uniform bound $\|Z^\varepsilon(s)\|_{\boldsymbol{Z}} \leq C$ for all $\varepsilon > 0$ and $s \in [0,S]$ implies that for each fixed s_* $\zeta^{\varepsilon_j}(s_*) \rightharpoonup \zeta_*$ in $\boldsymbol{Z}_T$. Since, we know $\zeta^\varepsilon(s_*) \to \zeta(s_*)$ in $\boldsymbol{V}_T$ we obtain (iii).

For $\xi^\varepsilon(s) = \mathrm{D}_z\mathcal{J}(\zeta^\varepsilon(s))$ we use (25) and the bounds on Z^ε to conclude that ξ^ε is bounded in $\mathrm{L}^\infty([0,T]; \boldsymbol{Z}^*)$. Choosing a further subsequence (not relabeled) we find $\xi^\varepsilon \overset{*}{\rightharpoonup} \xi$ in $\mathrm{L}^\infty([0,T], \boldsymbol{Z}^*)$. The semilinear structure (25) of $\mathrm{D}_z\mathcal{J}(t,\cdot) : \boldsymbol{Z} \to \boldsymbol{Z}^*$ implies weak continuity, which gives $\xi^{\varepsilon_n}(s) \rightharpoonup \xi(s) = \mathrm{D}_z(\zeta(s))$.

In the limit $\varepsilon \to 0$, we easily obtain the upper line of conditions in (90). To obtain the differential inclusion in the lower line, we use the equivalent formulation via the Legendre-Fenchel duality of the lower line in (86), namely

$$\mathcal{I}(\zeta^\varepsilon(S)) + \mathcal{M}_\varepsilon(\dot{Z}^\varepsilon, -\xi^\varepsilon) = \mathcal{I}(0, z_0) + \int_0^S \partial_t \mathcal{I}(\zeta^\varepsilon(s))\dot{\tau}^\varepsilon(s)\,\mathrm{d}s, \tag{93}$$
$$\text{where } \mathcal{M}_\varepsilon(V, \xi) = \int_0^S \widetilde{\mathcal{R}}_\varepsilon(V(s)) + \widetilde{\mathcal{R}}_\varepsilon^*(\xi(s))\,\mathrm{d}s.$$

We extend $\widetilde{\mathcal{R}}_\varepsilon^*$ for $\varepsilon \in [0,1]$ to $\boldsymbol{Z}^*$ by ∞ outside of $\boldsymbol{V}^*$. By the definition (92) and direct calculation we obtain the liminf estimates

$$V^\varepsilon \rightharpoonup V \in \boldsymbol{V} \implies \widetilde{\mathcal{R}}_0(V) \le \liminf_{\varepsilon \to 0} \widetilde{\mathcal{R}}_\varepsilon(V^\varepsilon),$$
$$\xi^\varepsilon \rightharpoonup \xi \in \boldsymbol{Z}^* \implies \widetilde{\mathcal{R}}_0^*(\xi) \le \liminf_{\varepsilon \to 0} \widetilde{\mathcal{R}}_\varepsilon^*(\xi^\varepsilon).$$

In fact, by [4, Sect. 3.3.1] this is equivalent to Mosco convergence of $\widetilde{\mathcal{R}}_\varepsilon$ to $\widetilde{\mathcal{R}}_0$. Hence, $\mathcal{M}_\varepsilon$ weakly Γ-converges to $\mathcal{M}_0$ on $\mathrm{L}^2([0,S];\boldsymbol{V}) \times \mathrm{L}^2([0,S];\boldsymbol{Z}^*)$, and we can pass to the limit $\varepsilon \to 0$ in (93) and obtain

$$\mathcal{I}(\zeta(S)) + \mathcal{M}_0(\dot{Z}, -\xi) \le \mathcal{I}(0, 0, z_0) + \int_0^S \partial_t \mathcal{I}(\zeta(s))\dot{\tau}(s)\,\mathrm{d}s,$$

where we use $\dot{\tau}^\varepsilon \overset{*}{\rightharpoonup} \dot{\tau}$ in $\mathrm{L}^\infty([0,S])$ and $\partial_t \mathcal{I}(\zeta^\varepsilon(\cdot)) \to \partial_t \mathcal{I}(\zeta(\cdot))$ in $\mathrm{L}^1([0,T])$ the linearity of $\partial_t \mathcal{I}(t,z) = -\langle \ell(t), z\rangle$.

Note that $\mathcal{M}_0(\dot{Z}, \xi) < \infty$ implies $\xi(\cdot) \in \mathrm{L}^1([0,S];\boldsymbol{V}^*)$, since $\widetilde{\mathcal{R}}_0^*(\xi(\cdot)) \in \mathrm{L}^1([0,S])$ and $\|\xi\|_{\boldsymbol{V}^*} \le \widetilde{\mathcal{R}}_0^*(\xi) + k_0$ with $k_0 = \sup_{\kappa \in \partial\Psi(0)} \|\kappa\|_{\boldsymbol{V}^*} < \infty$. Thus, exploiting $\|\dot{Z}(s)\|_{\boldsymbol{V}} \le 1$ a.e., we can use the chain-rule inequality (71) for the limit function ζ and find

$$\mathcal{M}_0(\dot{Z}, \xi) = \int_0^S \widetilde{\mathcal{R}}_0(\dot{Z}(s)) + \widetilde{\mathcal{R}}_0^*(-\xi(s))\,\mathrm{d}s \le \int_0^S \langle \dot{Z}(s), -\xi(s)\rangle\,\mathrm{d}s.$$

By the definition of the Legendre-Fenchel duality we have $\langle \dot{Z}(s), -\xi(s)\rangle \le \widetilde{\mathcal{R}}_0(\dot{Z}(s)) + \widetilde{\mathcal{R}}_0^*(-\xi(s))$ and conclude equalities. We obtain $0 \in \partial\widetilde{\mathcal{R}}_0(\dot{Z}(s)) + \xi(s)$, which gives the desired differential inclusion on the lower line in (90). □

Our main existence result for parametrized solutions uses the spaces

$$\boldsymbol{Z}_1 \Subset \boldsymbol{Z} \Subset \boldsymbol{V} \Subset \boldsymbol{Z}_{-1}, \quad \text{and} \quad \boldsymbol{V} \subset \boldsymbol{X},$$

as defined in Example 2.8.

Theorem 4.15 (Parametrized solutions). *Let all assumptions of Example 2.8 and* (25) *hold for* $(\boldsymbol{Z}, \mathcal{I}, \Psi, \mathbb{V})$ *and the spaces* $\boldsymbol{Z} \Subset \boldsymbol{V} \subset \boldsymbol{X}$. *Then, for each* $z_0 \in \boldsymbol{Z}_1$ *there exists a parametrized solution* $\zeta = (\tau, Z) : [0,S] \to \boldsymbol{Z}_T$ *with* $Z(0) = z_0$, *which further satisfies* $\zeta \in \mathrm{C_w}([0,S];\boldsymbol{Z}_T) \cap \mathrm{BV}([0,S],\boldsymbol{Z}_T)$.

The proof of this result relies on the convergence result established above together with an estimate on S^ε. The latter is based on higher-order a priori estimates. It is a surprising feature of rate-independent systems that certain a priori estimates are independent of the dissipation functional $\mathcal{R}$. We are in the case of translationally invariant dissipations $\mathcal{R}(z,v) = \Psi(v)$, where Ψ is assumed to be convex, 1-homogeneous, and continuous on $\boldsymbol{V}$. The three a priori estimates we will use derive from the following basic properties of Ψ:

(i) $\langle \partial\Psi(v), v\rangle = \Psi(v)$,
(ii) " $\langle \mathrm{D}^2\Psi(v)[w], v\rangle = \langle \mathrm{D}^2\Psi(v)[v], w\rangle = 0$ ",
(iii) $\langle \partial\Psi(v_1) - \partial\Psi(v_2), v_1 - v_2\rangle \geq 0$.

Here the first relation is 1-homogeneity, and the third is simply monotonicity. The middle relation was put into quotation marks, since $\mathrm{D}^2\Psi$ does not exist; however, by 0-homogeneity of $\partial\Psi$ the directional derivative $\partial\Psi(v)$ in the direction v is 0.

We first state the corresponding a priori estimates that are obtained from

$$0 \in \partial\Psi(\dot z(t)) + \varepsilon\mathbb{V}\dot z(t) + \mathrm{D}_z\mathcal{I}(t, z(t)) \tag{94}$$

by assuming smoothness and (i) by applying $\langle\cdot, \dot z\rangle$, (ii) by differentiation with respect to t and applying $\langle\cdot, \dot z\rangle$, and (iii) by differentiation with respect to t and applying $\langle\cdot, \ddot z\rangle$:

$$\Psi(\dot z) + \varepsilon\|\dot z\|^2_{\boldsymbol{V}} + \langle \mathrm{D}_z\mathcal{I}(t, z(t)), \dot z\rangle = 0, \tag{95a}$$

$$\varepsilon\langle \mathbb{V}\dot z, \ddot z\rangle + \langle \mathrm{D}_z^2\mathcal{I}(t,z)\dot z, \dot z\rangle + \langle \partial_t \mathrm{D}_z\mathcal{I}(t,z), \dot z\rangle = 0, \tag{95b}$$

$$\varepsilon\|\ddot z\|^2_{\boldsymbol{V}} + \langle \mathrm{D}_z^2\mathcal{I}(t,z)\dot z, \ddot z\rangle + \langle \partial_t \mathrm{D}_z\mathcal{I}(t,z), \ddot z\rangle \leq 0. \tag{95c}$$

The next result presents sufficient conditions on the solution such that these relations can be proved rigorously.

Lemma 4.16 (A priori estimates). *Let the assumptions* (25) *hold, then for each initial value* $z_0 \in \boldsymbol{Z}$ *there exists* $C > 0$ *such that for all* $\varepsilon \in \,]0,1[$ *and all solutions* z *of* (94) *with* $z(0) = z_0$ *we have:*

(A) If $z \in \mathrm{H}^1([0,T], \boldsymbol{V}) \cap \mathrm{C_w}([0,T], \boldsymbol{Z})$, *then* (95a) *holds a.e., and we have the energy balance*

$$\mathcal{I}(t, z(t)) + \int_s^t \Psi(\dot z(r)) + \varepsilon\|\dot z(r)\|^2_{\boldsymbol{V}}\,\mathrm{d}r = \mathcal{I}(s, z(s)) + \int_s^t \partial_r\mathcal{I}(r, z(r))\,\mathrm{d}r. \tag{96}$$

(B) If $z \in \mathrm{H}^1([0,T], \boldsymbol{Z}) \cap \mathrm{H}^2([0,T], \boldsymbol{Z}_{-1})$, *then* (95b) *holds a.e.,* $\dot z \in \mathrm{C}([0,T]; \boldsymbol{V})$, *and*

$$\varepsilon \max_{t\in[0,T]} \|\dot z(t)\|^2_{\boldsymbol{V}} + \int_0^T \|\dot z(r)\|^2_{\boldsymbol{Z}}\,\mathrm{d}r \leq C(1 + \varepsilon\|\dot z(0)\|^2_{\boldsymbol{V}}). \tag{97}$$

(C) If $z \in \mathrm{H}^1([0,T],\boldsymbol{Z}_1) \cap \mathrm{H}^2([0,T],\boldsymbol{V})$, *then* (95c) *holds a.e.,* $\dot z \in \mathrm{C}([0,T];\boldsymbol{Z})$, *and*

$$\max_{t\in[0,T]} \|\dot z(t)\|_{\boldsymbol{Z}}^2 + \varepsilon \int_0^T \|\ddot z(r)\|_{\boldsymbol{V}}^2\,\mathrm{d}r \le C(1+\|\dot z(0)\|_{\boldsymbol{Z}}^2). \tag{98}$$

Proof. Part (A) follows simply by using the chain rule.

For Part (B) we set $g(t) = \varepsilon\mathbb{V}\dot z(t) + \mathrm{D}_z\mathcal{I}(t,z(t))$. By (25) we have $g \in \mathrm{H}^1([0,T],\boldsymbol{Z}^*)$. From (94) we have $0 \in \partial\Psi(\dot z(t)) + g(t)$, and the characterization of $\partial\Psi(v)$ in Lemma 2.2 gives

$$\langle g(s), \dot z(t)\rangle \ge -\Psi(\dot z(t)) = \langle g(t), \dot z(t)\rangle \text{ for all } s,t \in \,]0,T]\,.$$

The left-hand side attains its global minimum at $s=t$. If additionally $s=t$ is a point of differentiability of g, then the left-hand side is differentiable with respect to s with derivative 0 at $s=t$, i.e., $0 = \langle \dot g(t), \dot z(t)\rangle$. This is the desired relation in (95b). The a priori estimate (97) follows by standard arguments using (25), cf. the proof of Proposition 4.17.

For part (C) we use the monotonicity of $\partial\Psi$ and obtain

$$\varepsilon\|\dot z(t)-\dot z(s)\|_{\boldsymbol{V}}^2 + \langle g(t)-g(s), \dot z(t)-\dot z(s)\rangle \le 0$$

for all $s,t \in [0,T]$. By the assumptions we have $\dot z, \mathbb{V}^{-1}g \in \mathrm{H}^1([0,T],\boldsymbol{V})$, thus we can divide by $(t-s)^2$ and pass to the limit a.e. This provides (95c), and the a priori estimate (98) again follows by standard arguments using (25). □

The following result relies on parabolic estimates, which essentially use the semilinear structure. For Galerkin approximations, we are able to exploit the above a priori estimates, which then survive in the limit. An essential novel feature is the derivation of an estimate that is invariant under rescaling. This estimate then allows us to derive upper bounds for S^ε. For more details and applications to quasilinear problems see [53].

Proposition 4.17. *Let assumptions of Example 2.8 and* (25) *hold for the RIS* $(\boldsymbol{Z},\mathcal{I},\Psi)$ *and the spaces* $\boldsymbol{Z} \Subset \boldsymbol{V} \subset \boldsymbol{X}$. *Then, for each* $z_0 \in \boldsymbol{Z}$ (94) *has a unique solution* $z^\varepsilon \in \mathrm{H}^1([0,T],\boldsymbol{V}) \cap \mathrm{C}_{\mathrm{w}}([0,T],\boldsymbol{Z})$ *with* $z(0) = z_0$, *which satisfies* $z^\varepsilon \in \mathrm{L}^2([0,T];\boldsymbol{Z}_1)$. *If additionally* $z_0 \in \boldsymbol{Z}_1$, *then there exists a constant* $C>0$ *independent of* ε *such that* z^ε *lies in* $\mathrm{H}^1([0,T],\boldsymbol{Z})$ *and satisfies the a priori estimate*

$$\int_0^T \|\dot z^\varepsilon(t)\|_{\boldsymbol{Z}}\,\mathrm{d}t \le C\Big(\varepsilon\|\dot z^\varepsilon(0)\|_{\boldsymbol{V}} + \int_0^T \Psi(\dot z^\varepsilon(t)) + \|\dot\ell(t)\|_{\boldsymbol{V}^*}\,\mathrm{d}t\Big). \tag{99}$$

Proof. We first construct solutions for $z_0 \in \boldsymbol{Z}_1$, which means $\mathbb{A}z_0 \in \boldsymbol{V}$. Then, from the equation we find $\dot z(0) \in \boldsymbol{V}$. We consider approximate solutions z_N by using some Galerkin projector P_N which commutes with $\mathbb{V}$ and $\mathbb{A}$, i.e., $\langle \mathbb{A}v, P_N w\rangle = \langle \mathbb{A}P_N v, w\rangle$ and $\langle \mathbb{V}v, P_N w\rangle = \langle \mathbb{V}P_N v, w\rangle$ and has finite-dimensional range $\boldsymbol{X}_N = P_N\boldsymbol{Z} = P_N\boldsymbol{V}$. (Such a P_N can be constructed using the eigenpairs (λ_j, ϕ_j) of $\mathbb{A}\phi = \lambda\mathbb{V}\phi$, which exist due to $\boldsymbol{Z} \Subset \boldsymbol{V}$.) Now take $z_N \in \mathrm{H}^1([0,T], P_N\boldsymbol{Z})$ as the unique solution of $z_N(0) = P_N z_0$ and

$$P_N^*\boldsymbol{Z}^* \ni 0 \in P_N^*\big(\partial\Psi(\dot z_N) + \varepsilon\mathbb{V}\dot z_N + \mathbb{A}z_N + \mathrm{D}\Phi(z_N) - \ell(t)\big). \tag{100}$$

This inclusion can be inverted to $\dot z_N = M_N(-P_N(\mathbb{A}z_N{+}\mathrm{D}\Phi(z_N){-}\ell))$, where $M_N : \boldsymbol{X}_N^* \to \boldsymbol{X}_N$ is the Lipschitz continuous inverse of the strictly monotone mapping $\boldsymbol{X}_N \ni v \mapsto P_N(\partial\Psi(v){+}\varepsilon\mathbb{V}v) \subset \boldsymbol{X}_N^*$. Since on $\boldsymbol{X}_N$ the norms of $\boldsymbol{Z}$ and $\boldsymbol{V}$ are equivalent, we conclude $z_N \in \mathrm{W}^{2,p}([0,T], \boldsymbol{X}_N)$. For these finite-dimensional approximations we are now able to exploit the assumption $z_0 \in \boldsymbol{Z}_1$, which gives $\|\dot z_N(0)\|_{\boldsymbol{V}} \to \|\dot z(0)\|_{\boldsymbol{V}} < \infty$. Thus, the a priori estimate (97) provides boundedness of $(z_N)_{N\in\mathbb{N}}$ in $\mathrm{H}^1([0,T],\boldsymbol{Z})$. We obtain a weakly converging subsequence (not relabeled) $z_N \rightharpoonup z$ in $\mathrm{H}^1([0,T],\boldsymbol{Z})$. It is easy to see that z is a solution of (94) with $z(0) = z_0$.

To show uniqueness we consider two solutions z_1 and z_2 in $\mathrm{H}^1([0,T],\boldsymbol{Z})$. Using (25f) and (25g) we find $z_j \in \mathrm{L}^2([0,T],\boldsymbol{Z}_1)$. Setting $w = z_1 - z_2$, the monotonicity of $\partial\Psi$ gives

$$0 \geq \varepsilon\|\dot w(t)\|_{\boldsymbol{V}}^2 + \frac{1}{2}\frac{\mathrm{d}}{\mathrm{d}t}\langle Aw(t), w(t)\rangle + \langle H(t)w(t), \dot w(t)\rangle$$

with $H(t) = \int_0^1 \mathrm{D}^2\Phi(z_2(t){+}rw(t))\,\mathrm{d}r$, where $H(t) \in \mathscr{L}(\boldsymbol{Z},\boldsymbol{V}^*)$ by (25f). Because of (25c) we may assume without loss of generality that the norm in $\boldsymbol{Z}$ is given via $\|z\|_{\boldsymbol{Z}}^2 = \langle \mathbb{A}z, z\rangle$. Thus, with (25f) we obtain

$$\varepsilon\|\dot w\|_{\boldsymbol{V}}^2 + \frac{1}{2}\frac{\mathrm{d}}{\mathrm{d}t}\|w\|_{\boldsymbol{Z}}^2 \leq C\|w\|_{\boldsymbol{Z}}\|\dot w\|_{\boldsymbol{V}} \leq \varepsilon\|\dot w\|_{\boldsymbol{V}}^2 + \tfrac{C}{\varepsilon}\|w\|_{\boldsymbol{Z}}^2.$$

Applying Gronwall's lemma we arrive at

$$\|z_2(t){-}z_1(t)\|_{\boldsymbol{Z}} = \|w\|_{\boldsymbol{Z}} \leq \mathrm{e}^{Ct/\varepsilon}\|w(0)\|_{\boldsymbol{Z}} = \mathrm{e}^{Ct/\varepsilon}\|z_2(0){-}z_1(0)\|_{\boldsymbol{Z}}, \tag{101}$$

which implies the desired uniqueness if $z_0 \in \boldsymbol{Z}_1$. Hence, we conclude that the full sequence z_N converges weakly to z.

To obtain existence and uniqueness for the general case $z_0 \in \boldsymbol{Z}$ we proceed as follows. Using the same Galerkin approach as above we obtain uniform a priori estimates for z_N in $\mathrm{H}^1([0,T],\boldsymbol{V}){\cap}\mathrm{C}_{\mathrm{w}}([0,T],\boldsymbol{Z})$. Note that the Gronwall estimate (101) holds for the finite-dimensional approximations as well and that the constants are independent of N. Thus, we can pass to the limit along subsequences, obtain solutions of (94) with $z(0) = z_0$, and that these solutions still satisfy the Gronwall estimate, which implies uniqueness.

It remains to establish the a priori estimate (99). For this return to the Galerkin approximation z_N, which satisfies

$$\tfrac{\mathrm{d}}{\mathrm{d}t}\tfrac{\varepsilon}{2}\|\dot z_N(t)\|_{\boldsymbol V}^2 + \|\dot z_N(r)\|_{\boldsymbol Z}^2 \le C\|\dot z_N(t)\|_{\boldsymbol Z}\|\dot z_N(t)\|_{\boldsymbol V} + \|\dot\ell(t)\|_{\boldsymbol V^*}\|\dot z_N(t)\|_{\boldsymbol V}$$

a.e. on $[0,T]$, where here and in the rest of this proof the constant C may take different values but is independent of ε and N. We let $\nu = \|\dot z_N\|_{\boldsymbol V}$, $\zeta = \|\dot z_N(t)\|_{\boldsymbol Z}$, $\psi = \Psi(\dot z_N)$, and $\lambda = \|\dot\ell(t)\|_{\boldsymbol V^*}$, then with (25b), (25d), and $\vartheta = 1/(1{+}\theta)$ we find

$$\varepsilon\nu\dot\nu + \zeta^2 \;\le\; C\nu^{1-\vartheta}(\psi^\theta\zeta^{1-\theta})^\vartheta\zeta + \lambda\nu \;\le\; \tfrac12\zeta^2 + (C\psi{+}\lambda)\nu.$$

Using $\nu \le C\zeta$ we find $\varepsilon\nu\dot\nu + \frac{1}{2C}\nu\zeta \le (C\psi{+}\lambda)\nu$. Without loss of generality we may assume $\nu > 0$ (otherwise take $\sqrt{\nu^2{+}\delta}$, which satisfies the same estimate, and let $\delta \to 0^+$ afterwards). Dividing by $\nu > 0$ and integrating gives

$$\tfrac{1}{2C}\int_0^T \|\dot z_N(r)\|_{\boldsymbol Z}\,\mathrm{d}r \le \varepsilon\|\dot z_N(0)\|_{\boldsymbol V} + \int_0^T C\Psi(\dot z_N(r)) + \|\dot\ell(r)\|_{\boldsymbol V^*}\,\mathrm{d}r.$$

Estimate (99) follows with $N \to \infty$ if we show $\lim_{N\to\infty}\int_0^T \Psi(\dot z_N(r))\,\mathrm{d}r \le \int_0^T \Psi(\dot z(r))\,\mathrm{d}r$. Indeed, this equality follows by passing to the limit in the energy balance (96), since convergence in $\mathrm{H}^1([0,T],\boldsymbol Z)$ implies $\mathcal{I}(t,z_N(t)) \to \mathcal{I}(t,z(t))$ and $\partial_t\mathcal{I}(t,z_N(t)) \to \partial_t\mathcal{I}(t,z(t))$ for all $t\in[0,T]$. Thus, we find

$$\int_0^T \Psi(\dot z_N(r))\,\mathrm{d}r + \int_0^T \varepsilon\|\dot z_N(r)\|_{\boldsymbol V}^2\,\mathrm{d}r \to \int_0^T \Psi(\dot z(r))\,\mathrm{d}r + \int_0^T \varepsilon\|\dot z(r)\|_{\boldsymbol V}^2\,\mathrm{d}r.$$

Together with the weak lower semicontinuity of each of the integrals we obtain the desired convergence and the proof is complete. □

We are now ready for stating the major existence result for parametrized solutions, which is obtained via the vanishing-viscosity approach. We emphasize that the result is not optimal, in the sense that we have to assume too much regularity for the initial condition. This is also seen in the a priori estimate (99), where we obtained the higher $\boldsymbol Z$-norm, but we actually only need the $\boldsymbol V$-norm. The existence result is obtained by combining Proposition 4.17 and the vanishing-viscosity theory of Proposition 4.14.

Proof (of Theorem 4.15). For each $\varepsilon \in\,]0,1[$ we can construct a solution $z^\varepsilon \in \mathrm{H}^1([0,T];\boldsymbol Z)$ of (94) satisfying the rate-independent a priori estimate (99). The latter implies that for $\sigma^\varepsilon : [0,T] \mapsto t + \int_0^t \|\dot z^\varepsilon(r)\|_{\boldsymbol V}\,\mathrm{d}r$, the total length $S^\varepsilon = \sigma^\varepsilon(T)$ of the graph of z^ε in $\boldsymbol V_T$ stays bounded. We let $S = \limsup_{\varepsilon\to0} S^\varepsilon$ and choose a subsequence ε_j such that $S^{\varepsilon_j} \to S$.

We parametrize the graph of z^ε in $\boldsymbol V_T$ by arclength to obtain the Lipschitz functions $\zeta^\varepsilon = (\tau^\varepsilon, Z^\varepsilon) : [0,S] \to \boldsymbol V_T$, where we set $\zeta^\varepsilon(s) = \zeta^\varepsilon(S^\varepsilon)$ for $s > S^\varepsilon$. Since all functions ζ^{ε_j} have Lipschitz constant 1, the Arzelà-Ascoli theorem allows us to choose a further subsequence ζ^{ε_n} as in Proposition 4.14. The limit $\widetilde\zeta$ satisfies (90) and additionally $\widetilde\tau(S) = T$, which follows from $T = \tau^{\varepsilon_n}(S^{\varepsilon_n})$ and $|\tau^{\varepsilon_n}(S^{\varepsilon_n}) - \tau^{\varepsilon_n}(S)| \le |S^{\varepsilon_n} - S|$.

Thus we are able to do the reparametrization (91) and the desired solution ζ is constructed. It remains to show that Z lies in $\mathrm{BV}([0,S],\boldsymbol{Z})$. For this we recall the rate-independent estimate (99) for each z^ε. The right-hand side there is bounded independently of ε. For $\int_0^T \Psi(\dot z^\varepsilon(s))\,\mathrm{d}s$ this follows from the energy balance and for $\varepsilon\dot z^\varepsilon(0)$ this follows from $0 = w^\varepsilon(0) + \varepsilon\mathbb{V}\dot z^\varepsilon(0) + \mathbb{A}z_0 + \mathrm{D}\Phi(z_0) - \ell(0)$, where $w^\varepsilon(0) \in \partial\Psi(\dot z^\varepsilon(0))$ is bounded in $\boldsymbol{V}^*$ by (25d). Thus, $z_0 \in \boldsymbol{Z}_1$ and (25f) imply $\varepsilon\|\dot z^\varepsilon(0)\|_{\boldsymbol{V}} \leq C$. Since the BV norm is scaling invariant, we have $\mathrm{Var}_{\boldsymbol{Z}}(Z^\varepsilon,[0,S^\varepsilon]) = \mathrm{Var}_{\boldsymbol{Z}}(z^\varepsilon,[0,T]) \leq C$. In the limit $\varepsilon_n \to 0$ this estimate still holds by lower semicontinuity of the variation. □

Remark 4.18. In [53] it is additionally shown that the limit $\widetilde{\zeta}$ obtained in the middle of the above proof is nondegenerate in the sense that there exists a $\alpha_0 > 0$ such that $\dot{\widetilde{\tau}}(s) + \|\dot{\widetilde{Z}}(s)\|_{\boldsymbol{V}} \geq \alpha_0$. It is still an open problem, whether the arclength parametrization remains preserved in the limit $\varepsilon \to 0$. A positive result was obtained in [22] under strong conditions. Moreover, the metric approach in Sect. 5.2 is such that the arclength parametrization is preserved.

4.5 BV Solutions and Optimal Jump Paths

The drawback of parametrized solutions is that we need to deal with functions in the extended state space $\boldsymbol{Z}_T$. Thus, it is not easy to compare this notion to all the other solution types, which are defined for functions $z : [0,T] \to \boldsymbol{Z}$ only. Thus, for each parametrized solution ζ we consider all associated projections $z : [0,T] \to \boldsymbol{Z}$ defined as follows. For $\zeta = (\tau, Z) : [0,S] \to \boldsymbol{Z}_T$ with $\tau(0) = 0$ and $\tau(S) = T$ and τ monotone, we define $\mathfrak{P}(\zeta)$ to be the set of functions $z : [0,T] \to \boldsymbol{Z}$ such that for all $t \in [0,T]$ there exists $s \in [0,S]$ such that $(t, z(t)) = \zeta(s) = (\tau(s), Z(s))$.

We first show that such projections lead to local solutions, see (77). Then, we derive the notion of *BV solutions* in such a way that we can show that all these projections are BV solutions. Thus, we are able to study convergence of the viscous approximations $z^\varepsilon \in \mathrm{H}^1([0,T];\boldsymbol{V})$ solving (84) towards BV solutions z. Recall the a priori estimate (85), which allows us to find a limit $z \in \mathrm{BV}([0,T];\boldsymbol{X}) \cap \mathrm{L}^\infty([0,T];\boldsymbol{Z})$. This limit passage was used in several applications already, see e.g. [12, 19, 34, 37, 69]. The aim of this section is to characterize these limits, also called *approximable solutions*, as good as possible. In particular, we need to derive conditions that characterize the jumps occurring in the vanishing-viscosity limit. We follow [59], where the finite-dimensional case is treated in full detail.

We first motivate our definition of BV solutions by referring to parametrized solutions. Then, we give some more motivation by doing the vanishing-viscosity limit directly in the energetic formulation, which leads to a new central object called *vanishing-viscosity contact potential* $\mathfrak{p}$. It leads to a supplemented dissipation distance Δ in a natural way, which includes the rate-independent contributions to the dissipation as well as the contributions from the vanishing viscosity. The associated dissipation functional $\mathrm{Diss}_{\mathfrak{p},\mathcal{J}}$ then leads to the notion of BV solutions that are defined to satisfy a local stability condition and the energy balance with the new dissipation functional. Finally, we present the convergence results (i) for the vanishing-viscosity limit $z^\varepsilon \to z$ and (ii) for the time-discrete incremental approximations given by

$$z_i^h \in \operatorname*{Arg\,min}_{z\in \boldsymbol{Z}} \Big(\mathcal{J}(hi,z) + \Psi(z-z_{i-1}^h) + \tfrac{\varepsilon}{2h}\|z-z_{i-1}^N\|_{\boldsymbol{V}}^2\Big).$$

In the latter case convergence of subsequences to BV solutions follows if ε, the time-step h, and the quotient h/ε tend to 0.

We start with a few facts about BV spaces. For a Banach space $\boldsymbol{Y}$ we let

$$\begin{aligned}&\mathrm{BV}([a,b];\boldsymbol{Y}) \overset{\mathrm{def}}{=} \{\, y:[a,b]\to \boldsymbol{Y} \mid \mathrm{Var}_{\boldsymbol{Y}}(y,[a,b])<\infty \,\} \quad \text{with}\\ &\mathrm{Var}_{\boldsymbol{Y}}(y,[a,b]) = \sup\Big\{ \textstyle\sum_{j=1}^N \|y(t_j)-y(t_{j-1})\|_{\boldsymbol{Y}} \;\Big|\; N\in\mathbb{N},\\ &\qquad\qquad\qquad\qquad\qquad\qquad a\le t_0<t_1<\cdots<t_N\le b\,\Big\}.\end{aligned}$$

As usual in evolutionary problems the functions in $\mathrm{BV}([a,b];\boldsymbol{Y})$ are defined everywhere on $[0,T]$ and the variation is sensitive to changing a function at a single point. Clearly, $\mathrm{BV}([a,b];\boldsymbol{Y})$ is a Banach space equipped with the norm $\|y\|_{\mathrm{BV}([a,b];\boldsymbol{Y})} = \|y(a)\|_{\boldsymbol{Y}} + \mathrm{Var}_{\boldsymbol{Y}}(y,[a,b])$. Moreover, $\mathrm{BV}([a,b];\boldsymbol{Y}) \subset \mathrm{L}^\infty([a,b];\boldsymbol{Y})$ with $\|y(t)\|_{\boldsymbol{Y}} \le \|y\|_{\mathrm{BV}([a,b];\boldsymbol{Y})}$ for all t and

$$\mathrm{Var}_{\boldsymbol{Y}}(y,[t_1,t_2]) + \mathrm{Var}_{\boldsymbol{Y}}(y,[t_2,t_3]) = \mathrm{Var}_{\boldsymbol{Y}}(y,[t_1,t_3]) \text{ for } y\in \mathrm{BV}([t_1,t_3];\boldsymbol{Y}).$$

Finally, for $y\in\mathrm{BV}([a,b];\boldsymbol{Y})$ and each $t\in[a,b]$ the right limit $y(t^+)$ and the left limit $y(t^-)$ (cf. (49)) exist in the strong norm topology of $\boldsymbol{Y}$.

For a given parametrized solution ζ we consider $z \in \mathfrak{P}(\zeta)$, then $\mathrm{Var}_{\boldsymbol{X}}(z,[0,T]) \le C\,\mathrm{Var}_{\boldsymbol{V}}(z,[0,T]) \le C\,\mathrm{Var}_{\boldsymbol{V}}([0,S],Z) \le CS < \infty$, since Z has Lipschitz constant less than 1. For $z\in\mathrm{BV}([0,T];\boldsymbol{X})$ we define the continuity set $C(z)$ and the jump set $J(z)$ of z by

$$C(z) = \{\, t\in[0,T] \mid z(t^-)=z(t)=z(t^+)\,\} \quad\text{and}\quad J(z)=[0,T]\setminus C(z),$$

where left and right limits exist in $\boldsymbol{X}$ and where $J(z)$ is countable.

For each $t\in J(z)$ the monotone function $\tau:[0,S]\to[0,T]$ may have plateaus $[a^t,b^t]$ with $b^t>a^t$, such that $\tau([a^t,b^t])=\{t\}$. Outside of all these intervals we are either in the sticking regime or in rate-independent slip. Hence, there exists $\xi(s)\in\overline{\partial}_z\mathcal{J}(\zeta(s))$ with $0\in\partial\Psi(0)+\xi(s)$. Thus, we obtain the local stability condition

$$0\in\partial\Psi(0)+\overline{\partial}_z\mathcal{J}(t,z(t)) \;\text{ for all } t\in C(z). \tag{102}$$

Using (89) and $\lambda(s) \geq 0$ we easily see that z also satisfies the energy inequality (77b). Thus, we have proved the following result.

Corollary 4.19. *If* (τ, Z) *is a parametrized solution of* $(\boldsymbol{Z}, \mathcal{I}, \Psi, \mathbb{V})$, *then each* $z \in \mathfrak{P}((\tau, Z))$ *is a local solution, see* (77).

The important addition to make local solutions into BV solutions is the careful analysis of the jumps. For each plateau $[a^t, b^t] \subset [0, S]$ associated with $t \in J(z)$, we denote by $y^t \in \mathrm{C}^{\mathrm{Lip}}([0,1]; \boldsymbol{V})$ the normalized jump curve $y^t(r) = Z(a^t + r(b^t - a^t))$. The point is that each such jump curve is an optimal curve in a specific sense. We first calculate the dissipation $D(t)$ along the jump curve associated with t, namely

$$D(t) = \int_{a^t}^{b^t} \Psi(\dot{Z}(s)) + M_\Psi^{\boldsymbol{V}}(-\mathrm{D}_z \mathcal{I}(t, Z(s)))\,\mathrm{d}s = \int_0^1 \mathfrak{p}(\dot{y}^t(r), -\mathrm{D}_z \mathcal{I}(t, y^t(r)))\,\mathrm{d}r$$

where $\mathfrak{p}(v, \xi) = \Psi(v) + \|v\|_{\boldsymbol{V}} M_\Psi^{\boldsymbol{V}}(\xi)$ with $M_\Psi^{\boldsymbol{V}}$ from (92). Here we have used that the arclength parametrization enforced $\|\dot{Z}(s)\|_{\boldsymbol{V}} = 1$, while the normalized jump curve satisfies $\|\dot{y}^t(r)\|_{\boldsymbol{V}} = b^t - a^t$. Thus, the factor in front of $M_\Psi^{\boldsymbol{V}}$ appeared because of the reparametrization only. However, the important effect is that the integrand now is 1-homogeneous in v, which reflects the rate independence nicely.

Another way to arrive at the same integrand gives the following definition.

Definition 4.20 (Vanishing-viscosity contact potential). Given a superlinear dissipation potential $\mathcal{R}_{\mathrm{sl}} : \boldsymbol{X} \to [0, \infty]$ we set $\mathcal{R}_\varepsilon(v) = \frac{1}{\varepsilon}\mathcal{R}_{\mathrm{sl}}(\varepsilon v)$. The *vanishing-viscosity contact potential* $\mathfrak{p} : \boldsymbol{X} \times \boldsymbol{X}^* \to \mathbb{R}_\infty$ is defined via

$$\mathfrak{p}(v, \xi) \stackrel{\mathrm{def}}{=} \inf_{\varepsilon > 0} \Big(\mathcal{R}_\varepsilon(v) + \mathcal{R}_\varepsilon^*(\xi)\Big) = \inf_{\varepsilon > 0} \Big(\tfrac{1}{\varepsilon}\mathcal{R}_{\mathrm{sl}}(v) + \tfrac{1}{\varepsilon}\mathcal{R}_{\mathrm{sl}}^*(\xi)\Big).$$

The *contact set* $\mathfrak{p}$ is given by $\mathfrak{C}_{\mathfrak{p}} \stackrel{\mathrm{def}}{=} \{\, (v, \xi) \mid \mathfrak{p}(v, \xi) = \langle \xi, v \rangle \,\}$.

For our choice $\mathcal{R}_{\mathrm{sl}}(v) = \Psi(v) + \frac{1}{2}\|v\|_{\boldsymbol{V}}^2$ we have $\mathcal{R}_\varepsilon^*(\xi) = \frac{1}{2\varepsilon} M_\Psi^{\boldsymbol{V}}(\xi)^2$ and find

$$\mathfrak{p}(v, \xi) = \Psi(v) + \|v\|_{\boldsymbol{V}} M_\Psi^{\boldsymbol{V}}(\xi).$$

The motivation for the definition of the vanishing-viscosity contact potential is obtained by the following lower bounds for the dissipation integrals

$$I_\varepsilon = \int_{t_1}^{t_2} \mathcal{R}_\varepsilon(\dot{z}) + \mathcal{R}_\varepsilon^*(\xi)\,\mathrm{d}t \geq \int_{t_1}^{t_2} \mathfrak{p}(\dot{z}, -\mathrm{D}\mathcal{I}(t, z(t)))\,\mathrm{d}t.$$

It turns out that this lower bound is sharp in the limit $\varepsilon \to 0$ along the jump curves.

We find the following important properties for $\mathfrak{p}$ and $\mathfrak{C}_\mathfrak{p}$:

$$\begin{aligned}
&\text{rate independence} && \mathfrak{p}(\lambda v,\xi)=\lambda\mathfrak{p}(v,\xi), && \text{(103a)}\\
&\text{lower bound} && \mathfrak{p}(v,\xi)\geq\langle\xi,v\rangle, && \text{(103b)}\\
&\text{separate convexity} && \mathfrak{p}(\cdot,\xi)\text{ is convex} && \text{(103c)}\\
& && \text{and }\mathfrak{p}(v,\cdot)\text{ has convex sublevels,} && \\
&\text{contact set} && (v,\xi)\in\mathfrak{C}_\mathfrak{p}\iff\xi\in\partial_v\mathfrak{p}(v,\xi) && \text{(103d)}\\
& && \iff\begin{cases} v=0\text{ or}\\ v\neq 0\text{ and }\xi\in\partial\Psi(v)+\frac{M_\Psi^{\boldsymbol{V}}(v)}{\|v\|_{\boldsymbol{V}}}\mathbb{V}v.\end{cases} &&
\end{aligned}$$

Using the contact potential $\mathfrak{p}$ we are now able to define a supplemented distance between points z_1 and z_2 involving both the dissipation due to Ψ and the possibly additional dissipation arising from fast viscous transitions:

$$\begin{aligned}
\Delta(t,z_1,z_2)\overset{\text{def}}{=}\inf\Big\{\ \int_0^1\mathfrak{p}(\dot{\widehat{z}}(r),-\xi(r))\,\mathrm{d}r\ \Big|\ \widehat{z}\in A_{\boldsymbol{V}}(z_1,z_2),\\
\xi(r)\in\overline{\partial}_z\mathcal{I}(t,\widehat{z}(r))\text{ a.e. in }[0,1]\ \Big\}\quad(104)\\
\text{where } A_{\boldsymbol{Y}}(z_1,z_2)\overset{\text{def}}{=}\{\,\widehat{z}\in\mathrm{C}^{\mathrm{Lip}}([0,1];\boldsymbol{Y})\mid\widehat{z}(0)=z_1,\ \widehat{z}(1)=z_2\,\}.
\end{aligned}$$

Note that Δ is defined with time t as a frozen parameter. Clearly, we have the triangle inequality $\Delta(t,z_0,z_2)\leq\Delta(t,z_0,z_1)+\Delta(t,z_1,z_2)$ and the lower estimate $\Delta(t,z_1,z_2)\geq\Psi(z_2{-}z_1)$.

The crucial observation is that the lower estimate (103b) and the classical chain rule $\frac{\mathrm{d}}{\mathrm{d}r}\mathcal{I}(t,z(r))=\langle\xi(r),\dot{z}(r)\rangle$ imply the estimate

$$\mathcal{I}(t,z_2)+\Delta(t,z_1,z_2)\geq\mathcal{I}(t,z_1)\ \text{ for all }z_1,z_2\in\boldsymbol{Z}.\qquad(105)$$

Thus, we define *optimal jump paths* by enforcing equality in this estimate:

$$\begin{aligned}
&\mathrm{OJP}(t,z_1,z_2)\overset{\text{def}}{=}\\
&\Big\{\,\widehat{z}\in A_{\boldsymbol{V}}(z_1,z_2)\ \Big|\ \Delta(t,z_1,z_2)=\mathcal{I}(t,z_1){-}\mathcal{I}(t,z_2)=\mathfrak{p}(\dot{\widehat{z}}(r),-\xi(r))\\
&\qquad\qquad\text{and }\xi(r)\in\overline{\partial}_z\mathcal{I}(t,\widehat{z}(r))\text{ for a.a. }r\in[0,1]\,\Big\}.
\end{aligned}$$

Clearly, these equalities imply that a.e. along the whole jump path the lower bound (103b) has to be an equality, i.e., the solution must lie in the contact set, viz. $(\dot{z}(r),-\xi(r))\in\mathfrak{C}_\mathfrak{p}$, which is again equivalent to $0\in\partial\mathfrak{p}(\cdot,-\xi(r))(\dot{z}(r))+\xi(r)$ and implies (see (103d))

$$0\in\partial\Psi(\dot{z}(r))+\lambda(r)\mathbb{V}\dot{z}(r)+\xi(r)\text{ with }\lambda(r)\geq 0\ \text{ and }\Psi^\circ(-\xi(r))\geq 1.\quad(106)$$

For the definition of BV solutions we use the associated supplemented dissipation functional $\mathrm{Diss}_{\mathfrak{p},\mathcal{J}}$ defined on functions $z\in\mathrm{BV}([0,T];\boldsymbol{X})\cap\mathrm{L}^\infty([0,T];\boldsymbol{Z})$. It takes into account the rate-independent friction via Ψ and the viscous friction at possible jumps:

$$\begin{aligned}\mathrm{Diss}_{\mathfrak{p},\mathcal{J}}(z,[t_1,t_2]) &= \mathrm{Cont}_{\Psi}(z,[t_1,t_2]) + \Delta(t_1,z(t_1),z(t_1^+)) + \Delta(t_2,z(t_2^-),z(t_2)) \\ &\quad + \textstyle\sum_{t\in J(z)} \big(\Delta(t,z(t^-),z(t)) + \Delta(t,z(t),z(t^+))\big), \\ \mathrm{Cont}_{\Psi}(z,[t_1,t_2]) &= \mathrm{Diss}_{\Psi}(z,[t_1,t_2]) - \Psi(z(t_1^+)-z(t_1)) - \Psi(z(t_2)-z(t_2^-)) \\ &\quad - \textstyle\sum_{t\in J(z)} \big(\Psi(z(t)-z(t^-)) + \Psi(z(t^+)-z(t))\big).\end{aligned} \tag{107}$$

Thus, $\mathrm{Diss}_{\mathfrak{p},\mathcal{J}}(z,[t_1,t_2])$ consists of the classical dissipation $\mathrm{Diss}_{\Psi}(z,[t_1,t_2])$ on continuous parts of z, while at jumps the integration of $\mathfrak{p}(\dot z,-\xi) = \Psi(\dot z)+\|\dot z\|_{\boldsymbol{V}} M_{\Psi}^{\boldsymbol{V}}(-\xi)$ along jump paths contains the rate-independent dissipation (which may be strictly larger than $\Psi(z_2-z_1)$) and the viscous contributions via $\|\dot z\|_{\boldsymbol{V}} M_{\Psi}^{\boldsymbol{V}}(-\xi)$.

Definition 4.21 (BV solutions). A function $z \in \mathrm{BV}([0,T];\boldsymbol{X})$ is called a *BV solution* of the RIS $(\boldsymbol{Z},\mathcal{J},\Psi,\mathbb{V})$, if $z \in \mathrm{L}^\infty([0,T];\boldsymbol{Z})$ and (108) hold:

$$\forall\, t \in C(z):\ z(t) \in \mathcal{S}_{\mathrm{loc}}(t) \overset{\mathrm{def}}{=} \{\, z \in \boldsymbol{Z} \mid 0 \in \partial\Psi(0) + \overline{\partial}_z \mathcal{J}(t,z)\,\}; \tag{108a}$$

$$\forall\, t{\in}[0,T]:\ \mathcal{J}(t,z(t)) + \mathrm{Diss}_{\mathfrak{p},\mathcal{J}}(z,\,[0,t]) = \mathcal{J}(0,z(0)) + \int_0^t \partial_\tau \mathcal{J}(\tau,z(\tau))\,\mathrm{d}\tau. \tag{108b}$$

The function z is called a *connectable BV solution*, if additionally the following holds:

$$\forall\, t \in J(z)\ \exists\, \widehat{z}^{\,t} \in \mathrm{OJP}(t,z(t^-),z(t^+))\ \exists\, r^t \in [0,1]:\ z(t) = \widehat{z}^{\,t}(r^t). \tag{108c}$$

Note that the energy inequality (77b) in the definition of local solutions differs from the energy identity (108b) exactly by replacing $\Delta(t,z(t^-),z(t^+))$ by $\Psi(z(t^+)-z(t^-))$. This also allows us to give precise formulae for the energy drop at jump points, which are in full analogy to (51) for energetic solutions:

$$\begin{aligned} &\mathcal{J}(t,z(t)) + \Delta(t,z(t^-),z(t)) = \mathcal{J}(t,z(t^-)), \\ &\mathcal{J}(t,z(t^+)) + \Delta(t,z(t),z(t^+)) = \mathcal{J}(t,z(t)), \\ &\mathcal{J}(t,z(t^-)) = \lim_{\tau\to t^-} \mathcal{J}(\tau,z(\tau)), \\ &\mathcal{J}(t,z(t^+)) = \lim_{\tau\to t^+} \mathcal{J}(\tau,z(\tau)), \\ &\Delta(t,z(t^-),z(t)) + \Delta(t,z(t),z(t^+)) = \Delta(t,z(t^-),z(t^+)). \end{aligned} \tag{109}$$

The existence of optimal jump paths is not needed for general BV solutions. If the RIS has the property that the infimum in the definition of $\Delta(t,z_1,z_2)$ is attained for all locally stable $z_1, z_2 \in \boldsymbol{Z}$ (i.e., $0 \in \partial\Psi(0) +$

$\overline{\partial}_z \mathcal{I}(t, z_j))$, then every BV solution is also connectable, as we may concatenate the optimal jump paths from $z(t^-)$ to $z(t)$ and from $z(t)$ to $z(t^+)$ to obtain an optimal jump path from $z(t^-)$ to $z(t^+)$.

The following corollary states that all BV solutions are local solutions, which is an easy consequence of the definitions, and that parametrized solutions give rise to connectable BV solutions. This also provides an existence result by employing Theorem 4.15.

Corollary 4.22. *Let the RIS $(\boldsymbol{Z}, \mathcal{I}, \Psi, \mathbb{V})$ be given.*

(A) If $(\tau, Z) \in \mathrm{C}^{\mathrm{Lip}}([0,T]; \boldsymbol{V}_T)$ is a parametrized solution, then every $z \in \mathfrak{P}((\tau, Z))$ is a connectable BV solution.
(B) Let the RIS $(\boldsymbol{Z}, \mathcal{I}, \Psi, \mathbb{V})$ be the standard Example 2.8 and $z_0 \in Z_1$. Then, there exists a connectable BV solution z with $z(0) = z_0$.
(C) Every BV solution is a local solution, cf. (77).

We now use the advantage that BV solutions are defined as functions from the time interval $[0, T]$ into the state space $\boldsymbol{Z}$ like the viscous approximations. Thus, the natural question is how the solutions z^ε converge to BV solutions. This question was first answered in [59] for the finite-dimensional setting. Here we give a similar result for our standard Example 2.8.

Theorem 4.23 (Pointwise convergence to BV solutions). *Let the RIS $(\boldsymbol{Z}, \mathcal{I}, \Psi, \mathbb{V})$ and the spaces $\boldsymbol{Z}_1 \Subset \boldsymbol{Z} \Subset \boldsymbol{V} \Subset Z_{-1} \subset \boldsymbol{X}$ be given as in Example 2.8. Choose any $z_0 \in \boldsymbol{Z}_1$ and consider viscous approximations $z^\varepsilon \in \mathrm{H}^1([0,T]; \boldsymbol{V}) \cap \mathrm{C}_{\mathrm{w}}([0,T]; \boldsymbol{Z})$ solving (24) with $z^\varepsilon(0) = z_0$. Then, there exists a subsequence $(z^{\varepsilon_n})_{n\in\mathbb{N}}$ with $\varepsilon_n \to 0$ and a limit function $z \in \mathrm{BV}([0,T]; \boldsymbol{V})$ such that*

$$\forall t \in [0,T]: \quad z^{\varepsilon_n}(t) \rightharpoonup z(t) \ \text{ in } \boldsymbol{Z};$$
$$z \text{ is a connectable BV solution to } (\boldsymbol{Z}, \mathcal{I}, \Psi, \mathbb{V}).$$

Moreover, any pointwise limit z of a subsequence of $(z^\varepsilon)_{\varepsilon>0}$ is a BV solution.

The last statement shows that in this case *all approximable solutions* are BV solutions.

Proof. The result follows by using the convergence of parametrized solutions once again. Given the sequence z^ε we define the associated parametrized solution $\zeta^\varepsilon \in \mathrm{C}^{\mathrm{Lip}}([0, S^\varepsilon]; \boldsymbol{V}_T)$. As in the proof of Theorem 4.15 we obtain a subsequence $(\zeta^{\varepsilon_j})_{j\in\mathbb{N}}$ that converges pointwise to a limit $\zeta = (\tau, Z)$ (which is possibly not arclength parametrized). For this ζ we consider all $\widetilde{z} : [0,T] \to \boldsymbol{Z}$ with $\widetilde{z} \in \mathfrak{P}(\zeta)$. The value of $\widetilde{z}$ is uniquely defined at all $t \in [0,T] \setminus P$, where the countable set P is the image under τ of the plateaus of τ.

We first claim that $z^{\varepsilon_j}(t_*) \rightharpoonup z(t_*)$ for all $t_* \in [0,T] \setminus P$. To show this we use that τ^{ε_j} converges to τ uniformly on $[0,T]$. Since t_* is not in the image of a plateau of τ, the pseudo-inverse $\sigma : t \mapsto \min\{\, s \mid \tau(s) = t \,\}$ is continuous in t_*. Moreover, there is a unique s_* with $\tau(s_*) = t_*$ and $s_* = \sigma(t_*)$. Thus, for each $\delta > 0$ there exists $\rho_\delta > 0$ such that $|\sigma(t)-\sigma(t_*)| \leq \delta$ for $|t-t_*| \leq \rho_\delta$. In particular, there is a j_δ such that $\|\tau^{\varepsilon_j}-\tau\|_{\mathrm{L}^\infty} \leq \rho_\delta$ for $j \geq j_\delta$. Now choose $s_j \in [0,S]$ such that $\tau^{\varepsilon_j}(s_j) = t_* = \tau(s_*)$, then $|\tau(s_j)-\tau(s_*)| = |\tau(s_j)-\tau^{\varepsilon_j}(s_j) \leq \rho_\delta$ and we obtain $|s_j-s_*| \leq \delta$ for $j \geq j_\delta$. With this we find

$$z(t_*)-z^{\varepsilon_j}(t_*) = Z(\sigma(t_*))-Z^{\varepsilon_j}(s_*)+Z^{\varepsilon_j}(s_*)-Z^{\varepsilon_j}(s_j) = Z(s_*)-Z^{\varepsilon_j}(s_*)+w_j,$$

where $\|w_j\|_{\boldsymbol{V}} \leq |s_j-s_*| \leq \delta$. Since $Z^{\varepsilon_j}(s_*) \rightharpoonup Z(s_*)$ in $\boldsymbol{Z}$, we obtain $\|z^{\varepsilon_j}(t_*)-z(t_*)\|_{\boldsymbol{V}} \to 0$, and the desired weak convergence follows by the a priori bound in the Hilbert space $\boldsymbol{Z}$.

It remains to establish pointwise convergence in the possible jump points. However, P is countable, so we may choose a further subsequence to obtain weak pointwise convergence for all $t \in [0,T]$. Proceeding as above, it is not difficult to show that the limit $z(t_\circ)$ for $t_\circ \in P$ satisfies $z(t_\circ) = Z(s_\circ)$, where $s_\circ$ lies in a plateau $[a^\circ, b^\circ]$ on which $\tau(s) = t_\circ$ holds. Clearly, the pointwise limit satisfies $z \in \mathrm{BV}([0,T];\boldsymbol{V})$ and it is a connectable BV solution.

To show that every pointwise limit point must be a BV solution, take any pointwise converging sequence (z^{ε_n}) and denote the limit by z. By choosing a further subsequence as above, we obtain a BV solution $\widehat{z}$ as limit of the further subsequence. Obviously, $z = \widehat{z}$, and we are done. □

We want to emphasize that in general the convergence result may fail in the infinite-dimensional setting. For this we refer to Example 4.4, where the pointwise limit z of the viscous approximations z^ε is a local solution but not a BV solution. In fact, the solution $z : [0,T] \to \boldsymbol{Z} = \boldsymbol{V}$ is strongly continuous, i.e. there are no jumps, but the energy balance does not hold because of missing viscous contributions. We heal this remedy in the next section by introducing the notion of weak BV solutions.

In general it is not clear whether BV solutions can be completed to parametrized solutions. We need an additional condition, which is certainly satisfied in the finite-dimensional setting discussed in [59], namely

$$\begin{aligned} &\exists C>0\ \forall t \in [0,T]\ \forall z_1, z_2 \in \boldsymbol{Z} \text{ with } 0 \in \partial\Psi(0) + \overline{\partial}_z \mathcal{I}(t,z_j) : \\ &y \in \mathrm{OJP}(t,z_1,z_2) \implies \mathrm{Var}_{\boldsymbol{V}}(y,[0,1]) \leq C\Delta(t,z_1,z_2). \end{aligned} \tag{110}$$

We conjecture that this condition also holds for the standard Example 2.8, which is suggested by (99), where the term involving $\dot{\ell}$ disappears and, at least formally, $\dot{z}(0) = 0$, since z_1 is locally stable.

Proposition 4.24. *If the RIS $(\boldsymbol{Z},\mathcal{I},\Psi,\mathbb{V})$ satisfies (110), then for each connectable BV solution $z \in \mathrm{BV}([0,T];\boldsymbol{V})$ there is a parametrized solution $(\tau, Z) \in \mathrm{C}^{\mathrm{Lip}}([0,T];\boldsymbol{V}_T)$ such that $z \in \mathfrak{P}((\tau,Z))$.*

Proof. The graph of z has finite length $T+\mathrm{Var}_{\boldsymbol{V}}(z,[0,T])$ in $\boldsymbol{V}_T$. For each $t \in J(z)$, we now add the graphs $Y^t \stackrel{\text{def}}{=} \{\,(t,y^t(r)) \mid r\in[0,1]\,\}$ of the optimal jump paths. By assumption (110) we know that the length of this curve is bounded by $C\Delta(t,z(t^-),z(t^+))$. Using the energy balance (108b) we see that the total length of all these added curves is finite. By construction $G(z)\cup\bigcup_{t\in J(z)}Y^t$ is a connected curve of finite length in $\boldsymbol{V}_T$ that can be reparametrized with respect to arclength, which provides the function $\zeta=(\tau,Z):[0,S]\to\boldsymbol{V}_T$.

We still have to show that ζ satisfies (87). The conditions in the upper line are trivial. For $s\in[0,S]$ such that $\zeta(s)\in Y^t$ we have $\dot\tau(s)=0$ and the differential inclusion in (87) holds because of (106). In the other case we have local stability, namely $0=\partial\Psi(0)+\xi(s)$ with $\xi(s)\in\overline{\partial}_z\mathcal{I}(\zeta(s))$. However, in these points the energy balance together with the chain rule provides $0=\Psi(\dot Z(s))+\langle\xi(s),\dot Z(s)\rangle$, which implies the desired differential inclusion by Lemma 2.2. □

We conclude this section with a result concerning time discretizations. Time-incremental minimization techniques are the central tools in generalized gradient flows as well as for energetic solutions, see Sect. 3.2. In our vanishing-viscosity approach, we are especially interested in the interaction between the smallness of the time steps and the smallness of the viscosity. It turns out that BV solutions are easily obtained by a joint limit. For simplicity we again study a RIS of the form $(\boldsymbol{Z},\mathcal{I},\Psi,\mathbb{V})$ for small viscosity ε. We also discretize the time interval in the form $\Pi=(t_0,t_1,\ldots,t_{N_\Pi})\in\mathrm{Part}([0,T])$ with fineness $\phi(\Pi)=\max\{\,t_k-t_{k-1}\mid k=1,\ldots,N_\Pi\,\}$, see (34). The incremental minimization problem for the viscous problem reads as follows:

$$z_k\in\mathrm{Arg\,min}\{\,\mathcal{I}(t_k,z)+\Psi(z-z_{k-1})+\frac{\varepsilon}{t_k-t_{k-1}}\|z-z_{k-1}\|^2_{\mathbb{V}}\mid z\in\boldsymbol{Z}\,\}.$$

We denote by $\underline{z}^{\Pi,\varepsilon}:[0,T]\to\boldsymbol{Z}$ the piecewise constant interpolant, see (39). Then the following result was proved in [59] for the finite-dimensional setting. A corresponding convergence result of incremental problems to parametrized solutions was also obtained in [22]. We expect that a similar result holds in infinite dimensions, in particular for our standard Example 2.8.

Theorem 4.25 (Convergence of viscous time discretizations). *Assume that all Banach spaces are finite-dimensional, that $\mathcal{I}\in\mathrm{C}^1(\boldsymbol{Z}_T)$ and that the RIS $(\boldsymbol{Z},\mathcal{I},\Psi,\mathbb{V})$ satisfies the standard coercivity assumptions. Take any sequence $(\Pi_n)_{n\in\mathbb{N}}$ of partitions and any sequence $(\varepsilon_n)_{n\in\mathbb{N}}$ of viscosities such that*

$$\phi(\Pi_n)\to 0,\quad \varepsilon_n\to 0,\quad \frac{\varepsilon_n}{\phi(\Pi_n)}\to\infty. \tag{111}$$

For an initial value $z_0\in\boldsymbol{Z}$ construct $\underline{z}^n\stackrel{\text{def}}{=}\underline{z}^{\Pi_n,\varepsilon_n}:[0,T]\to\boldsymbol{Z}$. Then, there exist a subsequence $(\underline{z}^{n_l})_{l\in N}$ and a BV solution z for $(\boldsymbol{Z},\mathcal{I},\Psi,\mathbb{V})$ such that

$$\forall t \in [0,T] : z^{n_l}(t) \rightharpoonup z(t) \text{ in } \boldsymbol{Z} \text{ for } l \to \infty.$$

Moreover, any pointwise limit of a subsequence of $(\underline{z}^n)_{n\in\mathbb{N}}$ *is a BV solution.*

4.6 Weak BV Solutions and Time-Varying Dissipation Distances

Here we introduce a slightly more complicated rate-independent dissipation which allows us to define a weaker version of BV solutions. As a generalization to the definition of $\Delta(t,z_0,z_1)$ in (104) we now define a dissipation distance $\widehat{\mathfrak{D}}_{\mathfrak{p}}^{\mathcal{J}}$ in the extended phase space $\boldsymbol{Z}_T$, i.e. $\widehat{\mathfrak{D}}_{\mathfrak{p}}^{\mathcal{J}} : \boldsymbol{Z}_T \times \boldsymbol{Z}_T \to [0,\infty]$. It measures the minimal dissipation needed for moving from a point z_0 to a point z_1 during the time interval $[t_0,t_1]$. For $t_1 < t_0$ we set $\widehat{\mathfrak{D}}_{\mathfrak{p}}^{\mathcal{J}}(t_0,z_1,t_1,z_1) = \infty$ and for $t_0 < t_1$ we let

$$\begin{aligned} &\widehat{\mathfrak{D}}_{\mathfrak{p}}^{\mathcal{J}}(t_0,z_0,t_1,z_1) \\ &\overset{\text{def}}{=} \inf\Big\{ \int_0^1 \mathfrak{p}(\dot{Z}(r),-\xi(r))\,\mathrm{d}r \;\Big|\; (\tau,Z) \in A_{\boldsymbol{V}_T}((t_0,z_0),(t_1,z_1)), \\ &\qquad\qquad \dot{\tau}(r) \geq 0 \text{ and } \xi(r) \in \overline{\partial}_z \mathcal{J}(\tau(r),Z(r)) \text{ a.e. in } [0,1] \Big\}. \end{aligned} \tag{112}$$

Clearly, we have $\widehat{\mathfrak{D}}_{\mathfrak{p}}^{\mathcal{J}}(t,z_0,t,z_1) = \Delta(t,z_0,z_1)$, $\widehat{\mathfrak{D}}_{\mathfrak{p}}^{\mathcal{J}}(t_0,z_0,t_1,z_1) \geq c\Psi(z_1-z_0)$ and the triangle inequality

$$\widehat{\mathfrak{D}}_{\mathfrak{p}}^{\mathcal{J}}(t_0,z_0,t_2,z_2) \leq \widehat{\mathfrak{D}}_{\mathfrak{p}}^{\mathcal{J}}(t_0,z_0,t_1,z_1) + \widehat{\mathfrak{D}}_{\mathfrak{p}}^{\mathcal{J}}(t_1,z_1,t_2,z_2).$$

The associated dissipation functional for curves $z : [0,T] \to \boldsymbol{Z}$ reads

$$\begin{aligned} \mathrm{Diss}_{\mathfrak{p}}^{\mathcal{J}}(z,[r,s]) = \sup\Big\{ \sum_{k=1}^{N} \widehat{\mathfrak{D}}_{\mathfrak{p}}^{\mathcal{J}}(t_{k-1},z(t_{k-1}),t_k,z(t_k)) \;\Big|\; N \in \mathbb{N}, \\ r \leq t_0 < t_1 < \cdots < t_{N-1} < t_N \leq s \Big\}. \end{aligned}$$

From the definitions, it is immediate that this dissipation is greater or equal to $\mathrm{Diss}_{\mathfrak{p},\mathcal{J}}$ defined in (107). With the new dissipation functional $\widehat{\mathfrak{D}}_{\mathfrak{p}}^{\mathcal{J}}$ we are now able to define a new notion of BV solution.

Definition 4.26 (Weak BV solutions). Let the RIS $(\boldsymbol{Z},\mathcal{J},\Psi,\mathbb{V})$ be given. A function $z \in \mathrm{BV}([0,T];\boldsymbol{X}) \cap \mathrm{L}^\infty([0,T];\boldsymbol{Z})$ is called *weak BV solution*, if

$$\forall t \in C(z): \ z(t) \in \mathcal{S}_{\mathrm{loc}}(t) \stackrel{\mathrm{def}}{=} \{ z \in \boldsymbol{Z} \mid 0 \in \partial\Psi(0) + \overline{\partial}_z \mathcal{I}(t,z) \}; \tag{113a}$$

$$\forall t \in [0,T]: \mathcal{I}(t,z(t)) + \mathrm{Diss}_{\mathfrak{p}}^{\mathcal{I}}(z,[0,t]) = \mathcal{I}(0,z(0)) + \int_0^t \partial_\tau \mathcal{I}(\tau,z(\tau))\,\mathrm{d}\tau. \tag{113b}$$

The following example shows that $\mathrm{Diss}_{\mathfrak{p}}^{\mathcal{I}}(z,[0,T])$ may be bigger than $\mathrm{Diss}_{\mathfrak{p},\mathcal{I}}(z,[0,T])$ even for continuous curves. Moreover, it states that the vanishing-viscosity limit in Example 4.4 provides a weak BV solution that is not a BV solution.

Example 4.27. We reconsider the setting of Example 4.4 where

$$\boldsymbol{X} = \mathrm{L}^1(\Omega), \quad \boldsymbol{V} = \boldsymbol{Z} = \mathrm{L}^2(\Omega) \text{ with } \Omega =]0,1[\,,$$
$$\mathfrak{p}(v,\xi) = \|v\|_{\mathrm{L}^1} + \|v\|_{\mathrm{L}^2} \|\max\{|\xi(\cdot)|-1,0\}\|_{\mathrm{L}^2}, \quad \mathcal{I}(t,z) = \int_\Omega \mathcal{U}(z) - (t+x)z\,\mathrm{d}x.$$

The vanishing-viscosity limit z obtained there jumps from -2 to 6 along the line $t+x=3$. It can be shown that for $2 \le r < t \le 3$ we have

$$\begin{aligned} 9(t-r) = \Psi(z(t)-z(r)) &= \mathrm{Diss}_{\mathfrak{p},\mathcal{I}}(z,[r,t]) \\ &< \widehat{\mathcal{D}}_{\mathfrak{p}}^{\mathcal{I}}(r,z(r),t,z(t)) = \mathrm{Diss}_{\mathfrak{p}}^{\mathcal{I}}(z,[r,t]) = 25(t-r). \end{aligned}$$

We should interpret the additional dissipation as a limiting effect of uncountably many infinitesimal jumps.

Considering the additional term ϱ in the energy balance in Example 4.4, we see that z fails to be a BV solution, however it is a weak BV solution.

The notion of weak BV solutions is close to the notion of energetic solutions, if we replace the standard dissipation distance $\mathcal{D}$ by the extended dissipation distance $\widehat{\mathcal{D}}_{\mathfrak{p}}^{\mathcal{I}}$. However, global stability with respect to $\widehat{\mathcal{D}}_{\mathfrak{p}}^{\mathcal{I}}$ is useless, since by construction of $\widehat{\mathcal{D}}_{\mathfrak{p}}^{\mathcal{I}}$ the points are globally stable. This is in agreement with the fact, that in principle every point z_0 should be possible as an initial condition. Nevertheless, weak BV solutions can be obtained via a similar incremental minimization problem. Given a partition $(0 = t_0, t_1, \ldots, t_N = T)$ and an initial state $z_0 \in \boldsymbol{Z}$, find $z_1, \ldots, z_N$ via

$$z_k \in \mathrm{Arg\,min}\{ \mathcal{I}(t_k,z) + \widehat{\mathcal{D}}_{\mathfrak{p}}^{\mathcal{I}}(t_{k-1},z_{k-1},t_k,z) \mid z \in \boldsymbol{Z} \} \cap \mathcal{S}_{\mathrm{loc}}(t_k).$$

The additional constraint $z_k \in \mathcal{S}_{\mathrm{loc}}(t_k)$ is added to avoid solutions that linger too long in jump paths. We expect that the intersection of Argmin and S_{loc} is always nonempty, since for any point in Argmin we can start an optimal jump path that ends in a point in the intersection and still lies in Argmin because of the definition of $\widehat{\mathcal{D}}_{\mathfrak{p}}^{\mathcal{I}}$.

5 Metric Formulations

Here we discuss the notions of parametrized and BV solutions in a metric setting by generalizing the ideas from Banach spaces to general metric spaces. Thus, the theory becomes more general in the sense that, as in Sect. 3 for energetic solutions, we dispose of the linear structure. However, in contrast to the previous section we have to restrict to the case that the viscous dissipations is proportional to the square of the rate-independent dissipation, i.e., the theory of this section includes all the results of the previous one, if we assume $\mathcal{R}_\varepsilon(v) = \Psi(v) + \frac{\varepsilon}{2}\Psi(v)^2$. Thus, the standard Example 2.8 can not be treated with the methods developed below.

The theory is based on the abstract approach to evolutionary problems in general metric spaces. We refer to [1,3,63] for general reading and to [60] for more details on the results presented here.

5.1 Metric Velocity, Slope, and Evolution

In this section we recall results presented in [1], starting from a complete metric space $(\mathcal{Z}, \mathcal{D})$, where we again use the letter $\mathcal{Z}$ (instead of $\boldsymbol{Z}$) to indicate that $\mathcal{Z}$ does not need a linear structure. Moreover, we now assume that $\mathcal{D} : \mathcal{Z}\times\mathcal{Z} \to [0,\infty[$ is a true metric distance, i.e. in addition to the assumptions above it is also symmetric and assumes only finite value. (It is expected that the theory can be generalized to extended quasi-distances, cf. [63].)

A curve $z : [0,T] \to \mathcal{Z}$ is called *absolutely continuous* (written as $z \in \mathrm{AC}([0,T];\mathcal{Z})$), if $\exists\ m \in \mathrm{L}^1([0,T])$ such that $\mathcal{D}(q(r), q(t)) \leq \int_r^t m(s)\,\mathrm{d}s$ for $0 \leq r < t \leq T$.

Theorem 5.1 (Metric velocity). *If* $z \in \mathrm{AC}([0,T],\mathcal{Z})$, *then for a.a.* $t \in [0,T]$ *the* metric velocity $|\dot{z}|(t) \overset{\mathrm{def}}{=} \lim_{h\to 0} \frac{1}{h}\mathcal{D}(z(t), z(t{+}h))$ *exists. Moreover,* $|\dot{z}|(t) \leq m(t)$ *a.e. and* $\mathrm{Diss}_\mathcal{D}(z,[r,t]) = \int_r^t |\dot{z}|(s)\,\mathrm{d}s$.

The dot $\dot{}$ and the norm $|\cdot|$ in the notation $|\dot{z}| \in \mathrm{L}^1([0,T])$ are used only to indicate that the metric velocity relates to the norm of a velocity in the classical case. In fact, if $\mathcal{Z} = \boldsymbol{Z}$, $\mathcal{D}$ is defined via a rate-independent dissipation potential $\mathcal{R}$ in the sense of (74), and $z \in \mathrm{W}^{1,1}([0,T),\boldsymbol{Z})$, then $|\dot{z}|(t) = \mathcal{R}(z(t), \dot{z}(t))$.

We emphasize that the metric concept is even useful in the case of non-reflexive Banach spaces $\boldsymbol{X}$ which we then equip with the distance induced by the norm. For instance, we may consider $\mathcal{Z} = \mathrm{L}^1(\mathbb{R})$ with $\mathcal{D}(z_0, z_1) = \|z_1{-}z_0\|_{\mathrm{L}^1}$. In Example 4.1 the curve z lies in $z \in \mathrm{AC}([0,T];\mathrm{L}^1(\mathbb{R}))$ but not in $\mathrm{W}^{1,1}([0,T];\mathrm{L}^1(\mathbb{R}))$. The metric velocity exists, namely $|\dot{z}|(t) = |\dot{\alpha}(t)| + |\dot{\beta}(t)|$.

For a functional $\mathcal{J} : \mathcal{Z} \to \mathbb{R}$ we define the *metric slope* $|\partial\mathcal{J}|_*(z)$ of $\mathcal{J}$ in the point z via

$$|\partial\mathcal{J}|_*(z) \overset{\text{def}}{=} \limsup_{\widetilde{z}\to z} \frac{\max\{\mathcal{J}(z)-\mathcal{J}(\widetilde{z}),0\}}{\mathcal{D}(z,\widetilde{z})}.$$

For functionals $\mathcal{I} : \mathcal{Z}_T \to \mathbb{R}$ we write with a slight abuse of notation

$$|\partial\mathcal{I}|_*(t,z) \overset{\text{def}}{=} |\partial\mathcal{I}(t,\cdot)|_*(z).$$

Again the sign ∂ and the dual norm $|\cdot|_*$ in the notation $|\partial\mathcal{J}|_*$ are formal only and should indicate that in the classical case the metric slope relates to the (dual) norm of (sub)gradient. In fact, if $\mathcal{J} \in \mathrm{C}^1(\mathcal{Z})$ and $\mathcal{D}$ is given via $\mathcal{R}$ as above, then $|\partial\mathcal{J}|_*(z) = \mathcal{R}^\circ(z,-\mathrm{D}\mathcal{J}(z))$, where $\mathcal{R}^\circ(z,\xi) = \sup\{\langle\xi,v\rangle \mid \mathcal{R}(z,v) \le 1\}$.

As a major assumption on our RIS $(\mathcal{Z},\mathcal{I},\mathcal{D})$ we impose the following chain-rule inequality:

$$\begin{aligned}&z\in\mathrm{AC}([0,T];\mathcal{Z}),\ t\mapsto|\partial\mathcal{I}|_*(t,z(t)) \text{ measurable},\ \int_0^T |\dot z|(t)|\partial\mathcal{I}|_*(t,z(t))\,\mathrm{d}t<\infty\\ &\implies\quad t\mapsto\mathcal{I}(t,z(t)) \text{ is absolutely continuous on } [0,T] \text{ and}\\ &\qquad\ \tfrac{\mathrm{d}}{\mathrm{d}t}\mathcal{I}(t,z(t))+|\dot z|(t)|\partial\mathcal{I}|_*(t,z(t)) \ge \partial_t\mathcal{I}(t,z(t)).\end{aligned}\tag{114}$$

Clearly, such a chain-rule inequality holds in the classical setting since

$$\tfrac{\mathrm{d}}{\mathrm{d}t}\mathcal{I}(t,z(t)) - \partial_t\mathcal{I}(t,z(t)) = \langle \mathrm{D}_z\mathcal{I}(t,z),\dot z\rangle \ge -\mathcal{R}(z,\dot z)\,\mathcal{R}^\circ(z,-\mathrm{D}\mathcal{I}(t,z)).$$

For a lower semicontinuous, convex function $\psi : [0,\infty[\to [0,\infty[$ we can now define metric solutions to the evolutionary systems $(\mathcal{Z},\mathcal{I},\mathcal{D},\psi)$ in analogy to the energetic formulations in Sect. 2.5.

Definition 5.2 (Metric evolution, ψ-gradient flow). Let $(\mathcal{Z},\mathcal{I},\mathcal{D})$ satisfy the chain-rule inequality (114). A function $z \in \mathrm{AC}([0,T],\mathcal{Z})$ is called a *metric solution* of the *ψ-gradient flow*, also called a *metric evolutionary system* (in the sense of De Giorgi) $(\mathcal{Z},\mathcal{I},\mathcal{D},\psi)$ if

$$\begin{aligned}&\mathcal{I}(T,z(T)) + \textstyle\int_0^T \psi(|\dot z|(t))+\psi^*(|\partial\mathcal{I}|_*(t,z(t)))\,\mathrm{d}t\\ &\le \mathcal{I}(0,z(0)) + \textstyle\int_0^T \partial_t\mathcal{I}(t,z(t))\,\mathrm{d}t.\end{aligned}\tag{115}$$

As in Sect. 2.5 the chain-rule inequality implies that (115) is equivalent to

$$\tfrac{\mathrm{d}}{\mathrm{d}t}\mathcal{I}(t,z(t)) + \psi(|\dot z|(t)) + \psi^*(|\partial\mathcal{I}|_*(t,z(t))) = \partial_t\mathcal{I}(t,z(t)) \quad \text{a.e. on } [0,T].$$

So far, the theory of metric evolutionary systems relies heavily on the absolute continuity of z. In the case of RIS we would like to choose $\psi(\nu) = \nu$, which led to $\psi^*(\xi) = \chi_{[0,1]}(\xi)$. Then, any approximation procedure leads to a priori estimates for $|\dot z|$ in $\mathrm{L}^1([0,T])$ only, which would not be enough to pass to the limit since jumps may develop.

5.2 Parametrized Metric Solutions

As in the Banach-space setting we introduce viscous approximations via $\psi_\varepsilon(\nu) = \nu + \frac{\varepsilon}{2}\nu^2$. The Legendre transform is $\psi_\varepsilon^*(\xi) = \frac{1}{2\varepsilon}\max\{\xi-1, 0\}^2$. The associated metric evolutionary system reads

$$\frac{\mathrm{d}}{\mathrm{d}t}\mathcal{I}(t, z^\varepsilon)+\psi_\varepsilon(|\dot{z}^\varepsilon|(t))+\psi_\varepsilon^*(|\partial\mathcal{I}|_*(t, z^\varepsilon(t))) \leq \partial_t\mathcal{I}(t, z^\varepsilon(t)), \quad z^\varepsilon(0) = z_0. \tag{116}$$

Using the standard energy bounds we find the a priori estimates

$$\mathcal{I}(t, z^\varepsilon(t)) \leq \mathrm{e}^{Ct}\mathcal{I}(0, z_0), \quad \int_0^T \psi_\varepsilon(|\dot{z}^\varepsilon|(t))+\psi_\varepsilon^*(|\partial\mathcal{I}|_*(t, z^\varepsilon(t)))\,\mathrm{d}t \leq \mathrm{e}^{CT}\mathcal{I}(0, z_0).$$

The classical existence theory in [1, 3] works for the cases $\psi(\nu) = \nu^p/p$, whereas [63] treats the case of general convex and superlinear ψ, covering our case $\psi_\varepsilon(\nu) = \nu + \nu^2/\varepsilon$. Thus, for each $\varepsilon > 0$ we obtain a solution $z^\varepsilon \in \mathrm{AC}([0, T]; \mathcal{Z})$ with the estimates

$$\| |\dot{z}^\varepsilon| \|_{\mathrm{L}^1([0,T])} \leq C \quad \text{and} \quad \| |\dot{z}^\varepsilon| \|_{\mathrm{L}^2([0,T])} \leq C/\sqrt{\varepsilon}. \tag{117}$$

As in Sect. 4.4 we introduce an arclength parametrization of the graph of z^ε, namely $\{\,(t, z^\varepsilon(t)) \mid t \in [0, T]\,\} \subset \mathcal{Z}_T$ via $s^\varepsilon(t) = t + \int_0^t |\dot{z}^\varepsilon|(r)\,\mathrm{d}r$ and the inverse functions $t = \tau^\varepsilon(s)$. The good message here is that we immediately have an upper bound for the total length S^ε by using (117). Thus, we define the arclength parametrized solutions as

$$\zeta^\varepsilon(s) = (\tau^\varepsilon(s), Z^\varepsilon(s)) \in \mathcal{Z}_T \quad \text{with } Z^\varepsilon(s) = z^\varepsilon(\tau^\varepsilon(s)).$$

We introduce the function

$$M_\varepsilon(\alpha, \nu, \xi) = \begin{cases} \alpha\psi_\varepsilon(\nu/\alpha) + \alpha\psi_\varepsilon^*(\xi) & \text{for } \alpha > 0, \\ \infty & \text{otherwise,} \end{cases}$$

which explicitly means $M_\varepsilon(\alpha, \nu, \xi) = \nu + \frac{\varepsilon}{2\alpha}\nu^2 + \frac{\alpha}{2\varepsilon}\max\{\xi-1, 0\}^2$ for $\alpha > 0$.

The rescaling of (116) leads to the relations

$$\left.\begin{array}{l} \tau^\varepsilon(0) = 0,\ \tau^\varepsilon(S^\varepsilon) = T,\ \dot{\tau}^\varepsilon(s) \geq 0,\ \dot{\tau}^\varepsilon(s) + |\dot{Z}^\varepsilon|(s) = 1, \\ \frac{\mathrm{d}}{\mathrm{d}s}\mathcal{I}(\zeta^\varepsilon(s)) + M_\varepsilon(\dot{\tau}^\varepsilon(s), |\dot{Z}^\varepsilon|(s), |\partial\mathcal{I}|_*(\zeta^\varepsilon(s)) \leq \partial_t\mathcal{I}(\zeta^\varepsilon(s))\dot{\tau}^\varepsilon(s) \end{array}\right\} \tag{118}$$

a.e. on $[0, S^\varepsilon]$. On $[0, \infty[^3$ it is easy to see that the functions M_ε converge to

$$M_0 : (\alpha, \nu, \xi) \mapsto \begin{cases} \nu\max\{\xi, 1\} & \text{for } \alpha = 0, \\ \nu + \chi_{[0,1]}(\xi) & \text{for } \alpha > 0, \end{cases}$$

in the sense of Γ-convergence. In fact, we have the following liminf estimate for the associated functionals, which is an application of Ioffe's theorem [32], see [60, Lem. 3.1]. It uses convexity in (α,ν) and monotonicity in ξ.

Proposition 5.3 (Lower semicontinuity). *Let* $\mathcal{M}_\varepsilon(\alpha,\nu,\xi)=\int_0^S M_\varepsilon(\alpha(s),\nu(s),\xi(s))\,\mathrm{d}s$ *for* $\varepsilon\geq 0$ *and assume that the sequence* $((\alpha^\varepsilon,\nu^\varepsilon,\xi^\varepsilon))_{\varepsilon>0}$ *satisfies* $\alpha^\varepsilon\rightharpoonup\widehat{\alpha}$ *and* $\nu^\varepsilon\rightharpoonup\widehat{\nu}$ *in* $\mathrm{L}^1([0,S])$ *and* $\widehat{\xi}(s)\leq\liminf_{\varepsilon\to0^+}\xi^\varepsilon(s)$ *a.e. on* $[0,S]$, *then we have* $\mathcal{M}_0(\widehat{\alpha},\widehat{\nu},\widehat{\xi})\leq\liminf_{\varepsilon\to0}\mathcal{M}_\varepsilon(\alpha^\varepsilon,\nu^\varepsilon,\xi^\varepsilon)$.

We now define parametrized metric solutions by using the functions M_0 in place of the small viscosity functions M_ε. Without loss of generality we only consider arclength parametrized solutions.

Definition 5.4 (Parametrized metric solutions). A function $\zeta=(\tau,Z):[0,S]\to\mathcal{Z}_T$ is called *parametrized metric solution* of the RIS $(\mathcal{Z},\mathcal{I},\mathcal{D})$, if $\zeta\in\mathrm{C}^{\mathrm{Lip}}([0,S];\mathcal{Z}_T)$ and if for a.a. $s\in[0,S]$ we have

$$\begin{aligned}&\tau(0)=0,\ \tau(S)=T,\ \dot\tau(s)\geq0,\ \dot\tau(s)+|\dot Z|(s)=1,\\&\tfrac{\mathrm{d}}{\mathrm{d}s}\mathcal{I}(\zeta(s))+M_0(\dot\tau(s),|\dot Z|(s),|\partial\mathcal{I}|_*(\zeta(s))\leq\partial_t\mathcal{I}(\zeta(s))\dot\tau(s).\end{aligned}\tag{119}$$

Since $M_0(\alpha,\nu,\xi)\geq\nu\xi$, the *contact set* Ξ of M_0 plays a central role:

$$\Xi\overset{\text{def}}{=}\{\,(\alpha,\nu,\xi)\mid M_0(\alpha,\nu,\xi)=\nu\xi\,\}$$

The explicit form of M_0 gives three distinct regimes: $\Xi=\Xi^{\text{stick}}\cup\Xi^{\text{slide}}\cup\Xi^{\text{jump}}$ with

$$\begin{aligned}\Xi^{\text{stick}}&=\{\,(\alpha,0,\xi)\mid\alpha\geq0,\ \xi\leq1\,\},\\\Xi^{\text{slide}}&=\{\,(\alpha,\nu,1)\mid\alpha,\nu\geq0\,\},\\\Xi^{\text{jump}}&=\{\,(0,\nu,\xi)\mid\nu\geq0,\ \xi\geq1\,\},\end{aligned}$$

see Fig. 4. Using the chain-rule inequality (114) and the last line of (119) we conclude that for a parametrized metric solution we have

$$(\dot\tau(s),|\dot Z|(s),|\partial\mathcal{I}|_*(\zeta(s)))\in\Xi\quad\text{for a.a. } s\in[0,S].$$

This leads to an alternative equivalent definition for parametrized metric solutions, which highlights these different regimes more clearly.

Lemma 5.5. *A function* $\zeta=(\tau,Z):[0,S]\to\mathcal{Z}_T$ *is a parametrized metric solution of the RIS* $(\mathcal{Z},\mathcal{I},\mathcal{D})$ *if and only if for a.a.* $s\in[0,S]$ *we have*

$$\begin{aligned}&\dot\tau(s)\geq0\quad\text{and}\quad\dot\tau(s)+|\dot Z|(s)=1;\\&\dot\tau(s)>0\implies|\partial\mathcal{I}|_*(\zeta(s))\leq1;\end{aligned}$$

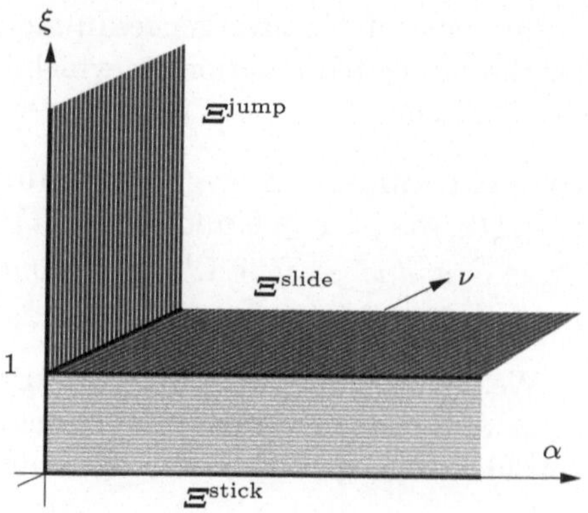

Fig. 4 The three different regimes for parametrized metric solutions

$$|\dot{Z}|(s) > 0 \implies |\partial\mathcal{I}|_*(\zeta(s)) \geq 1;$$
$$\tfrac{\mathrm{d}}{\mathrm{d}s}\mathcal{I}(\zeta(s)) + |\dot{Z}|(s)|\partial\mathcal{I}|_*(\zeta(s)) = \partial_s\mathcal{I}(\zeta(s))\dot{\tau}(s).$$

The last condition in the above lemma says that if a solution moves, then, it moves along a gradient flow curve. Moreover, integrating the last relation we obtain the energy balance

$$\begin{aligned}&\mathcal{I}(\zeta(s)) + \int_0^s |\dot{Z}|(s) + \gamma(s)\,\mathrm{d}s = \mathcal{I}(0,z_0) + \int_0^s \partial_r\mathcal{I}(\zeta(r))\,\mathrm{d}r\\ &\text{with } \gamma(s) = |\dot{Z}|(s)\max\{|\partial\mathcal{I}|_*(\zeta(s)){-}1, 0\},\end{aligned} \tag{120}$$

which is analogous to (89) for parametrized solutions in the Banach-space setting. Thus, we again see an explicit term arising in jump paths that is needed to lead to a correct energy balance.

The next result shows that under quite general assumptions we have convergence of parametrized solutions in the vanishing-viscosity limit and thus obtain existence of parametrized metric solutions. In contrast to the Banach-space setting we have much more general initial conditions and we are able to show that the arclength parametrization is inherited by the limit. In the additional conditions on $\mathcal{I}$ we use the topology on $\mathcal{Z}$ that is induced by the metric $\mathcal{D}$.

Theorem 5.6 (Vanishing viscosity, parametrized metric solutions). *Let the RIS $(\mathcal{Z}, \mathcal{I}, \mathcal{D})$ satisfy the assumptions from above and*

$$\begin{aligned}&\mathcal{I}(t,\cdot) : \mathcal{Z} \to \mathbb{R} \textit{ has sequentially compact sublevels,}\\ &\partial_t\mathcal{I} : \mathcal{Z}_T \to \mathbb{R} \textit{ is continuous,}\\ &|\partial\mathcal{I}|_* : \mathcal{Z}_T \to [0,\infty] \textit{ is lower semicontinuous.}\end{aligned} \tag{121}$$

Choose any $z_0 \in \mathcal{Z}$*. Then, for any family of parametrized metric solutions* $\zeta^\varepsilon : [0,S] \to \mathcal{Z}_T$ *of* (118) *with* $\zeta^\varepsilon(0) = (0, z_0)$ *there exist a subsequence* $(\varepsilon_k)_{k\in\mathbb{N}}$ *with* $\varepsilon_k \to 0$ *and a limit function* $\zeta = (\tau, Z) \in \mathrm{AC}([0,S]; \mathcal{Z}_T)$ *such that*

$$\zeta \text{ is a parametrized metric solution with } \zeta(0) = (0, z_0),$$

$$\zeta^{\varepsilon_k} \to \zeta \text{ in } \mathrm{C}^0([0,T]; \mathcal{Z}_T);$$

$$\dot\tau^{\varepsilon_k} \overset{*}{\rightharpoonup} \dot\tau, \ \ |\dot Z^{\varepsilon_k}| \overset{*}{\rightharpoonup} |\dot Z| \text{ in } \mathrm{L}^\infty([0,S]).$$

Proof. Since the functions ζ^ε have the uniform Lipschitz bound 1 and all ζ^ε lie in a sequentially compact sublevel of $\mathcal{I}$, we can apply the Arzelà-Ascoli theorem and obtain a sequence converging uniformly. Since $\dot\tau^\varepsilon$ and $|\dot Z^{\varepsilon_k}|$ are bounded by 1, we may also assume the weak* convergence to limits μ and η, respectively. Obviously, $\mu+\eta \equiv 1$ and it is easy to see that $\mu = \dot\tau$. From

$$\begin{aligned}\mathcal{D}(Z(s_1), Z(s_2)) &= \lim_k \mathcal{D}(Z^{\varepsilon_k}(s_1), Z^{\varepsilon_k}(s_2)) \\ &\le \liminf_k \textstyle\int_{s_1}^{s_2} |\dot Z^{\varepsilon_k}|(s)\,\mathrm{d}s = \int_{s_1}^{s_2} \eta(s)\,\mathrm{d}s,\end{aligned}$$

we find $|\dot Z|(s) \le \eta(s)$ a.e. on $[0,S]$. From the assumptions in (121) we obtain

$$\mathcal{I}(\zeta(s)) \le \liminf_k \mathcal{I}(\zeta^{\varepsilon_k}(s)), \ \ \partial_t\mathcal{I}(\zeta(s)) = \lim_k \partial_t\mathcal{I}(\zeta^{\varepsilon_k}(s)),$$

$$\xi(s) = |\partial\mathcal{I}|_*(\zeta(s)) \le \liminf_k \xi^k(s), \ \text{ where } \xi^k(s) = |\partial\mathcal{I}|_*(\zeta^{\varepsilon_k}(s)).$$

We can now pass to the limit in the integrated version of the last line in (118) and, using Proposition 5.3, we obtain

$$\mathcal{I}(\zeta(S)) + \mathcal{M}_0(\dot\tau, \eta, |\partial\mathcal{I}|_*(\zeta(\cdot))) \le \mathcal{I}(0, z_0) + \textstyle\int_0^S \partial_t\mathcal{I}(\zeta(s))\dot\tau(s)\,\mathrm{d}s. \tag{122}$$

For the convergence in the last integral note that the integrand is the product of a strongly and a weakly converging sequence.

In the following estimates we first use the chain-rule-inequality (114), then $\nu\xi \le M_0(\alpha, \nu, \xi)$, next $|\dot Z| \le \eta$ and monotonicity of $M_0(\alpha, \cdot, \xi)$, and finally (122):

$$\begin{aligned}&\mathcal{I}(\zeta(0)) - \mathcal{I}(\zeta(S)) + \textstyle\int_0^S \partial_t\mathcal{I}(\zeta(s))\dot\tau(s)\,\mathrm{d}s \le \int_0^S |\dot Z|(s)|\partial\mathcal{I}|_*(\zeta(s))\,\mathrm{d}s \\ &\le \mathcal{M}_0(\dot\tau, |\dot Z|, |\partial\mathcal{I}|_*(\zeta(\cdot))) \le \mathcal{M}_0(\dot\tau, \eta, |\partial\mathcal{I}|_*(\zeta(\cdot))) \\ &\le \mathcal{I}(\zeta(0)) - \mathcal{I}(\zeta(S)) + \int_0^S \partial_t\mathcal{I}(\zeta(s))\dot\tau(s)\,\mathrm{d}s.\end{aligned}$$

We conclude that all estimates are equalities. In particular, we find the second line of (119), and the strict monotonicity of $M_0(\alpha, \cdot, \xi)$ implies $|\dot{Z}|(s) = \eta(s)$ a.e., which implies $\dot{\tau} + |\dot{Z}|(s) = 1$ a.e. □

5.3 Metric BV Solutions

Like in the Banach-space setting it is desirable to define a solution concept that is closely associated with parametrized solutions, but that avoids the artificial arclength parametrization. Our definition follows the spirit of Sect. 4.5, which is simpler than the original approach in [60]. The function space $\mathrm{BV}([0,T];\mathcal{Z})$ contains functions $z : [0,T] \to \mathcal{Z}$ defined everywhere and such that $\mathrm{Var}_{\mathcal{D}}(z, [0,T]) = \mathrm{Diss}_{\mathcal{D}}(z, [0,T]) < \infty$. As before, left and right limits $z(t^-)$ and $z(t^+)$ exist for each $t \in [0,T]$. Moreover, the set of jump times $J(z)$ is well defined and countable.

We again define a supplemented dissipation measure $\Delta_M(t,\cdot) : \mathcal{Z}{\times}\mathcal{Z} \to [0,\infty]$ via

$$\begin{aligned} &\Delta_M(t,z_0,z_1) \overset{\text{def}}{=} \inf\{ \textstyle\int_0^1 M_0(0, |\dot{\widehat{z}}|(r), |\partial\mathcal{I}|_*(t,\widehat{z}(r)))\,\mathrm{d}r \mid \widehat{z} \in A(z_0,z_1) \}, \\ &\text{where } A(z_0,z_1) \overset{\text{def}}{=} \{ \widehat{z} \in \mathrm{AC}([0,1];\mathcal{Z}) \mid \widehat{z}(0) = z_0,\ \widehat{z}(1) = z_1 \}. \end{aligned}$$

Definition 5.7 (Metric BV solutions). Let the metric RIS $(\mathcal{Z},\mathcal{I},\mathcal{D})$ be given. Then a function $z \in \mathrm{BV}([0,T];\mathcal{Z})$ is called *metric BV solution* if (123a) and (123b) hold:

$$\forall t \in C(z) :\ |\partial\mathcal{I}|_*(t, z(t)) \leq 1; \tag{123a}$$

$$\forall t{\in}[0,T] : \mathcal{I}(t,z(t)) + \mathrm{Diss}_{\mathcal{D},\mathcal{I}}(z,[0,t]) = \mathcal{I}(0,z(0)) + \int_0^t \partial_\tau\mathcal{I}(\tau,z(\tau))\,\mathrm{d}\tau. \tag{123b}$$

The metric BV solution z is called *connectable*, if additionally

$$\forall t \in J(z)\ \exists \widehat{z}^t \in \mathrm{OJP}(t, z(t^-), z(t^+))\ \exists r^t \in [0,1] :\ z(t) = \widehat{z}^t(r^t). \tag{123c}$$

Again the set of optimal jump paths is defined via $\mathrm{OJP}(t,z_1,z_2) \overset{\text{def}}{=}$

$$\begin{aligned} \{ z \in A(z_1,z_2) \mid \Delta_M(t,z_1,z_2) &= \mathcal{I}(t,z_1) - \mathcal{I}(t,z_2) \\ &= M_0(0,|\dot{z}|(r),|\partial\mathcal{I}|_*(t,z(r))) \text{ for a.a. } r \in [0,1] \}. \end{aligned}$$

In [60] the following two results concerning existence, convergence and time discretization are derived.

Theorem 5.8 (Convergence and existence of metric BV solutions). *Assume that $(\mathcal{Z}, \mathcal{I}, \mathcal{D})$ satisfies the assumption of Theorem 5.6. Then, for each $z_0 \in \mathcal{Z}$ with $\mathcal{I}(0, z_0) < \infty$ there exists a metric BV solution z with $z(0) = z_0$.*

Moreover, if $z^\varepsilon \in \mathrm{AC}([0,T]; \mathcal{Z})$ are solutions of the viscous metric evolutionary system (116) *with $z^\varepsilon(0) = z_0$, then there exist a subsequence $\varepsilon_k \to 0$ and a metric BV solution z such that $z^{\varepsilon_k}(t) \to z(t)$ for all $t \in [0,T]$.*

For a partition $\Pi = (0 = t_0, t_1, \ldots, t_N = T)$, the time-incremental minimization problems with small viscosity read

$$z_j \in \operatorname{Arg\,min}\{\, \mathcal{I}(t_k, z) + \mathcal{D}(z_{k-1}, z) + \tfrac{\varepsilon}{2(t_k - t_{k-1})} \mathcal{D}(z_{k-1}, z)^2 \mid z \in \mathcal{Z} \}.$$

Defining the piecewise constant interpolants $\underline{z}^{\Pi_\varepsilon}$ as before, we obtain the following result.

Theorem 5.9 (Convergence of time discretizations). *Assume that the RIS $(\mathcal{Z}, \mathcal{I}, \mathcal{D})$ satisfies the assumption of Theorem 5.6. Assume further that for sequence Π_k and ε_k we have*

$$\varepsilon_k \to 0, \quad \phi(\Pi_k) \to 0, \quad \varepsilon_k / \phi(\Pi_k) \to \infty.$$

Then there exist a subsequence (not relabeled) and a metric BV solution z such that $\underline{z}^{\Pi_k, \varepsilon_k}(t) \to z(t)$ for all $t \in [0,T]$.

Acknowledgements The research was partially supported by DFG via Research Unit FOR 787 *MicroPlast* (Project Mie 459/5, Regularizations and relaxations of time continuous problems in plasticity). The author is indebted to Tomaáš Roubíček, Riccarda Rossi, Giuseppe Savaré, Marita Thomas, and Sergey Zelik for helpful and stimulating discussions. He thanks Olga Kuphal and Riccarda Rossi for their careful proofreading and for improving the English. Finally, he thanks Luigi Ambrosio, Giuseppe Savaré, and CIME for organizing the wonderful Summer School on *Nonlinear Partial Differential Equations and Applications* in Cetraro.

References

1. L. Ambrosio, N. Gigli, G. Savaré, *Gradient flows in metric spaces and in the space of probability measures.* Lectures in Mathematics ETH Zürich (Birkhäuser Verlag, Basel, 2005)
2. H.-D. Alber, *Materials with Memory,* vol. 1682, *Lecture Notes in Mathematics* (Springer, Berlin, 1998)
3. L. Ambrosio, Minimizing movements. Rend. Accad. Naz. Sci. XL Mem. Mat. Appl. **19**(5), 191–246 (1995)
4. H. Attouch, *Variational Convergence of Functions and Operators.* Pitman Advanced Publishing Program (Pitman, 1984)
5. J.-P. Aubin, H. Frankowska, *Set-valued Analysis* (Birkhäuser, Boston, 1990)

6. B. Bourdin, G.A. Francfort, J.-J. Marigo, The variational approach to fracture. J. Elasticity **91**, 5–148 (2008) (also printed as a book at Springer)
7. M. Brokate, P. Krejčí, H. Schnabel, On uniqueness in evolution quasivariational inequalities. J. Convex Anal. **11**, 111–130 (2004)
8. G. Bouchitté, A. Mielke, T. Roubíček, A complete-damage problem at small strains. Z. angew. Math. Phys. (ZAMP) **60**(2), 205–236 (2009)
9. G. Bonfanti, A vanishing viscosity approach to a two degree-of-freedom contact problem in linear elasticity with friction. Annali dell'Universita di Ferrara **42**(1), 127–154 (1996)
10. A. Braides, *Γ-Convergence for Beginners* (Oxford University Press, Oxford, 2002)
11. M. Brokate, J. Sprekels, *Hysteresis and Phase Transitions* (Springer, New York, 1996)
12. F. Cagnetti, A vanishing viscosity approach to fracture growth in a cohesive zone model with prescribed crack path. Math. Models Meth. Appl. Sci. (M^3AS) **18**(7), 1027–1071 (2009)
13. P. Colli, On some doubly nonlinear evolution equations in Banach spaces. Jpn. J. Indust. Appl. Math. **9**, 181–203 (1992)
14. S. Conti, M. Ortiz, Minimum principles for the trajectories of systems governed by rate problems. J. Mech. Phys. Solid. **56**(5), 1885–1904 (2008)
15. P. Colli, A. Visintin, On a class of doubly nonlinear evolution equations. Comm. Part. Differ. Equat. **15**(5), 737–756 (1990)
16. G. Dal Maso, *An Introduction to Γ-Convergence* (Birkhäuser Boston, Boston, MA, 1993)
17. G. Dal Maso, R. Toader, A model for quasi–static growth of brittle fractures: existence and approximation results. Arch. Rational Mech. Anal. **162**, 101–135 (2002)
18. G. Dal Maso, A. DeSimone, M.G. Mora, M. Morini, Time-dependent systems of generalized Young measures. Netw. Heterog. Media **2**, 1–36 (2007)
19. G. Dal Maso, A. DeSimone, M.G. Mora, M. Morini, A vanishing viscosity approach to quasistatic evolution in plasticity with softening. Arch. Rational Mech. Anal. **189**(3), 469–544 (2008)
20. E. De Giorgi, Γ-convergenza e G-convergenza. Boll. Unione Mat. Ital. V. A **14**, 213–220 (1977)
21. G. Dal Maso, G.A. Francfort, R. Toader, Quasistatic crack growth in nonlinear elasticity. Arch. Rational Mech. Anal. **176**, 165–225 (2005)
22. M. Efendiev, A. Mielke, On the rate–independent limit of systems with dry friction and small viscosity. J. Convex Anal. **13**(1), 151–167 (2006)
23. A. Fiaschi, A Young measures approach to quasistatic evolution for a class of material models with nonconvex elastic energies. ESAIM Control Optim. Calc. Var. **15**(2), 245–278 (2009)
24. G. Francfort, A. Garroni, A variational view of partial brittle damage evolution. Arch. Rational Mech. Anal. **182**, 125–152 (2006)
25. G.A. Francfort, C.J. Larsen, Existence and convergence for quasi-static evolution of brittle fracture. Comm. Pure Appl. Math. **56**, 1495–1500 (2003)
26. G. Francfort, A. Mielke, Existence results for a class of rate-independent material models with nonconvex elastic energies. J. Reine Angew. Math. **595**, 55–91 (2006)
27. A. Garroni, C.J. Larsen, Threshold-based quasi-static brittle damage evolution. Arch. Ration. Mech. Anal. **194**(2), 585–609 (2009)
28. A. Giacomini, M. Ponsiglione, A Γ-convergence approach to stability of unilateral minimality properties in fracture mechanics and applications. Arch. Ration. Mech. Anal. **180**, 399–447 (2006)
29. F. Gastaldi, M.D.P.M. Marques, J.A.C. Martins, Mathematical analysis of a two degree-of-freedom frictional contact problem with discontinuous solutions. Math. Comput. Model. **28**, 247–261 (1998)
30. K. Gröger, Evolution equations in the theory of plasticity. *Theory of nonlinear operators (Proc. Fifth Internat. Summer School, Central Inst. Math. Mech. Acad. Sci. GDR, Berlin, 1977)*, vol. 6, *Abh. Akad. Wiss. DDR, Abt. Math. Naturwiss. Tech., 1978* (Akademie-Verlag, Berlin, 1978), pp. 97–107

31. W. Han, B.D. Reddy, *Plasticity (Mathematical Theory and Numerical Analysis)*, vol. 9, *Interdisciplinary Applied Mathematics* (Springer, New York, 1999)
32. A.D. Ioffe, On lower semicontinuity of integral functionals, I. SIAM J. Contr. Optim. **15**(4), 521–538 (1977)
33. M. Kružík, A. Mielke, T. Roubíček, Modelling of microstructure and its evolution in shape-memory-alloy single-crystals, in particular in CuAlNi. Meccanica **40**, 389–418 (2005)
34. D. Knees, A. Mielke, C. Zanini, On the inviscid limit of a model for crack propagation. Math. Models Meth. Appl. Sci. (M^3AS) **18**, 1529–1569 (2008)
35. P. Krejčí, Evolution variational inequalities and multidimensional hysteresis operators. *Nonlinear differential equations (Chvalatice, 1998)*, vol. 404 *Chapman & Hall/CRC Res. Notes Math.* (Chapman & Hall/CRC, Boca Raton, FL, 1999), pp. 47–110
36. M. Kunze, M.D.P. Monteiro Marques, On parabolic quasi-variational inequalities and state-dependent sweeping processes. Topol. Methods Nonlinear Anal. **12**, 179–191 (1998)
37. D. Knees, C. Zanini, A. Mielke, Crack propagation in polyconvex materials. Phys. D **239**, 1470–1484 (2010)
38. A. Mainik, A. Mielke, Existence results for energetic models for rate–independent systems. Calc. Var. Part. Diff. Eqns. **22**, 73–99 (2005)
39. A. Mainik, A. Mielke, Global existence for rate-independent gradient plasticity at finite strain. J. Nonlinear Sci. **19**(3), 221–248 (2009)
40. A. Mielke, Energetic formulation of multiplicative elasto–plasticity using dissipation distances. Contin. Mech. Thermodyn. **15**, 351–382 (2003)
41. A. Mielke, Deriving new evolution equations for microstructures via relaxation of variational incremental problems. Comput. Methods Appl. Mech. Eng. **193**(48-51), 5095–5127 (2004)
42. A. Mielke, Evolution in rate-independent systems (Chap. 6), in *Handbook of Differential Equations, Evolutionary Equations, vol. 2*, ed. by C.M. Dafermos, E. Feireisl (Elsevier B.V., Amsterdam, 2005), pp. 461–559
43. A. Mielke, A mathematical framework for generalized standard materials in the rate-independent case, in *Multifield Problems in Solid and Fluid Mechanics*, vol. 28 *Lecture Notes in Applied and Computational Mechanics*, ed. by R. Helmig, A. Mielke, B.I. Wohlmuth (Springer, Berlin, 2006), pp. 351–379
44. A. Mielke, Complete-damage evolution based on energies and stresses. Discrete Cont. Dyn. Syst. Ser. S **4**, 423–439 (2011)
45. A. Mielke, M. Ortiz, A class of minimum principles for characterizing the trajectories of dissipative systems. ESAIM Contr. Optim. Calc. Var. **14**, 494–516 (2008)
46. A. Mielke, R. Rossi, Existence and uniqueness results for a class of rate-independent hysteresis problems. Math. Models Meth. Appl. Sci. (M^3AS) **17**, 81–123 (2007)
47. A. Mielke, T. Roubíček, Numerical approaches to rate-independent processes and applications in inelasticity. Math. Model. Numer. Anal. (M2AN) **43**, 399–428 (2009)
48. A. Mielke, T. Roubíček, *Rate-Independent Systems: Theory and Application* (in preparation, 2011)
49. A. Mielke, U. Stefanelli, A discrete variational principle for rate-independent evolution. *Advances Calculus Variations* **1**(4), 399–431 (2008)
50. A. Mielke, F. Theil, A mathematical model for rate-independent phase transformations with hysteresis, in *Proceedings of the Workshop on "Models of Continuum Mechanics in Analysis and Engineering"*, ed. by H.-D. Alber, R.M. Balean, R. Farwig (Shaker-Verlag, Aachen, 1999), pp. 117–129
51. A. Mielke, F. Theil, On rate–independent hysteresis models. Nonl. Diff. Eqns. Appl. (NoDEA) **11**, 151–189 (2004) (Accepted July 2001)
52. A. Mielke, A.M. Timofte, Two-scale homogenization for evolutionary variational inequalities via the energetic formulation. SIAM J. Math. Anal. **39**(2), 642–668 (2007)

53. A. Mielke, S. Zelik, On the vanishing-viscosity limit in parabolic systems with rate-independent dissipation terms. Submitted to Ann. Sc. Norm. Super. Pisa Cl. Sci. (5). 2010w (WIAS preprint 1500)
54. J.A.C. Martins, M.D.P. Monteiro Marques, F. Gastaldi, On an example of nonexistence of solution to a quasistatic frictional contact problem. Eur. J. Mech. A Solids **13**, 113–133 (1994)
55. J.-J. Moreau, Application of convex analysis to the treatment of elastoplastic systems, in *Applications of Methods of Functional Analysis to Problems in Mechanics*, ed. by P. Germain, B. Nayroles, Lecture Notes in Mathematics, vol. 503. (Springer, New York, 1976), pp. 56–89
56. J.-J. Moreau, Evolution problem associated with a moving convex set in a Hilbert space. J. Differ. Equat. **26**(3), 347–374 (1977)
57. A. Mielke, L. Paoli, A. Petrov, On the existence and approximation for a 3D model of thermally induced phase transformations in shape-memory alloys. SIAM J. Math. Anal. **41**(4), 1388–1414 (2009)
58. A. Mielke, T. Roubíček, U. Stefanelli, Γ-limits and relaxations for rate-independent evolutionary problems. Calc. Var. Part. Diff. Eqns. **31**, 387–416 (2008)
59. A. Mielke, R. Rossi, G. Savaré, BV solutions and viscosity approximations of rate-independent systems. ESAIM Control Optim. Calc. Var. 2011, in press. DOI:10.1051/cocv/2010054
60. A. Mielke, R. Rossi, G. Savaré, Modeling solutions with jumps for rate-independent systems on metric spaces. Discr. Cont. Dyn. Syst. A **25**(2), 585–615 (2009)
61. J.A.C. Martins, F.M.F. Simões, F. Gastaldi, M.D.P. Monteiro Marques, Dissipative graph solutions for a 2 degree-of-freedom quasistatic frictional contact problem. Int. J. Eng. Sci. **33**, 1959–1986 (1995)
62. A. Mielke, F. Theil, V.I. Levitas, A variational formulation of rate–independent phase transformations using an extremum principle. Arch. Ration. Mech. Anal. **162**, 137–177 (2002)
63. R. Rossi, A. Mielke, G. Savaré, A metric approach to a class of doubly nonlinear evolution equations and applications. Ann. Sc. Norm. Super. Pisa Cl. Sci. 5 **VII**, 97–169 (2008)
64. R. Rossi, G. Savaré, Gradient flows of non convex functionals in Hilbert spaces and applications. ESAIM Control Optim. Calc. Var. **12**, 564–614 (2006)
65. T. Roubíček, *Nonlinear Partial Differential Equations with Applications* (Birkhäuser Verlag, Basel, 2005)
66. U. Stefanelli, A variational characterization of rate-independent evolution. Mathem. Nach. **282**(11), 1492–1512 (2009)
67. M. Thomas, A. Mielke, Damage of nonlinearly elastic materials at small strain: existence and regularity results. Z. angew. Math. Mech. (ZAMM) **90**(2), 88–112 (2010). Submitted. WIAS preprint 1397
68. M. Thomas, *Rate-independent damage processes in nonlinearly elastic matherials.* Ph.D thesis, Institut für Mathematik, Humboldt-Universität zu Berlin, February 2010
69. R. Toader, C. Zanini, An artificial viscosity approach to quasistatic crack growth. Boll. Unione Matem. Ital. **2**(1), 1–36 (2009)
70. A. Visintin, *Differential Models of Hysteresis* (Springer, Berlin, 1994)

Optimal Transport and Curvature

Alessio Figalli and Cédric Villani

1 Introduction

These notes record the six lectures for the CIME Summer Course held by the second author in Cetraro during the week of June 23–28, 2008, with minor modifications. Their goal is to describe some recent developments in the theory of optimal transport, and their applications to differential geometry. We will focus on two main themes:

(a) Stability of lower bounds on Ricci curvature under measured Gromov–Hausdorff convergence.
(b) Smoothness of optimal transport in curved geometry.

The main reference for all the material covered by these notes (and much more) is the recent book of the second author [40].

These notes are organized as follows:

- In Sect. 2 we recall some classical facts of metric and differential geometry; then in Sect. 3 we study the optimal transport problem on Riemannian manifolds. These sections introduce the basic objects and the notation.
- In Sect. 4 we reformulate lower bounds on Ricci curvature in terms of the "displacement convexity" of certain functionals, and deduce the stability. Then in Sect. 5 we address the question of the smoothness of the optimal transport on Riemannian manifold. These two sections, focusing

A. Figalli (✉)
Department of Mathematics, The University of Texas at Austin, 1 University Station, C1200, Austin, TX 78712-1082, USA
e-mail: figalli@math.utexas.edu

C. Villani
Institut Henri Poincaré, Ecole Normale Supérieure, 11 rue Pierre et Marie Curie 75231 Paris Cedex 05, France
e-mail: cvillani@umpa.ens-lyon.fr

S. Bianchini et al., *Nonlinear PDE's and Applications*, Lecture Notes in Mathematics 2028, DOI 10.1007/978-3-642-21861-3_4,

on Problems (a) and (b) respectively, constitute the heart of these notes, and can be read independently of each other.

- Section 6 is devoted to a recap and the discussion of a few open problems; finally Sect. 7 gives a selection of the most relevant references.

2 Bits of Metric Geometry

The apparent redundancy in the title of this section is intended to stress the fact that we shall be concerned with geometry only from the metric point of view (rather than from the topological, or differential point of view), be it either in some possibly nonsmooth metric space, or in a smooth Riemannian manifold.

2.1 *Length*

Let (X, d) be a complete separable metric space. Given a Lipschitz curve $\gamma : [0, T] \to X$, we define its *length* by

$$L(\gamma) := \sup \left\{ \sum_{i=0}^{N} d(\gamma(t_i), \gamma(t_{i+1})) \,\middle|\, 0 = t_0 \leq t_1 \leq \ldots \leq t_{N+1} = T \right\}.$$

It is easily checked that the length of a curve is invariant by reparameterization.

In an abstract metric space the velocity $\dot{\gamma}(t)$ of a Lipschitz curve does not make sense; still it is possible to give a meaning to the "modulus of the velocity", or metric derivative of γ, or *speed:*

$$|\dot{\gamma}(t)| := \limsup_{h \to 0} \frac{d(\gamma(t+h), \gamma(t))}{|h|}$$

For almost all t, the above limsup is a true limit [1, Theorem 4.1.6], and the following formula holds:

$$L(\gamma) = \int_0^T |\dot{\gamma}(t)| \, dt. \tag{1}$$

2.2 *Length Spaces*

In the previous section we have seen how to write the length of curves in terms of the metric d. But once the length is defined, we can introduce a new distance on X:

$$d'(x,y) := \inf \left\{ L(\gamma) \,\middle|\, \gamma \in \mathrm{Lip}([0,1],X),\, \gamma(0)=x,\, \gamma(1)=y \right\}.$$

By triangle inequality, $d' \geq d$. If $d' = d$ we say that (X,d) is a *length space.* It is worth recording that (X,d') defined as above is automatically a length space.

Example 2.1. Take $X = \mathbb{S}^1 \subset \mathbb{R}^2$, and $d(x,y) = |x-y|$ the standard Euclidean distance in $\mathbb{R}^2$; then $d'(x,y) = 2\arcsin\frac{|x-y|}{2}$, so (X,d) is not a length space. More generally, if X is a closed subset of $\mathbb{R}^n$ then (X,d) is a length space if and only if X is convex.

2.3 Geodesics

A curve $\gamma : [0,1] \to X$ which minimizes the length, among all curves with $\gamma(0) = x$ and $\gamma(1) = y$, is called a *geodesic*, or more properly a *minimizing geodesic.*

The property of being a minimizing geodesic is stable by restriction: if $\gamma : [0,1] \to X$ is a geodesic, then for all $a < b \in [0,1]$, $\gamma|_{[a,b]} : [a,b] \to X$ is a geodesic from $\gamma(a)$ to $\gamma(b)$.

A length space such that any two points are joined by a minimizing geodesic is called a *geodesic space.*

Example 2.2. By the (generalized) Hopf–Rinow theorem, any locally compact complete length space is a geodesic space [4].

It is a general fact that a Lipschitz curve γ can be reparameterized so that $|\dot\gamma(t)|$ is constant [1, Theorem 4.2.1]. Thus, any geodesic $\gamma : [0,1] \to X$ can be reparameterized so that $|\dot\gamma(t)| = L(\gamma)$ for almost all $t \in [0,1]$. In this case, γ is called a *constant-speed minimizing geodesic.* Such curves are minimizers of the action functional

$$A(\gamma) := \frac{1}{2}\int_0^1 |\dot\gamma(t)|^2\,dt.$$

More precisely, we have:

Proposition 2.3. *Let (X,d) be a length space. Then*

$$d(x,y) = \inf_{\gamma(0)=x,\,\gamma(1)=y} \sqrt{\int_0^1 |\dot\gamma(t)|^2\,dt} \qquad \forall\, x,y \in X.$$

Moreover, if (X,d) is a geodesic space, then minimizers of the above functional are precisely constant-speed minimizing geodesics.

Sketch of the proof. By (1) we know that

$$d(x,y) = \inf_{\gamma(0)=x,\,\gamma(1)=y} \int_0^1 |\dot\gamma(t)|\,dt \qquad \forall\, x,y \in X.$$

By Jensen's inequality,

$$\int_0^1 |\dot\gamma(t)|\,dt \le \sqrt{\int_0^1 |\dot\gamma(t)|^2\,dt},$$

with equality if and only if $|\dot\gamma(t)|$ is constant for almost all t. The conclusion follows easily. □

2.4 Riemannian Manifolds

Given an n-dimensional C^∞ differentiable manifolds M, for each $x \in M$ we denote by T_xM the tangent space to M at x, and by $TM := \cup_{x\in M}(\{x\} \times T_xM)$ the whole tangent bundle of M. On each tangent space T_xM, we assume that is given a symmetric positive definite quadratic form $g_x : T_xM \times T_xM \to \mathbb{R}$ which depends smoothly on x; $g = (g_x)_{x\in M}$ is called a *Riemannian metric*, and (M,g) is a *Riemannian manifold.*

A Riemannian metric defines a scalar product and a norm on each tangent space: for each $v,w \in T_xM$

$$\langle v,w\rangle_x := g_x(v,w), \qquad |v|_x := \sqrt{g_x(v,v)}.$$

Let U be an open subset of $\mathbb{R}^n$ and $\Phi : U \to \Phi(U) = V \subset M$ a chart. Given $x = \Phi(x^1,\dots,x^n) \in V$, the vectors $\frac{\partial}{\partial x^i} := \frac{\partial \Phi}{\partial x_i}(x^1,\dots,x^n)$, $i = 1,\dots,n$, constitute a basis of T_xM: any $v \in T_xM$ can be written as $v = \sum_{i=1}^n v^i \frac{\partial}{\partial x^i}$. We can use this chart to write our metric g in coordinates inside V:

$$g_x(v,v) = \sum_{i,j=1}^n g_x\left(\frac{\partial}{\partial x^i}, \frac{\partial}{\partial x^j}\right) v^i v^j = \sum_{i,j=1}^n g_{ij}(x) v^i v^j,$$

where by definition $g_{ij}(x) := g_x\left(\frac{\partial}{\partial x^i}, \frac{\partial}{\partial x^j}\right)$. We also denote by g^{ij} the coordinates of the inverse of g: $g^{ij} = (g_{ij})^{-1}$; more precisely, $\sum_j g^{ij} g_{jk} = \delta^i_k$, where δ^i_k denotes Kronecker's delta:

$$\delta^i_k = \begin{cases} 1 \text{ if } i = k, \\ 0 \text{ if } i \neq k. \end{cases}$$

In the sequel we will use these coordinates to perform many computations. Einstein's convention of summation over repeated indices will be used systematically: $a_k b^k = \sum_k a_k b^k$, $g_{ij} v^i v^j = \sum_{i,j} g_{ij} v^i v^j$, etc.

2.5 Riemannian Distance and Volume

The notion of "norm of a tangent vector" leads to the definition of a distance on a Riemannian manifold (M, g), called *Riemannian distance*:

$$\begin{aligned} d(x,y) &= \inf_{\gamma(0)=x,\,\gamma(1)=y} \int_0^1 \sqrt{g_{\gamma(t)}\big(\dot\gamma(t),\dot\gamma(t)\big)}\,dt \\ &= \inf_{\gamma(0)=x,\,\gamma(1)=y} \sqrt{\int_0^1 g_{\gamma(t)}\big(\dot\gamma(t),\dot\gamma(t)\big)\,dt} \qquad \forall\, x,y \in X. \end{aligned}$$

This definition makes (M, d) a length space.

A Riemannian manifold (M, g) is also equipped with a natural reference measure, the *Riemannian volume*:

$$\mathrm{vol}(dx) = n\text{-dimensional Hausdorff measure on } (M,d) = \sqrt{\det(g_{ij})}\,dx^1 \dots dx^n.$$

This definition of the volume allows to write a change of variables formula, exactly as in $\mathbb{R}^n$ (see for instance [40, Chap. 1]).

2.6 Differential and Gradients

Given a smooth map $\varphi : M \to \mathbb{R}$, its differential $d\varphi : TM \to \mathbb{R}$ is defined as

$$d\varphi(x) \cdot v := \left.\frac{d}{dt}\right|_{t=0} \varphi(\gamma(t)),$$

where $\gamma : (-\varepsilon, \varepsilon) \to M$ is any smooth curve such that $\gamma(0) = x$, $\dot\gamma(0) = v$ (this definition is independent of the choice of γ). Thanks to the Riemannian metric, we can define the *gradient* $\nabla\varphi(x)$ at any point $x \in M$ by the formula

$$\big\langle \nabla\varphi(x), v \big\rangle_x := d\varphi(x) \cdot v.$$

Observe carefully that $\nabla\varphi(x)$ is a tangent vector (i.e. an element of $T_x M$), while $d\varphi(x)$ is a cotangent vector (i.e. an element of $T_x^* M := (T_x M)^*$, the dual space of T_xM). Using coordinates induced by a chart, we get

$$g_{ij}(\nabla\varphi)^i v^j = (d\varphi)_i v^i \quad \Longrightarrow \quad (\nabla\varphi)^i = g^{ij}(d\varphi)_j.$$

2.7 Geodesics in Riemannian Geometry

On a Riemannian manifold, constant-speed minimizing geodesics satisfy a second order differential equation:

$$\ddot{\gamma}^k + \Gamma_{ij}^k \dot{\gamma}^i \dot{\gamma}^j = 0, \tag{2}$$

where Γ_{ij}^k are the Christoffel symbols defined by

$$\Gamma_{ij}^k = \frac{1}{2} g^{k\ell} \left(\frac{\partial g_{j\ell}}{\partial x^i} + \frac{\partial g_{i\ell}}{\partial x^j} - \frac{\partial g_{ij}}{\partial x^\ell} \right).$$

Exercise 2.4. Prove the above formula.
Hint. Consider the action functional $A(\gamma) = \frac{1}{2}\int_0^1 g_{\gamma(t)}\big(\dot{\gamma}(t), \dot{\gamma}(t)\big)\, dt$ for a geodesic γ and, working in charts, make variations of the form $A(\gamma + \varepsilon h)$, with h vanishing at the end points. Then use $\frac{d}{d\varepsilon}|_{\varepsilon=0} A(\gamma + \varepsilon h) = 0$ and the arbitrariness of h.

2.8 Exponential Map and Cut Locus

From now on, by a geodesic we mean a solution of the geodesic equation (2), and we will explicitly mention whether it is or not minimizing. We also assume M to be complete, so that geodesics are defined for all times.

The *exponential map* $\exp : TM \to M$ is defined by

$$\exp_x(v) := \gamma_{x,v}(1),$$

where $\gamma_{x,v} : [0, +\infty) \to M$ is the unique solution of (2) starting at $\gamma_{x,v}(0) = x$ with velocity $\dot{\gamma}_{x,v}(0) = v$.

We observe that the curve $(\exp_x(tv))_{t\geq 0}$ is a geodesic defined for all times, but in general is not minimizing for large times (on the other hand, it is possible to prove that $\exp_x(tv)$ is always minimizing between x and $\exp_x(\varepsilon v)$ for $\varepsilon > 0$ sufficiently small). We define the *cut time* $t_c(x, v)$ as

$$t_c(x,v) := \inf\Big\{t > 0 \,|\, s \mapsto \exp_x(sv) \text{ is not minimizing between } x \text{ and } \exp_x(tv)\Big\}.$$

Example 2.5. On the sphere $\mathbb{S}^n$, the geodesics starting from the north pole $N = (0, \ldots, 0, 1)$ with unit speed describe great circles passing through the south pole $S = (0, \ldots, 0, -1)$. These geodesics are minimizing exactly until they reach S after a time π. Thus $t_c(N, v) = \pi$ for any $v \in T_N M$ with unit norm. By homogeneity of the sphere and time-rescaling, we get $t_c(x, v) = \frac{\pi}{|v|_x}$ for any $x \in \mathbb{S}^n$, $v \in T_x M \setminus \{0\}$.

Given two points $x, y \in M$, if there exists a unique minimizing geodesic $(\exp_x(tv))_{0 \le t \le 1}$ going from x to y in time 1, we will write (with a slight abuse of notation) $v = (\exp_x)^{-1}(y)$.

Given $x \in M$, we define the *cut locus* of x as

$$\mathrm{cut}(x) := \Big\{ \exp_x\big(t_c(x, \xi)\xi\big) \mid \xi \in T_x M,\ |\xi|_x = 1 \Big\}$$

We further define

$$\mathrm{cut}(M) := \{(x, y) \in M \times M \mid y \in \mathrm{cut}(x)\}.$$

Example 2.6. On the sphere $\mathbb{S}^n$, the cut locus of a point consists only of its antipodal point, i.e. $\mathrm{cut}(x) = \{-x\}$.

It is possible to prove that, if $y \notin \mathrm{cut}(x)$, then x and y are joined by a unique minimizing geodesic. The converse is close to be true: $y \notin \mathrm{cut}(x)$ if and only if there are neighborhoods U of x and V of y such that any two points $x' \in U$, $y' \in V$ are joined by a unique minimizing geodesic. In particular $y \notin \mathrm{cut}(x)$ if and only if $x \notin \mathrm{cut}(y)$.

2.9 First Variation Formula and (Super)Differentiability of Squared Distance

Exactly as in the computation for the geodesic equations (Exercise 2.4), one can compute the first variation of the action functional at a geodesic: let $\gamma : [0, 1] \to M$ be a constant-speed minimizing geodesic from x to y, and let $x' \simeq x + \delta x$ to $y' \simeq y + \delta y$ be perturbations of x and y respectively. (When $\delta x \in T_x M$ with $|\delta x| \ll 1$, $x + \delta x$ is an abuse of notation for, say, $\exp_x(\delta x)$, or for $h(1)$, where $h(s)$ is any smooth path with $h(0) = x$ and $\dot{h}(0) = \delta x$.) The *formulation of first variation* states that

$$A(\gamma') = A(\gamma) + \big(\langle \dot\gamma(1), \delta y\rangle_y - \langle \dot\gamma(0), \delta x\rangle_x\big) + O(|\delta y|^2) + O(|\delta x|^2).$$

Here γ' can be any curve from x' to y', geodesic or not, the important point is that γ' be a C^1 perturbation of γ. Below is a more rigorous statement:

Proposition 2.7. *Let $\gamma : [0, 1] \to M$ be a constant-speed minimizing geodesic from x to y. Consider a C^1 family of curves $\gamma^\varepsilon : [0, 1] \to M$,*

$\varepsilon \in (-\varepsilon_0, \varepsilon_0)$ with $\gamma^0 = \gamma$, and let X be the vector field along γ defined by $X(t) = (d/d\varepsilon)|_{\varepsilon=0}\gamma^\varepsilon(t)$. Then

$$\left.\frac{d}{d\varepsilon}\right|_{\varepsilon=0} A(\gamma^\varepsilon) = \langle \dot\gamma(1), X(1)\rangle_y - \langle \dot\gamma(0), X(0)\rangle_x.$$

The proof of this fact is analogous to the proof of the formula of the geodesic equation, with the only exception that now the variation we consider do not vanish at the boundary points, and so when doing integration by parts one has to take care of boundary terms.

We can now compute the (super)differential of the squared distance; the result is most conveniently expressed in terms of the gradient. Fix $x_0 \in M$, and consider the function $F(x) := \frac{1}{2}d(x_0, x)^2$.

Proposition 2.8. *If x_0 and x are joined by a unique minimizing geodesic, then F is differentiable at x and $\nabla F(x) = -(\exp_x)^{-1}(x_0)$.*

Sketch of the proof. Let $\gamma : [0,1] \to M$ be the unique constant-speed minimizing geodesic from x_0 to x, and let $x_\varepsilon \simeq x + \varepsilon w$ be a perturbation of x. Let γ_ε be a minimizing geodesic connecting x_0 to x_ε; so $\gamma_\varepsilon(t) = \exp_{x_0}(tv_\varepsilon)$ for some $v_\varepsilon \in T_{x_0}M$. Up to extraction of a subsequence, γ_ε converges to some minimizing geodesic, which is necessarily γ, and v_ε converges to $(\exp_{x_0})^{-1}(x)$. So $\gamma_\varepsilon(t)$ is a C^1 perturbation of γ, and the first variation formula yields

$$F(x_\varepsilon) = A(\gamma_\varepsilon) = A(\gamma) + \varepsilon\langle\dot\gamma(1), w\rangle_x + O(\varepsilon^2) = F(x) - \varepsilon\big\langle(\exp_x)^{-1}(x_0), w\big\rangle_x + O(\varepsilon^2).$$

□

In case x_0 and x are joined by several minimizing geodesics, the above argument fails. On the other hand, one still has *superdifferentiability*: there exists $p \in T_xM$ such that

$$F(x') \le F(x) + \langle p, \delta x\rangle_x + O(|\delta x|^2).$$

Proposition 2.9. *For any $x \in M$, F is superdifferentiable at x.*

Sketch of the proof. Let $\gamma : [0,1] \to M$ be a constant-speed minimizing geodesic from x_0 to x. Then, for any perturbation $x' \simeq x + \delta x$ of x, we can perturb γ into a smooth path γ' (not necessarily minimizing!) connecting x_0 to x'. The first variation formula yields

$$F(x') \le A(\gamma') = A(\gamma) + \langle\dot\gamma(1), \delta x\rangle_x + O(|\delta x|^2),$$

so $\dot\gamma(1)$ is a supergradient for F at x. □

By the above proposition we deduce that, although the (squared) distance is not smooth, its only singularities are upper crests. By the above proof we also see that F is differentiable at x if and only if x_0 and x are joined by a unique minimizing geodesic. (Indeed, a superdifferentiable function F is differentiable at x if and only if F has only one supergradient at x.)

2.10 Hessian and Second Order Calculus

Let $\varphi : M \to \mathbb{R}$ be a smooth function. The *Hessian* $\nabla^2\varphi(x) : T_xM \to T_xM$ is defined by

$$\langle \nabla^2\varphi(x) \cdot v, v\rangle_x := \frac{d^2}{dt^2}\bigg|_{t=0} \varphi(\gamma(t)),$$

where $\gamma(t) = \exp_x(tv)$. Observe that, when we defined the differential of a function, we could use any curve starting from x with velocity v. In this case, as the second derivative of $\varphi(\gamma(t))$ involves $\ddot{\gamma}(0)$, we are not allowed to choose an arbitrary curve in the definition of the Hessian.

2.11 Variations of Geodesics and Jacobi Fields

Let us consider a family $(\gamma_\theta)_{-\varepsilon\le\theta\le\varepsilon}$ of constant-speed geodesics $\gamma_\theta : [0,1] \to M$. Then, for each $t \in [0,1]$, we can consider the vector field

$$J(t) := \frac{\partial}{\partial\theta}\bigg|_{\theta=0} \gamma_\theta(t) \in T_{\gamma(t)}M.$$

The vector field J is called a *Jacobi field* along $\gamma = \gamma_0$. By differentiating the geodesic equations with respect to θ, we get a second order differential equation for J:

$$\frac{\partial}{\partial\theta}\left(\ddot{\gamma}_\theta^k + \Gamma_{ij}^k(\gamma_\theta)\dot{\gamma}_\theta^i\dot{\gamma}_\theta^j\right) = 0$$

gives

$$\ddot{J}^k + \frac{\partial\Gamma_{ij}^k}{\partial x^\ell} J^\ell \dot{\gamma}^i\dot{\gamma}^j + 2\Gamma_{ij}^k \dot{J}^i\dot{\gamma}^j = 0.$$

This complicated equation takes a nicer form if we choose time-dependent coordinates determined by a moving orthonormal basis $\{e_1(t), \ldots, e_n(t)\}$ of $T_{\gamma(t)}M$, such that

$$\dot{e}_\ell^k(t) + \Gamma_{ij}^k(\gamma(t))e_\ell^i(t)\dot{\gamma}^j(t) = 0$$

(in this case, we say that the basis is *parallel transported* along γ). With this choice of the basis, defining $J^i(t) := \langle J(t), e_i(t)\rangle_{\gamma(t)}$ we get

$$\ddot{J}_i(t) + R_i^j(t)J_j(t) = 0.$$

For our purposes it suffices to know that R_i^j is a symmetric matrix; in fact one can show that $R_i^j(t) = \langle \mathrm{Riem}(\dot\gamma, e_i)\cdot\dot\gamma, e_j\rangle$, where Riem denotes the *Riemann tensor* of (M, g).

We now write the *Jacobi equation* in matrix form: let $\boldsymbol{J}(t) = (J_1(t), \ldots, J_n(t))$ be a matrix of Jacobi fields, and define $J_{ij}(t) := \langle J_i(t), e_j(t)\rangle_{\gamma(t)}$, with $\{e_1(t), \ldots, e_n(t)\}$ parallel transported as before. Then

$$\ddot{\boldsymbol{J}}(t) + R(t)\boldsymbol{J}(t) = 0,$$

where $R(t)$ is a symmetrix matrix involving derivatives of the metric $g_{ij}(\gamma(t))$ up to the second order, and such that (up to identification) $R(t)\dot\gamma(t) = 0$.

2.12 Sectional and Ricci Curvatures

The matrix R appearing in the Jacobi fields equation allows to define the *sectional curvature* in a point x along a plane $P \subset T_xM$: let $\{e_1, e_2\}$ be an orthonormal basis of P, and consider γ the geodesic starting from x with velocity e_1. We now complete $\{e_1, e_2\}$ into an orthonormal basis of T_xM, and we construct $\{e_1(t), \ldots, e_n(t)\}$ as above. Then the sectional curvature at x along P is given by

$$\mathrm{Sect}_x(P) := R_{22}(0)$$

Remark 2.10. The sectional curvature has the following geometric interpretation: given $v, w \in T_xM$ unit vectors, with (non-oriented) angle θ,

$$d\big(\exp_x(tv), \exp_x(tw)\big) = \sqrt{2(1-\cos\theta)}\,t\left(1 - \frac{\sigma\cos^2(\theta/2)}{6}t^2 + O(t^4)\right), \quad (3)$$

where σ is the sectional curvature at x along the plane generated by v and w. Thus the sectional curvatures infinitesimally measure the tendency of geodesics to converge ($\sigma > 0$) or diverge ($\sigma < 0$). We observe that formula (3) implies Gauss's Theorema Egregium, namely that the sectional curvature is invariant under local isometry.

The *Ricci curvature* at point $x \in M$ is a quadratic form on the tangent space defined as follows: fix $\xi \in T_xM$, and complete ξ to an orthonormal basis $\{\xi = e_1, e_2, \ldots, e_n\}$. Denoting by $[e_i, e_j]$ the plane generated by e_i and e_j for $i \neq j$, we define

$$\mathrm{Ric}_x(\xi, \xi) := \sum_{j=2}^{n} \mathrm{Sect}_x([e_1, e_j]).$$

Another equivalent definition consists in considering the geodesic starting from x with velocity ξ, take $\{e_1(t), \ldots, e_n(t)\}$ obtained by parallel transport, and define $\mathrm{Ric}_x(\xi, \xi) = \mathrm{tr}\big(R(0)\big)$.

2.13 Interpretation of Ricci Curvature Bounds

In this paragraph we give a geometric interpretation of the Ricci curvature. For more details we refer to [40, Chap. 14] and references therein.

Let ξ be a C^1 vector field (i.e. $\xi(x) \in T_xM$) defined in a neighborhood of $\{\gamma(t) \,|\, 0 \le t \le 1\}$, and consider the map

$$T_t(x) := \exp_x\big(t\xi(x)\big).$$

We want to compute the Jacobian of this map.

Think of $d_xT_t : T_xM \to T_{T_t(x)}M$ as an array $(J_1(t), \ldots, J_n(t))$ of Jacobi fields. Expressed in an orthonormal basis $\{e_1(t), \ldots, e_n(t)\}$ obtained by parallel transport, we get a matrix $\boldsymbol{J}(t)$ which solves the Jacobi equations $\ddot{\boldsymbol{J}} + R\boldsymbol{J} = 0$. Moreover, since $T_0(x) = x$ and $\dot{T}_0(x) = \xi(x)$, we have the system

$$\begin{cases} \ddot{\boldsymbol{J}} + R\boldsymbol{J} = 0, \\ \boldsymbol{J}(0) = I_n, \\ \dot{\boldsymbol{J}}(0) = \nabla\xi, \end{cases}$$

where $\nabla\xi$ is defined as

$$\langle \nabla\xi \cdot e_i, e_j \rangle := \frac{d}{ds}\bigg|_{s=0} \Big\langle \xi\big(\exp_x(se_i)\big), e_j\big(\exp_x(se_i)\big) \Big\rangle_{\exp_x(se_i)},$$

and $e_j\big(\exp_x(se_i)\big)$ is obtained by parallel transport along $s \mapsto \exp_x(se_i)$.

We now define $\mathscr{J}(t) := \mathrm{Jac}_x T_t = \det \boldsymbol{J}(t)$. Then

$$\frac{d}{dt} \log \mathscr{J}(t) = \mathrm{tr}\left(\dot{\boldsymbol{J}}(t)\boldsymbol{J}(t)^{-1}\right)$$

as long as $\det \boldsymbol{J}(t) > 0$. Let $U(t) := \dot{\boldsymbol{J}}(t)\boldsymbol{J}(t)^{-1}$. Using the Jacobi equation for $\boldsymbol{J}$, we get

$$\dot{U}(t) = -R(t) - U(t)^2.$$

Taking the trace, we deduce the important formula

$$\frac{d}{dt}\mathrm{tr}\big(U(t)\big) + \mathrm{tr}\left(U(t)^2\right) + \mathrm{Ric}_{\gamma(t)}(\dot\gamma(t), \dot\gamma(t)) = 0.$$

Remark 2.11. The above formula is nothing else than the Lagrangian version of the celebrated *Bochner formula*, which is Eulerian in nature:

$$-\nabla\cdot\big((\xi\cdot\nabla)\xi\big)+\xi\cdot\nabla(\nabla\cdot\xi)+\mathrm{tr}\,\big((\nabla\xi)^2\big)+\mathrm{Ric}(\xi,\xi)=0.$$

Assume now that $U(0)$ is symmetric (this is the case for instance if $\xi=\nabla\psi$ for some function ψ, as $U(0)=\nabla^2\psi$). In this case, since R is symmetric, U and U^* solves the same differential equation with the same initial condition, and by uniqueness $U(t)$ is symmetric too. We can therefore apply the inequality

$$\mathrm{tr}\,\big(U(t)^2\big)\geq\frac{1}{n}\big[\mathrm{tr}(U(t))\big]^2$$

(a version of the Cauchy–Schwarz inequality). Combining all together we arrive at:

Proposition 2.12. *If $T_t(x)=\exp_x\big(t\nabla\psi(x)\big)$, then $\mathcal{J}(t)=\mathrm{Jac}_x T_t$ satisfies*

$$\frac{d^2}{dt^2}\log\mathcal{J}(t)+\frac{1}{n}\left(\frac{d}{dt}\log\mathcal{J}(t)\right)^2+\mathrm{Ric}_{\gamma(t)}(\dot\gamma(t),\dot\gamma(t))\leq 0.$$

From this proposition, we see that lower bounds on the Ricci curvature estimate the *averaged* tendency of geodesics to converge (in the sense of the Jacobian determinant).

The above proposition can be reversed, and one can prove for instance

$$\mathrm{Ric}\geq 0 \text{ throughout } M \qquad\Longleftrightarrow\qquad \frac{d^2}{dt^2}\log\mathrm{Jac}_x\big(\exp_x(t\nabla\psi(x))\big)\leq 0\quad\forall\psi, \tag{4}$$

where ψ is arbitrary in the class of semiconvex functions defined in the neighborhood of x, such that $\mathrm{Jac}_x\big(\exp_x(t\nabla\psi)\big)$ remains positive on $[0,1]$. A more precise and more general discussion can be found in [40, Chap. 14].

2.14 Why Look for Curvature Bounds?

Sectional upper and lower bounds, and Ricci lower bounds, turn out to be very useful in many geometric applications. For instance, Ricci bounds appears in inequalities relating gradients and measures, such as:

- Sobolev inequalities
- Heat kernel estimates
- Compactness of families of manifolds
- Spectral gap
- Diameter control.

For example, the Bonnet–Myers theorem states

$$\mathrm{Ric}_x \geq K g_x,\ K>0 \quad \Longrightarrow \quad \mathrm{diam}(M) \leq \pi\sqrt{\frac{n-1}{K}},$$

while Sobolev's inequalities on n-dimensional compact manifolds say that, if $\mathrm{Ric}_x \geq K g_x$ for some $K \in \mathbb{R}$, then

$$\|f\|_{L^{(n-1)/n}(d\mathrm{vol})} \leq C(n,K,\mathrm{diam}(M))\left(\|f\|_{L^1(d\mathrm{vol})} + \|\nabla f\|_{L^1(d\mathrm{vol})}\right) \qquad \forall f.$$

2.15 Stability Issue and (Measured) Gromov–Hausdorff Convergence

Sectional and Ricci curvatures are nonlinear combinations of derivatives of the metric g up to the second order. Therefore it is clear that if a sequence of Riemannian manifolds (M_k, g_k) converges (in charts) in C^2-topology to a Riemannian manifold (M,g), then both sectional and Ricci curvatures pass to the limit.

However much more is true: *lower bounds on these quantities pass to the limit under much weaker notions of convergence* (which is an indication of the stability/robustness of these bounds).

To make an analogy, consider the notion of convexity: a C^2 function $\phi : \mathbb{R}^n \to \mathbb{R}$ is convex if and only if $\nabla^2\phi \geq 0$ everywhere. In particular if a sequence of convex functions ϕ_k converges to a ϕ is C^2-topology, then ϕ is convex. However it is well-known that convexity pass to the limit under much weaker notions of convergence (for instance, pointwise convergence).

In a geometric context, a powerful weak notion of convergence is the *Gromov–Hausdorff convergence*:

Definition 2.13. A sequence $(X_k, d_k)_{k\in\mathbb{N}}$ of compact length spaces is said to converge in the Gromov–Hausdorff topology to a metric space (X,d) if there are functions $f_k : X_k \to X$ and positive numbers $\varepsilon_k \to 0$, such that f_k is an ε_k-approximate isometry, i.e.

$$\begin{cases} |d(f_k(x), f_k(y)) - d_k(x,y)| \leq \varepsilon_k \ \forall\, x,y \in X_k, \\ \\ \mathrm{dist}_d(f(X_k), X) \leq \varepsilon_k. \end{cases}$$

Here $\mathrm{dist}_d(A,B)$ denotes the distance between two sets $A, B \subset X$ measured with respect to d, namely

$$\mathrm{dist}_d(A,B) = \max(\sup_{x\in A}\inf_{y\in B} d(x,y),\ \sup_{y\in B}\inf_{x\in A} d(x,y)).$$

One pleasant feature of the Gromov–Hausdorff convergence is that a limit of length spaces is a length space, and a limit of geodesic spaces is a geodesic space. The key property is summarized in the following:

Exercise 2.14. Let γ_k be a geodesic in X_k for each $k \in X_k$. Prove that, although $f_k \circ \gamma_k$ is a priori a discontinuous curve in X, up to extraction $f_k \circ \gamma_k$ converges to a geodesic in X.
Hint. Argue by contradiction.

In a Riemannian context, by abuse of notation one will say that a sequence of compact Riemannian manifolds (M_k, g_k) converges to a Riemannian manifold (M, g) if (M_k, d_k) converges to (M, d), where d_k (resp. d) is the geodesic distance on (M_k, g_k) (resp. (M, g)). It is remarkable that sectional curvature bounds do pass to the limit under Gromov–Hausdorff convergence [4]), however weak the latter notion:

Theorem 2.15. *Let (M_k, g_k) be a sequence of Riemannian manifolds converging in the Gromov–Hausdorff topology to a Riemannian manifold (M, g). If all sectional curvatures of (M_k, g_k) are all bounded from below by some fixed number $\kappa \in \mathbb{R}$, then also the sectional curvatures of (M, g) are bounded from below by κ.*

A slightly weaker notion of convergence takes care not only of the distances, but also of the measures; it is the *measured Gromov–Hausdorff convergence*, introduced in the present form by Fukaya (related notions were studied by Gromov):

Definition 2.16. A sequence $(X_k, d_k, \mu_k)_{k\in\mathbb{N}}$ of compact length spaces, equipped with reference Borel measures μ_k, is said to converge in the measured Gromov–Hausdorff topology to a measured metric space (X, d, μ) if there are functions $f_k : X_k \to X$, and positive numbers $\varepsilon_k \to 0$, such that f_k is an ε_k-approximate isometry, and $(f_k)_\#\mu_k$ converge in the weak topology to μ.

By abuse of notation, we shall say that a sequence of compact Riemannian manifolds (M_k, g_k) converges to another Riemannian manifold (M, g) in the measured Gromov–Hausdorff topology if $(M_k, d_k, \mathrm{vol}_k)$ converges to (M, d, vol), where d_k and vol_k (resp. d and vol) are the geodesic distance and volume measure associated to (M_k, g_k) (resp. (M, g)). After these preparations, we can state the stability result on Ricci lower bounds, which will be proved in Sect. 4 using elements of calculus of variations, optimal transport, and its relation to Ricci curvature. The following statement is a particular case of more general results proved independently in [25] and [35,36]:

Theorem 2.17. *Let (M_k, g_k) be a sequence of compact Riemannian manifolds converging in the measured Gromov–Hausdorff topology to a compact Riemannian manifold (M, g). If the Ricci curvature of (M_k, g_k) is bounded below by $K\, g_k$, for some number $K \in \mathbb{R}$ independent of k, then also the Ricci curvature of (M, g) is bounded from below by $K\, g$.*

3 Solution of the Monge Problem in Riemannian Geometry

3.1 The Monge Problem with Quadratic Cost

Let (M, g) be a Riemannian manifold, d its geodesic distance, and let $P(M)$ denote the space of probability measures on M. The *Monge problem with quadratic cost* on M is the following: given $\mu, \nu \in P(M)$, consider the minimization problem

$$\inf_{T_{\#}\mu=\nu} \int_M d(x, T(x))^2 \, d\mu(x).$$

Here the infimum is taken over all measurable maps $T : M \to M$ such that the *push-forward* $T_{\#}\mu$ of μ by T (i.e. the Borel probability measure defined by $T_{\#}\mu(A) := \mu(T^{-1}(A))$ for all Borel subsets A of M) coincides with ν.

This problem has a nice *engineering interpretation*: if we define $c(x, y) = \frac{1}{2}d(x, y)^2$ to be the cost to move a unit mass from x to y, then the above minimization problem simply consists in minimizing the total cost (=work) by choosing the destination $T(x)$ for each x.

It is also possible to give an equivalent *probabilistic interpretation* of the problem:

$$\inf\Big\{\mathbb{E}[c(X, Y)] \,|\, \mathrm{law}(X) = \mu,\ \mathrm{law}(Y = \nu)\Big\},$$

so that we are minimizing a sort of correlation of two random variables, once their law is given.

Example 3.1. If $c(x, y) = -x \cdot y$ in $\mathbb{R}^n \times \mathbb{R}^n$, then we are just maximizing the correlation of the random variables X and Y, in the usual sense.

3.2 Existence and Uniqueness on Compact Manifolds

In [29], McCann generalized Brenier's theorem [3] to compact Riemannian manifolds (see [11] or [40, Chap. 10] for the case of more general cost functions on arbitrary Riemannian manifolds). McCann proved:

Theorem 3.2. *Let (M, g) be a compact connected Riemannian manifold, let $\mu(dx) = f(x)\mathrm{vol}(dx)$ and $\nu(dy) = g(y)\mathrm{vol}(dy)$ be probability measures on M, and consider the cost $c(x, y) = \frac{1}{2}d(x, y)^2$. Then*

(1) There exists a unique solution T to the Monge problem.
(2) T is characterized by the structure $T(x) = \exp_x(\nabla\psi(x))$ for some $\frac{d^2}{2}$-convex function $\psi : M \to \mathbb{R}$.

(3) *For μ_0-almost all x, there exists a unique minimizing geodesic from x to $T(x)$, which is given by $t \mapsto \exp_x\big(t\nabla\psi(x)\big)$.*
(4) $\mathrm{Jac}_x T = \frac{f(x)}{g(T(x))}$ *μ-almost everywhere.*

In the sequel of this section we shall explain this statement, and provide a sketch of the proof; much more details are in [40, Part I].

3.3 c-Convexity and c-Subdifferential

We recall that a function $\varphi : \mathbb{R}^n \to \mathbb{R} \cup \{+\infty\}$ is convex and lower semicontinuous convex if and only if

$$\varphi(x) = \sup_{y\in\mathbb{R}^n} \big[x \cdot y - \varphi^*(y)\big],$$

where

$$\varphi^*(x) := \sup_{x\in\mathbb{R}^n} \big[x \cdot y - \varphi(x)\big].$$

This fact is the basis for the definition of *c-convexity*, where $c : X \times Y \to \mathbb{R}$ is an arbitrary function:

Definition 3.3. A function $\psi : X \to \mathbb{R} \cup \{+\infty\}$ is c-convex if

$$\forall x \qquad \psi(x) = \sup_{y\in Y}\big[\psi^c(y) - c(x,y)\big],$$

where

$$\forall y \qquad \psi^c(y) := \inf_{x\in X}\big[\psi(x) + c(x,y)\big].$$

Moreover, for a c-convex function ψ, we define its *c-subdifferential* at x as

$$\partial^c\psi(x) := \big\{y \in Y \,|\, \psi(x) = \psi^c(y) - c(x,y)\big\}$$

With this general definition, when $c(x,y) = -x \cdot y$, the usual convexity coincides with the c-convexity and the usual subdifferential coincides with the c-subdifferential.

Remark 3.4. In the case of the Euclidean $\mathbb{R}^n$, a function ψ is $\frac{d^2}{2}$-convex if and only if $\psi(x) + \frac{|x|^2}{2}$ is convex.

The following facts are useful (see [40, Chap. 13]):

Proposition 3.5. *Let M be a compact Riemannian manifold. Then*

(a) *If $\psi : M \to \mathbb{R}$ is $\frac{d^2}{2}$-convex, then ψ is semiconvex (i.e. in any chart can be written as the sum of a convex and a smooth function).*

(*b*) *There exists a small number* $\delta(M) > 0$ *such that any function* $\psi : M \to \mathbb{R}$ *with* $\|\psi\|_{C^2(M)} \leq \delta(M)$ *is* $\frac{d^2}{2}$*-convex.*

(*c*) *If* $\psi : M \to \mathbb{R}$ *is a* $\frac{d^2}{2}$*-convex function of class* C^2*, then*

$$\nabla^2\psi(x) + \frac{\nabla_x^2 d\big(x, \exp_x(\nabla\psi(x))\big)^2}{2} \geq 0,$$

where $\nabla_x^2 d(x,y)^2$ *denotes the second derivative of the function* $d^2(x,y)$ *with respect to the* x *variable.*

Remark 3.6. A natural question is whether condition (*c*) is also sufficient for $\frac{d^2}{2}$-convexity (at least for C^2 functions). As we will see in Sect. 5.5, this is the case under a suitable (forth order) condition on the cost function $\frac{d^2}{2}$.

3.4 Sketch of the Proof Theorem 3.2

There are several ways to establish Theorem 3.2. One possibility is to go through the following five steps:

3.4.1 Step 1: Solve the Kantorovich Problem

In [17, 18], Kantorovich introduced a notion of weak solution of the optimal transport problem: look for *transport plans* instead of transport maps. A transport map between two probability measures μ and ν is a measurable map T such that $T_{\#}\mu = \nu$; while a transport plan is a probability measure π on $M \times M$ whose marginals are μ and ν, i.e.

$$\int_{M\times M} h(x)\, d\pi(x,y) = \int_M h(x)\, d\mu(x), \qquad \int_{M\times M} h(y)\, d\pi(x,y) = \int_M h(y)\, d\nu(y),$$

for all $h : M \to \mathbb{R}$ bounded continuous. Denoting by $\Pi(\mu,\nu)$ the set of transport plans between μ and ν, the new minimization problem becomes

$$\inf_{\pi\in\Pi(\mu,\nu)} \left\{ \int_{M\times M} d(x,y)^2\, d\pi(x,y) \right\}. \tag{5}$$

A solution of this problem is called an *optimal transport plan.* The connection between the formulation of Kantorovich and that of Monge is the following: any transport map T induces the plan defined by $(\mathrm{Id}_X \times T)_{\#}\mu$ which is concentrated on the graph of T. Conversely, if a transport plan is concentrated on the graph of a measurable function T, then it is induced by this map.

By weak compactness of the set $\Pi(\mu,\nu)$ and continuity of the function $\pi \mapsto \int d(x,y)^2\,d\pi$, it is simple to prove the existence of an optimal transport plan $\bar{\pi}$; so to prove the existence of a solution to the Monge problem it suffices to show that $\bar{\pi}$ is automatically concentrated on the graph of a measurable map T, i.e.

$$y = T(x) \qquad \text{for } \bar{\pi}\text{-almost every } (x,y).$$

Once this fact is proved, the uniqueness of optimal maps will follow from the observation that, if T_1 and T_2 are optimal, then $\pi_1 := (\mathrm{Id}_X \times T_1)_\# \mu$ and $\pi_2 := (\mathrm{Id}_X \times T_2)_\# \mu$ are both optimal plans, so by linearity $\bar{\pi} = \frac{1}{2}(\pi_1 + \pi_2)$ is optimal. If it is concentrated on a graph, this implies $T_1 = T_2$ μ-almost everywhere.

3.4.2 Step 2: The support of $\bar{\pi}$ is c-Cyclically Monotone

A set $S \subset M \times M$ is called *c-cyclically monotone* if, for all $N \in \mathbb{N}$, for all $\{(x_i, y_i)\}_{0 \le i \le N} \subset S$, one has

$$\sum_{i=0}^{N} c(x_i, y_i) \le \sum_{i=0}^{N} c(x_i, y_{i+1}),$$

where by convention $y_{N+1} = y_0$.

The above definition heuristically means that, sending the point x_i to the point y_i for $i = 0, \dots, N$ is globally less expensive than sending the point x_i to the point y_{i+1}. It is therefore intuitive that, since $\bar{\pi}$ is optimal, its support is c-cyclically monotone (see [14] or [40, Chap. 5] for a proof).

3.4.3 Step 3: Any c-Cyclically Monotone Set Is Contained in the c-Subdifferential of a c-Convex Function

A proof of this fact (which is due to Rockafellar for $c(x,y) = -x \cdot y$, and Rüschendorf for the general case) consists in constructing explicitly a c-convex function which does the work: given S c-cyclically monotone, we define

$$\psi(x) := \sup_{N \in \mathbb{N}} \sup_{\{(x_i,y_i)\}_{1 \le i \le N} \subset S} \Big\{ \big[c(x_0, y_0) - c(x_1, y_0)\big] +$$
$$\big[c(x_1, y_1) - c(x_2, y_1)\big] + \ldots + \big[c(x_N, y_N) - c(x, y_N)\big] \Big\},$$

where (x_0, y_0) is arbitrarily chosen in S. We leave as an exercise for the reader to check that with this definition ψ is c-convex, and that $S \subset \partial^c \psi(x) := \cup_{x \in M}(\{x\} \times \partial^c \psi(x))$.

3.4.4 Step 4: $\bar{\pi}$ Is Concentrated on a Graph

Applying Steps 2 and 3, we know that the support of $\bar{\pi}$ is contained in the c-subdifferential of a c-convex function $\bar{\psi}$. Moreover, as said in Proposition 3.5, c-convex functions with $c = \frac{d^2}{2}$ are semiconvex. In particular $\bar{\psi}$ is Lipschitz, and so it is differentiable vol-almost everywhere. Since μ is absolutely continuous with respect to vol, we deduce that $\bar{\psi}$ is differentiable μ-almost everywhere. This further implies that, for $\bar{\pi}$-almost every (x, y), $\bar{\psi}$ is differentiable at x.

Now, let us fix a point $(\bar{x}, \bar{y}) \in \mathrm{supp}(\pi)$ such that $\bar{\psi}$ is differentiable at $\bar{x}$. To prove that $\bar{\pi}$ is concentrated on a graph, it suffices to prove that $\bar{y}$ is uniquely determined as a function of $\bar{x}$. To this aim, we observe that:

(a) Since the support of $\bar{\pi}$ is contained in the c-subdifferential of $\bar{\psi}$, we have $\bar{y} \in \partial^c\bar{\psi}(\bar{x})$, and this implies that the function $x \mapsto \bar{\psi}(x) + c(x, \bar{y})$ is subdifferentiable at $\bar{x}$ and 0 belongs to the subdifferential .
(b) As shown in Sect. 2.9, $c(x, \bar{y}) = \frac{1}{2}d(x, \bar{y})^2$ is superdifferentiable everywhere.
(c) $\bar{\psi}$ is differentiable at $\bar{x}$.

The combination of (a), (b) and (c) implies that $c(x, \bar{y})$ is both upper and lower differentiable at $x = \bar{x}$, hence it is differentiable at $\bar{x}$. Since $\bar{x}$ was an arbitrary point where $\bar{\psi}$ is differentiable, this proves that

$$\nabla\bar{\psi}(x) + \nabla_x c(x, y) = 0 \qquad \text{for } \bar{\pi}\text{-almost every } (x, y).$$

By the first variation formula and the discussion in Sect. 2.9, this implies that there exists a unique geodesic joining x to y, and $\nabla\bar{\psi}(x) = (\exp_x)^{-1}(y)$. Thus we conclude that, for $\bar{\pi}$-almost every (x, y),

$$\begin{cases} y = \exp_x\big(\nabla\bar{\psi}(x)\big) \text{ (in particular } y \text{ is a function of } x), \\ \\ t \mapsto \exp_x\big(t\nabla\bar{\psi}(x)\big) \text{ is the unique minimizing geodesic between } x \text{ and } y. \end{cases}$$

3.4.5 Step 5: Change of Variable Formula

Here we give just a formal proof of the Jacobian equation, and we refer to [40, Chap. 11] for a rigorous proof.

Since π has marginals μ and ν, and is concentrated on the graph of T, for all bounded continuous functions $\zeta : M \to \mathbb{R}$ we have

$$\int_M \zeta(y)\, d\nu(y) = \int_{M\times M} \zeta(y)\, d\pi(x, y) = \int_M \zeta(T(x))\, d\mu(x),$$

that is $T_{\#}\mu = \nu$. Recalling that $\mu = f\text{vol}$ and $\nu = g\text{vol}$, we get, by change of variables,

$$\int_M \zeta(T(x))f(x)\,d\text{vol}(x) = \int_M \zeta(y)g(y)\,d\text{vol}(y)$$
$$= \int_M \zeta(T(x))g(T(x))|\det(d_xT)|\,d\text{vol}(x). \quad (6)$$

By the arbitrariness of ζ we conclude that $f(x) = g(T(x))|\det(d_xT)|$ μ-almost everywhere.

3.5 *Interpretation of the Function $\bar{\psi}$*

The function $\bar{\psi}$ appearing in the formula for the optimal transport map has an interpretation as the solution of a dual problem:

$$\sup_{\psi}\left[\int_M \psi^c(y)\,d\nu(y) - \int_M \psi(x)\,d\mu(x)\right].$$

The above maximization problem has the following economical interpretation: $\psi(x)$ is the price at which a "shipper" buys material at x, while $\psi^c(y)$ is the price at which he sells back the material at y. Then, since

$$\psi^c(y) = \inf_x\left[\psi(x) + c(x,y)\right] = \sup\Big\{\varphi(y)\,|\,\varphi(y) \le \psi(x) + c(x,y)\Big\},$$

this means that $\psi^c(y)$ is the maximum selling price which is below the sum "buy price + transportation cost", that is the maximum price to be "competitive". In other words, the shipper is trying to maximize his profit.

To prove that $\bar{\psi}$ solves the above maximization problem, we observe that:

(1) $\bar{\psi}^c(y) - \bar{\psi}(x) = c(x,y)$ on $\text{supp}(\bar{\pi})$ for an optimal plan $\bar{\pi}$.
(2) For any c-convex function ψ, $\psi^c(y) - \psi(x) \le c(x,y)$ on $M \times M$.

Combining these two facts, we get

$$\int_M \bar{\psi}^c(y)\,d\nu - \int_M \bar{\psi}(x)\,d\mu = \int_{M\times M}\left[\bar{\psi}^c(y) - \bar{\psi}(x)\right]d\bar{\pi}(x,y)$$
$$= \int_{M\times M} c(x,y)\,d\bar{\pi}(x,y)$$

$$
\begin{aligned}
&\geq \int_{M\times M} [\psi^c(y) - \psi(x)]\, d\bar{\pi}(x,y) \\
&= \int_M \psi^c\, d\nu - \int_M \psi\, d\mu,
\end{aligned}
$$

and so $\bar{\psi}$ is a maximizer.

4 Synthetic Formulation of Ricci Bounds

As we have seen in the last section, the optimal transport allows to construct maps $T(x) = \exp_x\big(\nabla\psi(x)\big)$, with $\psi : M \to \mathbb{R}$ globally defined. Moreover it involves a Jacobian formula for T. For this reason, it turns out to be a natural candidate for a global reformulation of Ricci curvature bounds. More precisely, Ricci curvature bounds can be reformulated in terms of *convexity inequalities* for certain nonlinear functionals of the density, along geodesics of optimal transport. This fact will provide the stability of such bounds, and other applications.

4.1 The 2-Wasserstein Space

Let (M,g) be a compact Riemannian manifold, equipped with its geodesic distance d and its volume measure vol. We denote with $P(M)$ the set of probability measures on M. Let

$$
W_2(\mu,\nu) := \min_{\pi\in\Pi(\mu,\nu)} \left\{ \int_{M\times M} d^2(x,y)\, d\pi(x,y) \right\}^{\frac{1}{2}}.
$$

The quantity $W_2(\mu,\nu)$ is called the *Wasserstein distance of order 2* between μ and ν. It is well-known that it defines a finite metric on $P(M)$, and so one can speak about geodesic in the metric space $P_2(M) := (P(M), W_2)$. This space turns out to be a geodesic space (see e.g. [40, Chap. 7]). We denote with $P_2^{ac}(M)$ the subset of $P_2(M)$ that consists of the Borel probability measures on M that are absolutely continuous with respect to vol.

4.2 Geodesics in $P_2(M)$

Given $\mu_0, \mu_1 \in P_2(M)$, we want to construct a geodesic between them. In general the geodesic is not unique, as can be seen considering $\mu_0 = \delta_x$ and

$\mu_1 = \delta_y$, where x and y can be joined by several minimizing geodesics. Indeed if $\gamma : [0,1] \to M$ is a geodesic form x to y, then $\mu_t := \delta_{\gamma(t)}$ is a Wasserstein geodesic from δ_x to δ_y. (Check it!)

On the other hand, if μ_0 (or equivalently μ_1) belongs to $P_2^{ac}(M)$, this problem has a simple answer (see [40, Chap. 7] for a general treatment):

Proposition 4.1. *Assume $\mu_0 \in P_2^{ac}(M)$, and let $T(x) = \exp_x\big(\nabla\psi(x)\big)$ be the optimal map between μ_0 and μ_1. Then the unique geodesic from μ_0 to μ_1 is given by $\mu_t := (T_t)_{\#}\mu_0$, with $T_t(x) := \exp_x\big(t\nabla\psi(x)\big)$.*

Proof. To prove that μ_t is a geodesic, we observe that

$$\begin{aligned} W_2(\mu_s,\mu_t)^2 &\le \int_M d\big(\exp_x\big(s\nabla\psi(x)\big), \exp_x\big(t\nabla\psi(x)\big)\big)\,d\mu_0(x) \\ &= (s-t)^2\int_M |\nabla\psi(x)|_x^2\,d\mu_0(x) = (t-s)^2 W_2(\mu_0,\mu_1)^2. \end{aligned}$$

This implies that $W_2(\mu_s,\mu_t) \le |t-s| W_2(\mu_0,\mu_1)$ for all $s,t \in [0,1]$, so the length of the path $(\mu_t)_{0\le t\le 1}$ satisfies

$$L\big((\mu_t)_{0\le t\le 1}\big) \le W_2(\mu_0,\mu_1).$$

But since the converse inequality is always true, we get $L\big((\mu_t)_{0\le t\le 1}\big) = W_2(\mu_0,\mu_1)$, so that μ_t is a geodesic.

The fact that μ_t is the unique geodesic is a consequence Theorem 3.2(c), together with the general fact that any Wasserstein geodesic takes the form $\mu_t = (e_t)_{\#}\Pi$, where Π is a probability measure on the set Γ of minimizing geodesics, and $e_t : \Gamma \to M$ is the evaluation map at time t: $e_t(\gamma) := \gamma(t)$ (see [40, Theorem 7.21 and Corollary 7.23]). □

Remark 4.2. One can write down the geodesic equations for $P_2(M)$, which has to be understood in a suitable weak sense (see [40, Chap. 13]):

$$\begin{cases} \dfrac{\partial\mu_t}{\partial t} + \operatorname{div}(\mu_t\nabla\psi_t) = 0, \\[2ex] \dfrac{\partial\psi_t}{\partial t} + \dfrac{|\nabla\psi_t|^2}{2} = 0. \end{cases}$$

4.3 Approximate Geodesics in Wasserstein Space

A key property of the Wasserstein space is that it depends continuously on the basis space, when the topology is the Gromov–Hausdorff topology. The following statement is proven in [25, Proposition 4.1]: *If $f_k : M_k \to M$ are*

ε_k-approximate isometries, with $\varepsilon_k \to 0$, then $(f_k)_\# : P_2(M_k) \to P_2(M)$ are $\tilde{\varepsilon}_k$-approximate isometries, with $\tilde{\varepsilon}_k \to 0$.

4.4 Reformulation of Ric ≥ 0

Let

$$H(\mu) := H_{\mathrm{vol}}(\mu) = \begin{cases} \int_M \rho \log(\rho)\, d\mathrm{vol} \text{ if } \mu = \rho \mathrm{vol}, \\ +\infty \qquad\qquad\qquad \text{otherwise.} \end{cases}$$

This is the *Boltzmann H functional*, or negative of the *Boltzmann entropy.* As shown in [34] (as a development of the works in [9, 33]), the inequality Ric ≥ 0 can be reformulated in terms of the convexity of H along Wasserstein geodesics:

Theorem 4.3. *Let (M, g) be a compact Riemannian manifold. Then* Ric ≥ 0 *if and only if $t \mapsto H(\mu_t)$ is a convex function of $t \in [0,1]$ for all Wasserstein geodesics $(\mu_t)_{0\leq t\leq 1}$.*

More generally, Ric $\geq K\, g$ if and only if, for all $\mu_0, \mu_1 \in P_2^{ac}(M)$,

$$H(\mu_t) \leq (1-t)H(\mu_0) + tH(\mu_1) - K\,\frac{t(1-t)}{2} W_2(\mu_0,\mu_1)^2 \qquad \forall t \in [0,1], \quad (7)$$

where $(\mu_t)_{0\leq t\leq 1}$ is the Wasserstein geodesic between μ_0 and μ_1.

Sketch of the proof. Let us consider just the case $K = 0$. By using the Jacobian equation (Theorem 3.2(d)), we get

$$\begin{aligned} H(\mu_t) &= \int_M \rho_t(x) \log\big(\rho_t(x)\big)\, d\mathrm{vol}(x) \\ &= \int_M \rho_t(T_t(x)) \log\big(\rho_t(T_t(x))\big) \mathrm{Jac}_x T_t \, d\mathrm{vol}(x) \\ &= \int_M \rho_0(x) \log\left(\frac{\rho_0(x)}{\mathrm{Jac}_x T_t}\right) d\mathrm{vol}(x) \\ &= H(\mu_0) - \int_M \log \mathrm{Jac}_x\big(\exp_x(t\nabla\psi)\big)\, d\mu_0. \end{aligned}$$

Then the direct implication follows from (4). The converse implication is obtained also from (4), using Proposition 3.5(b) to explore all tangent directions by minimizing geodesics in Wasserstein space. Details appear e.g. in [40, Chap. 17]. □

4.5 Application: Stability

Let us give a sketch of the proof of Theorem 2.17; we refer to [25] and [40, Chaps. 28 and 29] for details.

First of all, we reformulate the inequality $\mathrm{Ric}(M_k) \geq K\, g_k$ in terms of the convexity inequality (7); the goal is to prove that the inequality (7) holds on the limit manifold M.

Let $\mu_0 = \rho_0 \mathrm{vol}, \mu_1 = \rho_1 \mathrm{vol} \in P_2^{ac}(M)$, and define on M_k the probability measures

$$\mu_0^k := \frac{\rho_0 \circ f_k \mathrm{vol}_k}{\displaystyle\int_{M_k} \rho_0 \circ f_k \, d\mathrm{vol}_k}, \qquad \mu_1^k := \frac{\rho_1 \circ f_k \mathrm{vol}_k}{\displaystyle\int_{M_k} \rho_1 \circ f_k \, d\mathrm{vol}_k},$$

where $f_k : M_k \to M$ are the approximate isometries appearing in the definition of the measured Gromov–Hausdorff convergence. Let $(\mu_t^k)_{0 \leq t \leq 1}$ be the Wasserestein geodesic between μ_0^k and μ_1^k. Up to extraction of a subsequence, $(f_k)_\# \mu_t^k$ converges, uniformly in $t \in [0,1]$, to a geodesic μ_t in $P_2(M)$ between μ_0 and μ_1 (recall Exercise 2.14). It remains to show that the inequality (7) passes to the limit. Let us consider separately the three terms in this inequality.

4.5.1 The Term $H_k(\mu_0^k)$ Passes to the Limit

By an approximation argument, it suffices to consider the case $\rho_0 \in C(M)$. Then

$$Z_k := \int_{M_k} \rho_0 \circ f_k \, d\mathrm{vol}_k = \int_M \rho_0 \, d(f_k)_\# \mathrm{vol}_k \longrightarrow \int_M \rho_0 \, d\mathrm{vol} = 1,$$

so that (with obvious notation)

$$\begin{aligned} H_k(\mu_0^k) &= \int_{M_k} \left(\frac{\rho_0 \circ f_k}{Z_k}\right) \log\left(\frac{\rho_0 \circ f_k}{Z_k}\right) d\mathrm{vol}_k \\ &\simeq \int_{M_k} \rho_0 \circ f_k \log(\rho_0 \circ f_k) \, d\mathrm{vol}_k \\ &= \int_M \rho_0 \log(\rho_0) \, d(f_k)_\# \mathrm{vol}_k \longrightarrow \int_M \rho_0 \log(\rho_0) \, d\mathrm{vol} = H(\mu_0). \end{aligned}$$

The case of $H_k(\mu_1^k)$ is analogous.

4.5.2 The Term $W_2(\mu_0^k, \mu_1^k)^2$ Passes to the Limit

This follows from the fact that $(f_k)_\#$ are $\tilde{\varepsilon}_k$-approximate isometries, and the Wasserstein distance on a compact manifold metrizes the weak convergence: so

$$W_2(\mu_0^k, \mu_1^k) \simeq W_2\big((f_k)_\#\mu_0^k, (f_k)_\#\mu_1^k\big) \longrightarrow W_2(\mu_0, \mu_1).$$

4.5.3 The Term $H_k(\mu_t^k)$ Is Lower Semicontinuous Under Weak Convergence

This comes from the following general property: If $U : \mathbb{R}^+ \to \mathbb{R}$ is convex and continuous, then

- $P_2(M) \times P_2(M) \ni (\mu, \nu) \longmapsto \int U\left(\frac{d\mu}{d\nu}\right)\, d\nu$ is a convex lower semicontinuous functional.
- $\displaystyle\int U\left(\frac{d(f_\#\mu)}{d(f_\#\nu)}\right)\, d(f_\#\nu) \le U\left(\frac{d\mu}{d\nu}\right)\, d\nu$ for any function $f : M \to M$.

Combining these two facts, we get

$$H(\mu_t) \le \liminf_{k\to\infty} H_{(f_k)_\#\mathrm{vol}_k}\big((f_k)_\#\mu_t^k\big) \le \liminf_{k\to\infty} H_{\mathrm{vol}_k}(\mu_t^k),$$

which is the desired inequality.

5 The Smoothness Issue

Let (M, g) be a compact connected Riemannian manifold, let $\mu(dx) = f(x)\mathrm{vol}(dx)$ and $\nu(dy) = g(y)\mathrm{vol}(dy)$ be probability measures on M, and consider the cost $c(x, y) = \frac{1}{2}d(x, y)^2$. Assume f and g are C^∞ and strictly positive on M. Is the optimal map T smooth?

A positive answer to this problem has been given in the Euclidean space [5–7, 10, 38, 39] and in the case of the flat torus [8], but the general question of Riemannian manifolds remained open until the last years. Only recently, after two key papers of Ma, Trudinger and Wang [27] and Loeper [21], did specialists understand a way to attack this problem; see [40, Chap. 12] for a global picture and references.

There are several motivations for the investigation of the smoothness of the optimal map:

- It is a typical PDE/analysis question.
- It is a step towards a qualitative understanding of the optimal transport map.

- If it is a general phenomenon, then nonsmooth situations may be treated by regularization, instead of working directly on nonsmooth objects.

Moreover, as we will see, the study of this regularity issue allows to understand some geometric properties of the Riemannian manifold itself.

5.1 The PDE

Starting from the Jacobian equation

$$\left|\det(d_xT)\right| = \frac{f(x)}{g(T(x))},$$

and the relation $T(x) = \exp_x\big(\nabla\psi(x)\big)$, we can write a PDE for ψ. Indeed, since

$$\nabla\psi(x) + \nabla_x c(x, T(x)) = 0,$$

differentiating with respect to x and using the Jacobian equation we get

$$\det\Big[\nabla^2\psi(x) + \nabla_x^2 c\big(x, \exp_x\big(\nabla\psi(x)\big)\big)\Big] = \frac{f(x)}{g(T(x))\big|\det(d_{\nabla\psi(x)}\exp_x)\big|}.$$

(Observe that, since ψ is c-convex, the matrix appearing in the left-hand side is positive definite by Proposition 3.5(c).)

We see that ψ solves a Monge–Ampère type equation with a perturbation $\nabla_x^2 c\big(x, \exp_x\big(\nabla\psi(x)\big)\big)$ which is of first order in ψ. Unfortunately, for Monge–Ampère type equations lower order terms do matter, and it turns out that it is exactly the term $\nabla_x^2 c\big(x, \exp_x\big(\nabla\psi(x)\big)\big)$ which can create obstructions to the smoothness.

5.2 Obstruction I: Local Geometry

We now show how negative sectional curvature is an obstruction to regularity (indeed even to continuity) of optimal maps. We refer to [40, Theorem 12.4] for more details on the construction given below.

Let $M = \mathbb{H}^2$ be the hyperbolic plane (or a compact quotient thereof). Fix a point O as the origin, and fix a local system of coordinates in a neighborhood of O such that the maps $(x_1, x_2) \mapsto (\pm x_1, \pm x_2)$ are local isometries (it suffices for instance to consider the model of the Poincaré disk, with O equal to the origin in $\mathbb{R}^2$). Then define the points

$$A^{\pm} = (0, \pm\varepsilon), \quad B^{\pm} = (\pm\varepsilon, 0) \qquad \text{for some } \varepsilon > 0.$$

Take a measure μ symmetric with respect to 0 and concentrated near $\{A^+\} \cup \{A^-\}$ (say 3/4 of the total mass belongs to a small neighborhood of $\{A^+\} \cup \{A^-\}$), and a measure ν symmetric with respect to 0 and concentrated near $\{B^+\} \cup \{B^-\}$. Moreover assume that μ and ν are absolutely continuous, and have strictly positive densities everywhere. We denote by T the unique optimal transport map, and we assume by contradiction that T is continuous. By symmetry, we deduce that $T(O) = O$. Then, by counting the total mass, we deduce that there exists a point A' close to A^+ which is sent to a point B' near, say, B^+.

But, by negative curvature (if A' and B' are close enough to A and B respectively), Pythagore's Theorem becomes an inequality: $d(O, A')^2 + d(O, B')^2 < d(A', B')^2$, and this contradicts the c-cyclical monotonicity of the support of an optimal plan (see Step 2 of Theorem 3.2).

5.3 *Obstruction II: Topology of the c-Subdifferential*

Let $\varphi : \mathbb{R}^n \to \mathbb{R}$ be a convex function; its differential $\partial\varphi(x)$ is given by

$$\begin{aligned}\partial\varphi(x) &= \{y \in \mathbb{R}^n \mid \varphi(x) + \varphi^*(y) = x \cdot y\} \\ &= \{y \in \mathbb{R}^n \mid \varphi(z) - z \cdot y \geq \varphi(x) - x \cdot y \quad \forall z \in \mathbb{R}^n\}.\end{aligned}$$

Then $\partial\varphi(x)$ is a convex set, a fortiori connected.

If we now consider $\psi : M \to \mathbb{R}$ a c-convex function, $c = \frac{d^2}{2}$, then (see Sect. 3.3)

$$\begin{aligned}\partial^c\psi(x) &= \{y \in M \mid \psi(x) = \psi^c(y) - c(x, y)\} \\ &= \{y \in M \mid \psi(z) + c(z, y) \geq \psi(x) + c(x, y) \quad \forall z \in M\}.\end{aligned}$$

In this generality there is no reason for $\partial^c\psi(x)$ to be connected – and in general, this is not the case!

Following the construction given in [27, Sect. 7.3], Loeper showed that under adequate assumptions the connectedness of the c-subdifferential is a necessary condition for the smoothness of optimal transport [21] (see also [40, Theorem 12.7]):

Theorem 5.1. *Assume that there exist $\bar{x} \in M$ and $\psi : M \to \mathbb{R}$ c-convex such that $\partial^c\psi(\bar{x})$ is not (simply) connected. Then one can construct f and g, C^∞ strictly positive probability densities on M, such that the optimal map T from $f$vol to $g$vol is discontinuous.*

5.4 Conditions for the Connectedness of $\partial^c\psi$

We now wish to find some simple enough conditions implying the connectedness of sets $\partial^c\psi$.

5.4.1 First Attempt

Let us look at the simplest c-convex functions:

$$\psi(x) := -c(x, y_0) + a_0.$$

Assume that $\bar{x} \notin \mathrm{cut}(y)$, and let $\bar{y} \in \partial^c\psi(\bar{x})$. Then the function $\psi(x)+c(x,\bar{y})$ achieves its minimum at $x = \bar{x}$, so $\bar{x} \notin \mathrm{cut}(\bar{y})$ (see the argument in Step 4 of Theorem 3.2) and

$$-\nabla_x c(\bar{x}, y_0) + \nabla_x c(\bar{x}, \bar{y}) = 0.$$

Thus $(\exp_{\bar{x}})^{-1}(y_0) = (\exp_{\bar{x}})^{-1}(\bar{y})$ (see Sect. 2.9), which implies $\bar{y} = y_0$. In conclusion $\partial^c\psi(\bar{x}) = \{y_0\}$ is a singleton, automatically connected – so we do not learn anything!

5.4.2 Second Attempt

The second simplest example of c-convex functions are

$$\psi(x) := \max\bigl\{-c(x, y_0) + a_0, -c(x, y_1) + a_1\bigr\}.$$

Take a point $\bar{x} \notin \mathrm{cut}(y)$ belonging to the set $\{-c(x, y_0)+a_0 = -c(x, y_1)+a_1\}$, and let $\bar{y} \in \partial^c\psi(\bar{x})$. Since $\psi(x) + c(x, \bar{y})$ attains its minimum at $x = \bar{x}$, we get

$$0 \in \nabla^-_{\bar{x}}\bigl(\psi + c(\cdot, \bar{y})\bigr),$$

or equivalently

$$-\nabla_x c(\bar{x}, \bar{y}) \in \nabla^-\psi(\bar{x})$$

(recall that by Proposition 3.5(a) ψ is a semiconvex function, so that its subgradient $\nabla^-\psi(\bar{x})$, which is defined in charts as the set $\{p \,|\, \psi(\bar{x} + \delta x) \geq \psi(\bar{x}) + \langle p, \delta x\rangle + o(|\delta x|)\}$, is convex and non-empty). From the above inclusion we deduce that $\bar{y} \in \exp_{\bar{x}}\bigl(\nabla^-\psi(\bar{x})\bigr)$ (see Sect. 2.9). Moreover, it is not difficult to see that

$$\nabla^-\psi(\bar{x}) = \{(1-t)v_0 + tv_1 \,|\, t \in [0, 1]\}, \quad v_i := \nabla_x c(\bar{x}, y_i) = (\exp_{\bar{x}})^{-1}(y_i), \quad i = 0, 1.$$

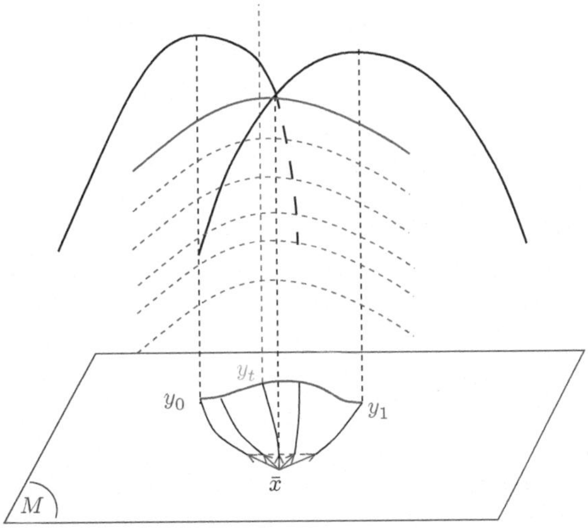

Fig. 1 Property (3): the mountain grown from y_t emerges exactly at the pass between the mountains centered at y_0 and y_1

Therefore, denoting by $[v_0, v_1]$ the segment joining v_0 and v_1, we have proved the inclusion

$$\partial^c\psi(\bar{x}) \subset \exp_{\bar{x}}\big([v_0, v_1]\big).$$

The above formula suggests the following definition of *c-segment*:

Definition 5.2. Let $\bar{x} \in M$, $y_0, y_1 \notin \mathrm{cut}(\bar{x})$. Then we define the c-segment from y_0 to y_1 with base $\bar{x}$ as

$$[y_0, y_1]_{\bar{x}} := \big\{y_t = \exp_{\bar{x}}\big((1-t)(\exp_{\bar{x}})^{-1}(y_0) + t(\exp_{\bar{x}})^{-1}(y_1)\big) \,|\, t \in [0,1]\big\}.$$

In [21], Loeper proved (a slightly weaker version of) the following result (see [40, Chap. 12] for the general result):

Theorem 5.3. *The following conditions are equivalent:*

(1) For any ψ c-convex, for all $\bar{x} \in M$, $\partial^c\psi(\bar{x})$ is connected.
(2) For any ψ c-convex, for all $\bar{x} \in M$, $(\exp_{\bar{x}})^{-1}\big(\partial^c\psi(\bar{x}) \setminus \mathrm{cut}(\bar{x})\big)$ is convex.
(3) For all $\bar{x} \in M$, for all $y_0, y_1 \notin \mathrm{cut}(\bar{x})$, if $[y_0, y_1]_{\bar{x}} = \{y_t\}_{t\in[0,1]}$ does not meet $\mathrm{cut}(\bar{x})$, then

$$d(x, y_t)^2 - d(\bar{x}, y_t)^2 \geq \min\big[d(x, y_0)^2 - d(\bar{x}, y_0)^2, d(x, y_1)^2 - d(\bar{x}, y_1)^2\big] \tag{8}$$

for all $x \in M$, $t \in [0,1]$ (Fig. 1).
(4) For all $\bar{x} \in M$, for all $y \notin \mathrm{cut}(\bar{x})$, for all $\eta, \xi \in T_{\bar{x}}M$ with $\xi \perp \eta$,

$$\frac{d^2}{ds^2}\bigg|_{s=0} \frac{d^2}{dt^2}\bigg|_{t=0} d\big(\exp_{\bar{x}}(t\xi), \exp_{\bar{x}}(p + s\eta)\big)^2 \leq 0,$$

where $p = (\exp_{\bar{x}})^{-1}(y)$.

Moreover, if these conditions are not satisfied, C^1 c-convex functions are not dense in Lipschitz c-convex functions.

Sketch of the proof. We give here only some elements of the proof.
(2) ⇒ (1): since $(\exp_{\bar{x}})^{-1}\big(\partial^c\psi(\bar{x}) \setminus \mathrm{cut}(\bar{x})\big)$ is convex, it is connected, and so its image by $\exp_{\bar{x}}$ is connected too.
(1) ⇒ (2): for $\psi_{\bar{x},y_0,y_1} := \max\big\{-c(\cdot,y_0)+c(\bar{x},y_0), -c(\cdot,y_1)+c(\bar{x},y_1)\big\}$ we have $(\exp_{\bar{x}})^{-1}\big(\partial^c\psi(\bar{x})\big) \subset [(\exp_{\bar{x}})^{-1}(y_0), \exp_{\bar{x}})^{-1}(y_1)]$, which is a segment. Since in this case connectedness is equivalent to convexity, if (1) holds we obtain $\partial^c\psi_{\bar{x},y_0,y_1} = [y_0,y_1]_{\bar{x}}$.

In the general case, we fix $y_0, y_1 \in \partial^c\psi(\bar{x})$. Then it is simple to see that

$$\partial^c\psi(\bar{x}) \supset \partial^c\psi_{\bar{x},y_0,y_1}(\bar{x}) = [y_0,y_1]_{\bar{x}},$$

and the result follows.
(2) ⇔ (3): condition (8) is equivalent to $\partial^c\psi_{\bar{x},y_0,y_1} = [y_0,y_1]_{\bar{x}}$. Then the equivalence between (2) and (3) follows arguing as above.
(3) ⇒ (4): fix $\bar{x} \in M$, and let $y := \exp_{\bar{x}}(p)$. Take ξ, η orthogonal and with unit norm, and define

$$y_0 := \exp_{\bar{x}}(p - \varepsilon\eta), \quad y_1 := \exp_{\bar{x}}(p + \varepsilon\eta) \qquad \text{for some } \varepsilon > 0 \text{ small.}$$

Moreover, let

$$h_0(x) := c(\bar{x},y_0) - c(x,y_0), \qquad h_1(x) := c(\bar{x},y_1) - c(x,y_1),$$

$$\psi := \max\{h_0,h_1\} = \psi_{\bar{x},y_0,y_1}.$$

We now define $\gamma(t)$ as a curve contained in the set $\{h_0 = h_1\}$ such that $\gamma(0) = \bar{x}$, $\dot{\gamma}(0) = \xi$. (See Fig. 2.)

Since $y \in [y_0,y_1]_{\bar{x}}$, by (3) we get $y \in \partial^c\psi(\bar{x})$, so that

$$\begin{aligned}
&\frac{1}{2}\big[h_0(\bar{x}) + h_1(\bar{x})\big] + c(\bar{x},y) = \psi(\bar{x}) + c(\bar{x},y)\\
&\le \psi(\gamma(t)) + c(\gamma(t),y) = \frac{1}{2}\big[h_0(\gamma(t)) + h_1(\gamma(t))\big] + c(\gamma(t),y),
\end{aligned}$$

where we used that $h_0 = h_1$ along γ. Recalling the definition of h_0 and h_1, we deduce

$$\frac{1}{2}\big[c(\gamma(t),y_0) + c(\gamma(t),y_1)\big] - c(\gamma(t),y) \le \frac{1}{2}\big[c(\bar{x},y_0) + c(\bar{x},y_1)\big] - c(\bar{x},y),$$

so the function $t \mapsto \frac{1}{2}\big[c(\gamma(t),y_0) + c(\gamma(t),y_1)\big] - c(\gamma(t),y)$ achieves its maximum at $t = 0$. This implies

$$\frac{d^2}{dt^2}\bigg|_{t=0}\left[\frac{1}{2}\big(c(\gamma(t),y_0)+c(\gamma(t),y_1)\big)-c(\gamma(t),y)\right]\leq 0,$$

i.e.

$$\left\langle\left[\frac{1}{2}\big(\nabla_x^2c(\bar{x},y_0)+\nabla_x^2c(\bar{x},y_1)\big)-\nabla_x^2c(\bar{x},y)\right]\cdot\xi,\xi\right\rangle\leq 0$$

(here we used that $\nabla_x c(\bar{x},y)=\frac{1}{2}\big[\nabla_x c(\bar{x},y_0)+\nabla_x c(\bar{x},y_1)\big]$). Thus the function

$$\eta\mapsto\left\langle\nabla_x^2 c\big(\bar{x},\exp_{\bar{x}}(p+\eta)\big)\cdot\xi,\xi\right\rangle$$

is concave, and proves (4). □

The above theorem leads to the definition of the *regularity* property:

Definition 5.4. The cost function $c=\frac{d^2}{2}$ is said to be regular if the properties listed in Theorem 5.3 are satisfied.

To understand why the above properties are related to smoothness, consider properties (3) in Theorem 5.3. It says that, if we take the function $\psi_{\bar{x},y_0,y_1}=\max\{-c(\cdot,y_0)+c(\bar{x},y_0),-c(\cdot,y_1)+c(\bar{x},y_1)\}$, then we are able to touch the graph of this function from below at $\bar{x}$ with the family of functions $\{-c(\cdot,y_t)+c(\bar{x},y_t)\}_{t\in[0,1]}$. This suggests that we could use this family to regularize the cusp of $\psi_{\bar{x},y_0,y_1}$ at the point $\bar{x}$, by slightly moving above the graphs of the functions $-c(\cdot,y_t)+c(\bar{x},y_t)$. (See Fig. 1.) On the other hand, if (3) does not holds, it is not clear how to regularize the cusp preserving the condition of being c-convex.

By what we said above, we see that the regularity property seems mandatory to develop a theory of smoothness of optimal transport. Indeed, if it is not satisfied, we can construct C^∞ strictly positive densities f,g such that the optimal map is not continuous. The next natural question is: when is it satisfied?

5.5 *The Ma–Trudinger–Wang Tensor*

As we have seen in Theorem 5.3, the regularity of $c=\frac{d^2}{2}$ is equivalent to

$$\frac{d^2}{ds^2}\bigg|_{s=0}\frac{d^2}{dt^2}\bigg|_{t=0}c\big(\exp_x(t\xi),\exp_x(p+s\tilde{\eta})\big)\leq 0, \qquad (9)$$

for all $p,\xi,\tilde{\eta}\in T_xM$, with ξ and $\tilde{\eta}$ orthogonal, $p=(\exp_x)^1(y)$ for some $y\notin\mathrm{cut}(x)$.

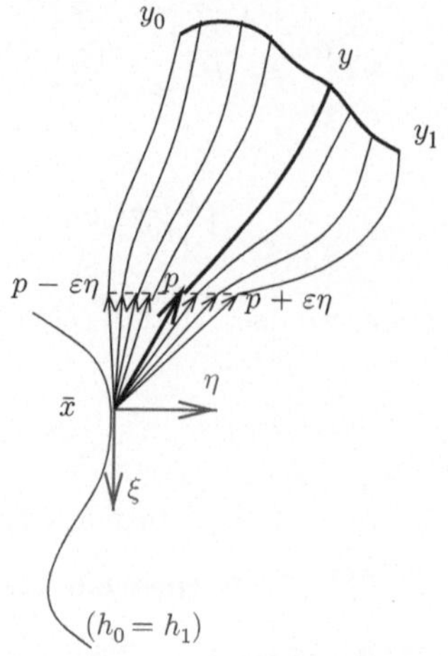

Fig. 2 Proof of (3) $\Rightarrow$ (4); y belongs to the c-segment with base $\overline{x}$ and endpoints y_0, y_1; ξ is tangent to the local hypersurface $(h_0 = h_1)$

Introduce a local system of coordinates $(x^1, \ldots, x^n)$ around x, and a system $(y^1, \ldots, y^n)$ around y. We want to express the above condition only in terms of c, using the relation $\nabla_x c(x, y) + (\exp_x)^{-1}(y) = 0$. By the definition of gradient (see Sect. 2.6), this relation is also equivalent to

$$- d_x c(x, y) = \langle (\exp_x)^{-1}(y), \cdot \rangle_x. \tag{10}$$

We now start to write everything in coordinates. We will write $c_j = \frac{\partial c}{\partial x^j}$, $c_{jk} = \frac{\partial^2 c}{\partial x^j \partial x^k}$, $c_{i,j} = \frac{\partial^2 c}{\partial x^i \partial y^j}$, and so on; moreover $(c^{i,j})$ will denote the coordinates of the inverse matrix of $(c_{i,j})$. Then (10) becomes

$$-c_i \xi^i = g_{jk} p^j \xi^k \qquad \forall \xi \in T_x M.$$

Differentiating at y in a direction $\eta \in T_y M$ we get

$$-c_{i,j} \xi^i \eta^j = g_{ij} [(d_p \exp_x)^{-1}(\eta)]^j \xi^i.$$

Thus we get a formula for $-d^2_{x,y} c : T_x M \times T_y M \to \mathbb{R}$:

$$- c_{i,j}(x, y) = g_{ik}(x) [(d_p \exp_x)^{-1}(\eta)]^k_j, \qquad y = \exp_x(p). \tag{11}$$

As shown in [20], $-d^2_{x,y} c$ defines a pseudo-metric on $T_x M \times T_y M$, which coincides with g along the diagonal $\{x = y\}$, and it is possible to interpret the regularity condition for the cost in terms of this pseudo-metric.

5.5.1 Rewriting the Orthogonality Condition

The operator $-d^2_{x,y}c$ can be used to transport tangent vectors at y in cotangent vectors at x, and viceversa. In particular, if we consider the covector $\tilde{\eta}_i := -c_{i,j}\eta^j$ ($\eta \in T_yM$), the orthogonality condition $g_x(\xi, \tilde{\eta}) = 0$ appearing in the definition of the regularity of the cost is equivalent to $\tilde{\eta}_i\xi^i = 0$, i.e.

$$0 = -c_{i,j}\xi^i\eta^j.$$

Thanks to (11), we have the formula $\eta = d_p \exp_x(\tilde{\eta})$. (Note: η is *not* the parallel transport of $\tilde{\eta}$ along the geodesic $\exp_x(tp)$!) In particular, if the sectional curvature of the manifold in non-negative everywhere, then the exponential map is 1-Lipschitz, and so $|\eta|_y \leq |\tilde{\eta}|_x$.

5.5.2 Rewriting the Ma–Trudinger–Wang Condition

Equation (9) can also be written as

$$\frac{\partial^2}{\partial p_\eta^2}\frac{\partial^2}{\partial x_\xi^2}\, c(x,y) \leq 0. \tag{12}$$

The meaning of the left-hand side in (12) is the following: first freeze y and differentiate $c(x,y)$ twice with respect to x in the direction $\xi \in T_xM$. Then, considering the result as a function of y, parameterize y by $p = -\nabla_x c(x,y)$, and differentiate twice with respect to p in the direction $\eta \in T_yM$ (see the discussion in the next paragraph). By the relation $p_i = -c_i(x,y)$ we get $\frac{\partial p_i}{\partial y^j} = -c_{i,j}$, which gives $\frac{\partial y^k}{\partial p_\ell} = -c^{k,\ell}$. Then, using $-c_{i,j}$ and $-c^{i,j}$ to raise and lower indices ($\eta^k = -c^{k,l}\eta_i$, etc.), it is just a (tedious) exercise to show that the expression in (12) is equal to

$$\sum_{ijklrs} \left(c_{ij,kl} - c_{ij,r}c^{r,s}c_{s,kl}\right)\xi^i\xi^j\eta^k\eta^l,$$

where we used the formula $d(M^{-1}) \cdot H = -M^{-1}HM^{-1}$.

We can now define the *Ma–Trudinger–Wang tensor* (in short *MTW tensor*):

$$\mathfrak{S}_{(x,y)}(\xi,\eta) := \frac{3}{2}\sum_{ijklrs} \left(c_{ij,r}c^{r,s}c_{s,kl} - c_{ij,kl}\right)\xi^i\xi^j\eta^k\eta^l.$$

In terms of this tensor, the regularity condition for the cost functions becomes

$$\mathfrak{S}_{(x,y)}(\xi,\eta) \geq 0 \qquad \text{whenever } (x,y) \in (M\times M)\setminus \mathrm{cut}(M),\; -c_{i,j}\xi^i\eta^j = 0.$$

5.6 Invariance of $\mathfrak{S}$

By the computations of the last paragraph, we have seen that $\mathfrak{S}$ is constructed by the expression in (12). Since that expression involves second derivatives (which are not intrinsic and depend on the choice of the coordinates), it is not a priori clear whether $\mathfrak{S}$ depends or not on the choice of coordinates. On the other hand, we can hope it does not, because of the (intrinsic) geometric interpretation of the regularity.

To see that $\mathfrak{S}$ is indeed independent of any choice of coordinates (so that one does not even need to use geodesic coordinates, as in (9)), we observe that, if we do a change of coordinates and compute first the second derivatives in x, we get some additional terms of the form

$$\Gamma_{ij}^k(x)c_k(x,y) = -\Gamma_{ij}^k(x)p_k(x,y) = \Gamma_{ij}^k(x)g_{k\ell}(x)p^\ell(x,y).$$

But when we differentiate twice with respect to p, this additional term disappears.

5.7 Relation to Curvature

Let $\xi, \eta \in T_xM$ two orthogonal unit vectors, and consider the functions

$$F(t,s) := \frac{d\big(\exp_x(t\xi), \exp_x(s\eta)\big)^2}{2}.$$

As $\frac{\partial}{\partial s}|_{s=0}F(t,s)$ and $\frac{\partial}{\partial t}|_{t=0}F(t,s)$ identically vanish, we have the Taylor expansion

$$F(t,s) \simeq At^2 + Bs^2 + Ct^4 + Dt^2s^2 + Es^4 + \dots$$

Since $F(t,0) = t^2$ and $F(0,s) = s^2$ we deduce $A = B = 1$ and $C = E = 0$. Hence by (3) we recover the identity

$$\mathfrak{S}_{(x,x)}(\xi,\eta) = -\frac{3}{2}\frac{\partial^2}{\partial s^2}\bigg|_{s=0}\frac{\partial^2}{\partial t^2}\bigg|_{t=0} F(t,s) = \mathrm{Sect}_x([\xi,\eta]),$$

first proved by Loeper [21]. This fact shows that the MTW tensor is a non-local version of the sectional curvature. In fact, as shown by Kim and McCann [20], $\mathfrak{S}$ is the sectional curvature of the manifold $M \times M$, endowed with the pseudo-metric $-d^2_{xy}c$. Combining the above identity with Theorems 5.1 and 5.3, we get the following important negative result:

Theorem 5.5. *Let (M,g) be a (compact) Riemannian manifolds, and assume that there exist $x \in M$ and a plane $P \subset T_xM$ such that* $\mathrm{Sect}_x(P) < 0$.

Then there exist C^∞ strictly positive probability densities f and g such that the optimal map is discontinuous.

After this negative result, one could still hope to develop a regularity theory on any manifold with non-negative sectional curvature. But such is not the case: as shown by Kim [19], the regularity condition is strictly stronger than the condition of nonnegativity of sectional curvatures.

5.8 The Ma–Trudinger–Wang Condition

Definition 5.6. We say that (M,g) satisfies the MTW(K) condition if, for all $(x,y) \in (M \times M) \setminus \mathrm{cut}(M)$, for all $\xi \in T_xM$, $\eta \in T_yM$,

$$\mathfrak{S}_{(x,y)}(\xi,\eta) \geq K|\xi|_x^2|\tilde{\eta}|_x^2 \qquad \text{whenever } -c_{i,j}(x,y)\xi^i\eta^j = 0,$$

where $\tilde{\eta}^i = -g^{i,k}(x)c_{k,j}(x,y)\eta^j \in T_xM$.

Some example of manifolds satisfying the Ma–Trudinger–Wang condition are given in [12,20–22]:

- $\mathbb{R}^n$ and $\mathbb{T}^n$ satisfy MTW(0).
- $\mathbb{S}^n$ and its quotients satisfy MTW(1).
- Products of spheres satisfy MTW(0).

We observe that the MTW condition is a nonstandard curvature condition, as it is fourth order and nonlocal. Therefore an important open problem is whether this condition is stable under perturbation. More precisely, we ask for the following:
Question: assume that (M,g) satisfies the MTW(K) condition for $K > 0$, and let g_ε be a C^4-perturbation of g. Does (M, g_ε) satisfy the MTW(K') condition for some $K' > 0$?

The answer to this question is easily seen to be affirmative for manifolds with nonfocal cut-locus like the projective space $\mathbb{RP}^n$ (see [23]). Moreover, as proven by Rifford and the first author [12], the answer is affirmative also for the 2-dimensional sphere $\mathbb{S}^2$.

The next property, called *Convexity of Tangent Injectivity Loci*, or (*CTIL*) in short, is useful to prove regularity and stability results [12,23]:

Definition 5.7. We say that (M,g) satisfies CTIL if, for all $x \in M$, the set

$$\mathrm{TIL}(x) := \{tv \in T_xM \mid 0 \leq t < t_c(x,v)\} \subset T_xM$$

is convex.

As shown by the second author [41], if CTIL is satisfied, then the MTW condition is stable under Gromov–Hausdorff convergence:

Theorem 5.8. *Let* (M_k, g_k) *be a sequence of Riemannian manifolds converging in the Gromov–Hausdorff topology to a Riemannian manifold* (M, g)*. If* (M_k, g_k) *satisfy* MTW(0) *and* CTIL*, then also* (M, g) *satisfies* MTW(0)*.*

The proof of this result uses that, under CTIL, MTW(0) is equivalent to the connectedness of the c-subdifferential of all c-convex functions ψ which solve the dual Kantorovich problem (see Sect. 3.5).

5.9 Local to Global

Under CTIL, one can prove that the MTW(0) condition is equivalent to the regularity condition (8) (see [41]).

Here we want to show that an "improved" MTW condition allows to prove an "improved" regularity condition, which in turns implies (Hölder) continuity of the optimal map.

Definition 5.9. Let $K, C \geq 0$. We say that (M, g) satisfies the MTW(K, C) condition if for all $(x, y) \in (M \times M) \setminus \mathrm{cut}(M)$, for all $\xi \in T_xM$, $\eta \in T_yM$,

$$\mathfrak{S}_{(x,y)}(\xi, \eta) \geq K|\xi|_x^2|\tilde{\eta}|_x^2 - C\big|\langle \xi, \tilde{\eta}\rangle_x\big||\xi|_x|\tilde{\eta}|_x,$$

where $\tilde{\eta}^i = -g^{i,k}(x)c_{k,j}(x, y)\eta^j \in T_xM$.

We observe that the second term appearing in the right hand side vanishes if $-c_{i,j}(x, y)\xi^i\eta^j = 0$. Moreover, by the Cauchy–Schwarz inequality and $\mathfrak{S}_{(x,x)}(\xi, \xi) = 0$, we must have $C \geq K$. Next, if MTW(K, C) holds for some $K > 0$, then the sectional curvatures are bounded from below by K, and so by Bonnet–Myers's Theorem the manifold is compact. We finally remark that, in a subset of $M \times M$ where c is smooth, a compactness argument shows that MTW(K) implies MTW(K, C) for some $C > 0$ [23, Lemma 2.3]. So the refinement from MTW(K) to MTW(K, C) is interesting only when the cost function loses its smoothness, i.e. close to the cut locus.

Example 5.10. As proved in [12], the sphere $\mathbb{S}^n$ and its quotients satisfy MTW(K, K) for some $K > 0$, and C^4-perturbations of $\mathbb{S}^2$ satisfy MTW(K, C) for some $K, C > 0$.

We now show that MTW(K, C) with $K > 0$ implies an "improved" regularity inequality. For simplicity, here we give a simpler version of the lemma, where we assume that (M, g) satisfies CTIL (otherwise one would need to apply an approximation lemma proved by the authors in [13]). (See Fig. 3.)

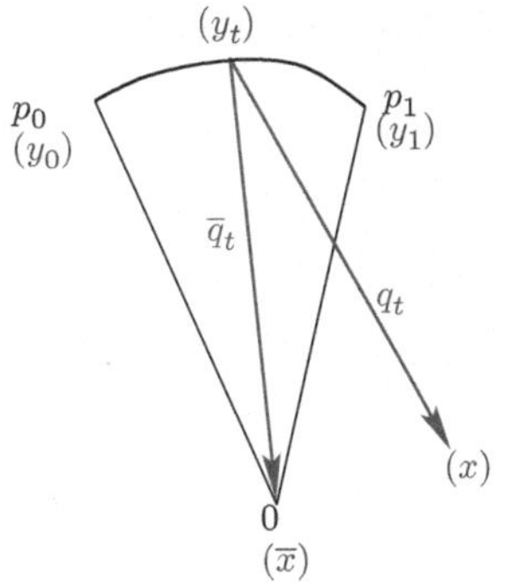

Fig. 3 Lemma 5.11: p_0, p_1 are tangent vectors at $\bar{x}$; q_t, $\bar{q}_t$ are tangent at $y_t = \exp_{\bar{x}} p_t$

Lemma 5.11. *Let (M, g) satisfies* CTIL *and* MTW(K, C) *with $K > 0$. For any $\bar{x} \in M$, let $(p_t)_{0 \le t \le 1}$ be a C^2 curve drawn in* TIL$(\bar{x})$, *and let $y_t = \exp_{\bar{x}}(p_t)$; let further $x \in M$. If*

$$|\ddot{p}_t|_{\bar{x}} \le \varepsilon_0 \, d(\bar{x}, x) |\dot{p}_t|^2_{\bar{x}}, \tag{13}$$

then there exists $\lambda = \lambda(K, C, \varepsilon_0) > 0$ such that, for any $t \in (0, 1)$,

$$d(x, y_t)^2 - d(\bar{x}, y_t)^2 \ge \min\Big(d(x, y_0)^2 - d(\bar{x}, y_0)^2,\ d(x, y_1)^2 - d(\bar{x}, y_1)^2\Big) \\ + 2\lambda t(1-t) d(\bar{x}, x)^2 |p_1 - p_0|^2_{\bar{x}}. \tag{14}$$

This result, first proved by Loeper and the second author [23], and then slightly modified by Rifford and the first author [12], is a refinement of the proof given by Kim and McCann [20] for the implication (4) ⇒ (3) in Theorem 5.3.

Sketch of the proof. Define the function

$$h(t) := -c(x, y_t) + c(\bar{x}, y_t) + \delta t(1-t),$$

with $c = \frac{d^2}{2}$. We want to prove that, for $\delta = \lambda d(\bar{x}, x)^2 |p_1 - p_0|^2_{\bar{x}}$, if λ is small enough then $h(t) \le \max\big(h(0), h(1)\big)$. The idea of the proof is by the maximum principle: if we show that $\ddot{h} > 0$ whenever $\dot{h} = 0$, this will imply the result.

Define $q_t := (\exp_{y_t})^{-1}(x)$, $\bar{q}_t := (\exp_{y_t})^{-1}(\bar{x})$, $\eta := q_t - \bar{q}_t$. Then, since

$$\dot{y}^i_t = c^{i,j} \dot{p}_j,$$

$$\ddot{y}^i_t = -c^{i,k} c_{k,\ell j} c^{\ell,r} c^{j,s} \dot{p}_r \dot{p}_s - c^{i,r} \ddot{p}_r,$$

(everything being evaluated at $(\bar{x}, y_t)$), after some computations one obtains

$$\dot{h}(t) = -c_{i,j}(\bar{x}, y_t)\eta^i \dot{y}_t^j + \delta(1-2t),$$

$$\ddot{h}(t) = -\big([c_{,ij}(x,y_t) - c_{,ij}(\bar{x},y_t)] - \eta^k c_{k,ij}(\bar{x},y_t)\big)\dot{y}_t^i\dot{y}_t^j + c_{i,j}\eta^i c^{j,r}\ddot{p}_r - 2\delta. \quad (15)$$

We now observe that the first term in the right hand side can be written as

$$\Phi(q_t) - \Phi(\bar{q}_t) - d_{\bar{q}_t}\Phi \cdot (q_t - \bar{q}_t),$$

with $\Phi(q) := c_{,ij}\big(\exp_{y_t}(q), y_t\big)\dot{y}_t^i\dot{y}_t^j$; therefore it is equal to

$$\int_0^1 \frac{d^2}{ds^2}\Phi(sq_t + (1-s)\bar{q}_t)\,ds = -\frac{2}{3}\int_0^1 \mathfrak{S}_{(y_t,x_s)}(\dot{y}_t, \eta)\,ds$$

with $x_s := \exp_{y_t}\big(sq_t + (1-s)\bar{q}_t\big)$, and we get

$$\begin{aligned}\ddot{h}(t) &= \frac{2}{3}\int_0^1 \mathfrak{S}_{(y_t,x_s)}(\dot{y}_t,\eta)\,ds + c_{i,j}\eta^i c^{j,r}\ddot{p}_r - 2\delta \\ &\geq \frac{2}{3}K|\tilde{\eta}|_{y_t}^2|\dot{y}_t|_{y_t}^2 - \frac{2}{3}C\big|\langle\tilde{\eta},\dot{y}_t\rangle_{y_t}\big||\dot{y}_t|_{y_t}|\tilde{\eta}|_{y_t} + c_{i,j}(\bar{x},y_t)\eta^i c^{j,r}\ddot{p}_r - 2\delta,\end{aligned}$$

where $\tilde{\eta} := (d_{p_t}\exp_{y_t})^{-1}(\eta) = (d_{p_t}\exp_{y_t})^{-1}(q_t - \bar{q}_t)$. To understand now why the result is true, we remark that for $\delta = 0$ the condition $\dot{h} = 0$ means $\langle\tilde{\eta},\dot{y}_t\rangle_{y_t} = -c_{i,j}(\bar{x},y_t)\eta^i\dot{y}_t^j = 0$, which gives

$$\ddot{h}(t) \geq \frac{2}{3}K|\tilde{\eta}|_{y_t}^2|\dot{y}_t|_{y_t}^2 + c_{i,j}\eta^i c^{j,r}\ddot{p}_r,$$

and thanks to the assumption of smallness on $|\ddot{p}|_{\bar{x}}$ one gets $\ddot{h} > 0$.

In the general case $\delta > 0$ small, the condition $\dot{h} = 0$ gives $|c_{i,j}(\bar{x},y_t)\eta^i\dot{y}_t^j| \leq \delta$, and using that $|\tilde{\eta}|_{\bar{x}} \geq |\eta|_{y_t} \geq d(x,\bar{x})$ (since the sectional curvature of the manifold is non-negative) one obtains the desired result. □

Remark 5.12. This local-to-global argument can also be used to give a differential characterization of c-convex functions (see Proposition 3.5). More precisely one has: assume that (M,g) satisfies MTW(0) and CTIL. Then $\psi \in C^2(M)$ is c-convex if and only if

$$\nabla^2\psi(x) + \frac{\nabla_x^2 d\big(x, \exp_x(\nabla\psi(x))\big)^2}{2} \geq 0 \qquad \forall x \in M.$$

5.10 A Smoothness Result

Theorem 5.13. *Let (M, g) be a (compact) Riemannian manifold satisfying* MTW(K, C) *with $K > 0$. Assume moreover that all* TIL(x) *are uniformly convex, and let f and g be two probability densities on M such that $f \leq A$ and $g \geq a$ for some $A, a > 0$. Then the optimal map between $\mu = f \,\mathrm{vol}$ and $\nu = g \,\mathrm{vol}$, with cost function $c = d^2/2$, is continuous.*

As shown by Loeper [21], this theorem can be refined into a Hölder regularity for the transport map. The argument of the proof, originally due to Loeper, has been simplified first by Kim and McCann [20], and then by Loeper and the second author [23]. The argument we present is borrowed from [23].

Proof. By Theorem 3.2, we know that the optimal map T can be written as $\exp_x(\nabla\psi(x))$, so it suffices to prove that ψ is C^1. Since ψ is semiconvex, we need to show that the subgradient $\nabla^-\psi(x)$ is a singleton for all $x \in M$. The proof is by contradiction.

Assume that there is $\bar{x} \in M$ and $p_0, p_1 \in \nabla^-\psi(\bar{x})$. Let $y_0 = \exp_{\bar{x}} p_0$, $y_1 = \exp_{\bar{x}} p_1$. Since the cost is regular, we have $y_i \in \partial^c\psi(\bar{x})$ for $i = 0, 1$, that is

$$\psi(\bar{x}) + c(\bar{x}, y_i) = \min_{x\in M}\Big[\psi(x) + c(x, y_i)\Big], \qquad i = 0, 1.$$

In particular

$$c(x, y_i) - c(\bar{x}, y_i) \geq \psi(\bar{x}) - \psi(x), \qquad i = 0, 1. \tag{16}$$

For $\varepsilon \in (0, 1)$, we define $D_\varepsilon \subset \overline{\mathrm{TIL}}(\bar{x})$ as follows: D_ε consists of the set of points $p \in T^*_{\bar{x}}M$ such that there exists a path $(p_t)_{0\leq t\leq 1} \subset \overline{\mathrm{TIL}}(\bar{x})$ from p_0 to p_1 such that, if we define $y_t = \pi_1 \circ \phi_1^H(\bar{x}, p_t)$, we have $\ddot{p}_t = 0$ for $t \notin [1/4, 3/4]$, $|\ddot{p}_t|_{y_t} \leq \varepsilon\eta_0|\dot{y}_t|^2_{y_t}$ for $t \in [1/4, 3/4]$, and $p = p_t$ for some $t \in [1/4, 3/4]$ (this is like a "sausage", see Fig. 4).

By uniform convexity of $\overline{\mathrm{TIL}}(\bar{x})$, if η_0 is sufficiently small then D_ε lies a positive distance σ away from the tangent cut locus $\mathrm{TCL}(\bar{x}) = \partial\big(\mathrm{TIL}(\bar{x})\big)$, with $\sigma \sim |p_0 - p_1|^2_{\bar{x}}$. Thus all paths $(p_t)_{0\leq t\leq 1}$ used in the definition of D_ε satisfy

$$|\dot{y}_t|_{y_t} \geq c|p_0 - p_1|_{\bar{x}} \qquad \forall t \in [1/4, 3/4].$$

Moreover condition (13) is satisfied if $\eta_0 \leq \varepsilon_0$ and $d(\bar{x}, x) \geq \varepsilon$. By simple geometric consideration, we see that D_ε contains a parallelepiped E_ε centered at $(p_0+p_1)/2$ with one side of length $\sim |p_0 - p_1|_{\bar{x}}$, and the other sides of length $\sim \varepsilon|p_0 - p_1|^2_{\bar{x}}$, such that all points y in this parallelepiped can be written as

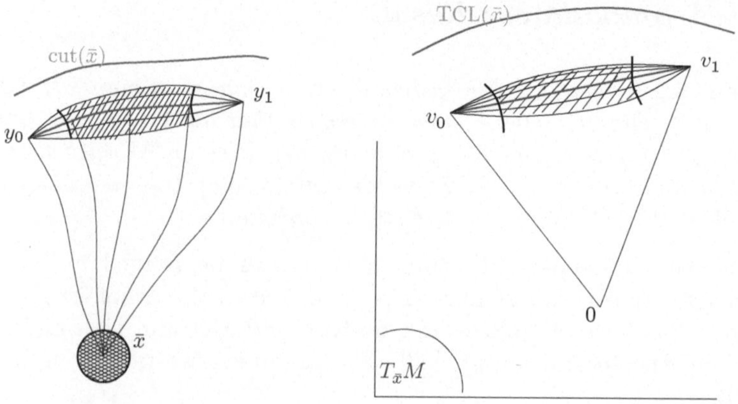

Fig. 4 Proof of Theorem 5.13: the volume of the ball $B(\bar{x}, \varepsilon)$ is much smaller than the volume of the "sausage" with base $\bar{x}$, endpoints y_0, y_1 and width $O(\varepsilon)$; TCL$(\bar{x})$ is the tangent cut locus at $\bar{x}$

y_t for some $t \in [1/3, 2/3]$, with y_t as in the definition of D_ε. Therefore

$$\mathscr{L}^n(E_\varepsilon) \geq c(M, \eta_0, |p_0 - p_1|_{\bar{x}})\varepsilon^{n-1},$$

with $\mathscr{L}^n$ denoting the Lebesgue measure on $T_{\bar{x}}M$. Since E_ε lies a positive distance from $\partial\big(\mathrm{TIL}(\bar{x})\big)$, we obtain

$$\mathrm{vol}(Y_\varepsilon) \sim \mathscr{L}^n(E_\varepsilon) \geq c(M, \eta_0, |p_0 - p_1|_{\bar{x}})\varepsilon^{n-1}, \qquad Y_\varepsilon := \exp_{\bar{x}}(E_\varepsilon).$$

We wish to apply Theorem 5.13 to the paths $(p_t)_{0\leq t\leq 1}$ used in the definition of D_ε. Since p_0, p_1 belong to $\overline{\mathrm{TIL}}(\bar{x})$ but not necessarily to $\mathrm{TIL}(\bar{x})$, we first apply the theorem with $(\theta p_t)_{0\leq t\leq 1}$ with $\theta < 1$, and then we let $\theta \to 1$; in the end, for any $y \in Y_\varepsilon$ and $x \in M \setminus B_\varepsilon(\bar{x})$,

$$d(x,y)^2 - d(\bar{x},y)^2 \geq \min\Big(d(x,y_0)^2 - d(\bar{x},y_0)^2,\ d(x,y_1)^2 - d(\bar{x},y_1)^2\Big) + \lambda\varepsilon^2|p_0 - p_1|^2_{\bar{x}},$$

for some $\lambda > 0$. Combining this inequality with (16), we conclude that

$$\text{for any } y \in Y_\varepsilon, \qquad y \notin \partial^c\psi(x) \qquad \forall x \in M \setminus B_\varepsilon(\bar{x}).$$

This implies that all the mass brought into Y_ε by the optimal map comes from $B_\varepsilon(\bar{x})$, and so

$$\mu(B_\varepsilon(\bar{x})) \geq \nu(Y_\varepsilon).$$

Since $\mu(B_\varepsilon(\bar{x})) \leq A\mathrm{vol}(B_\varepsilon(\bar{x})) \sim \varepsilon^n$ and $\nu(Y_\varepsilon) \geq a\mathrm{vol}(Y_\varepsilon) \gtrsim \varepsilon^{n-1}$, we obtain a contradiction as $\varepsilon \to 0$. $\square$

6 Recap and Perspectives

In these notes, we have seen two different connections of optimal transport to curvature:

1. *Ricci curvature:* the optimal transport is a way to give a synthetic formulation of lower Ricci curvature bounds.
2. *Sectional curvature:* a regularity theory for optimal transport on a manifold depends on the MTW condition, which reinforces non-negative sectional curvature.

We remark that, in both cases, the optimal transport goes well with lower bounds only.

A good thing is that in both cases there is a "soft" reformulation in terms of optimal transport:

$$\mathrm{Ric} \geq 0 \quad \Longleftrightarrow \quad t \mapsto H(\mu_t) \text{ is convex, } \forall\, (\mu_t)_{0 \leq t \leq 1} \text{ geodesic in } P_2(M);$$

$\mathfrak{S} \geq 0 \Longleftrightarrow \partial^c\psi(x)$ is connected, $\forall\psi$ solution of the dual Kantorovich problem. Observe that these reformulations have the advantage of being very stable, and at the same time can be used to generalize some differential concepts out of the Riemannian setting.

6.1 The Curvature-Dimension Condition

As we already said, while the sectional curvature gives a pointwise control on distances, the Ricci curvature gives an averaged control, and is related to Jacobian estimates with respect to a reference measure (which in Sect. 4 was always the volume measure). For this reason, a natural general setting where one can study Ricci bounds is the one of measured metric spaces (X, d, ν) (see Sect. 2.15).

As an example, consider the measured metric spaces $(M^n, g, e^{-V}\mathrm{vol})$, with $V \in C^2(M)$. Modify the classical Ricci tensor into $\mathrm{Ric}_{N,\nu} := \mathrm{Ric} + \nabla^2 V - \frac{\nabla V \otimes \nabla V}{N-n}$ for $N \geq n$ (where by convention $\frac{0}{0} = 0$, and N plays the role of an "effective" dimension). Then the *curvature-dimension* condition $\mathrm{CD}(K, N)$, classically used in probability theory and geometry, consists in $\mathrm{Ric}_{N,\nu} \geq K$. Exactly as $\mathrm{Ric} \geq K$, also this more general condition can be reformulated in terms of optimal transport. Up to minor variations, the following definition was introduced independently by Sturm [35, 36] and by Lott and the second author [25, 26] (recall the definition of H from Sect. 4.4):

Definition 6.1. A compact measured metric space (X, d, ν) is said to satisfy $\mathrm{CD}(K, \infty)$ if, for all $\mu_0, \mu_1 \in P_2(X)$, there exists a Wasserstein geodesic

$(\mu_t)_{0\le t\le 1}$ between μ_0 and μ_1 such that

$$H_\nu(\mu_t) \le (1-t)H_\nu(\mu_0) + tH_\nu(\mu_1) - K\frac{t(1-t)}{2}W_2(\mu_0,\mu_1)^2 \qquad \forall t \in [0,1].$$

A similar definition for $\mathrm{CD}(0,N)$ is obtained by choosing $K = 0$ and replacing the nonlinearity $r\log r$ by $-r^{1-1/N}$. There is also a more complicated definition which works for the general $\mathrm{CD}(K,N)$ criterion [36, 40].

Example 6.2. $(\mathbb{R}^n, |\cdot|, e^{-|x|^2/2}dx)$ satisfies $\mathrm{CD}(1,\infty)$; $(\mathbb{R}^n, \|\cdot\|, dx)$ satisfies $\mathrm{CD}(0,N)$ for any norm $\|\cdot\|$.

Ohta [30] recently performed some exploration of Finsler geometry along these lines.

6.2 Open Problem: Locality

A natural question is whether the above definition of $\mathrm{Ric}_{\infty,\nu} \ge K$ is local or not. As shown in [35, 36] and [40, Chap. 30], this question has an affirmative answer if geodesics are non-branching, but it is open in the general case. The answer would however be affirmative if the following (elementary but tricky) conjecture from [40] were true:

Conjecture 6.3. Let $0 < \theta < 1$, $0 \le \alpha \le \pi$, and assume that $f : [0,1] \to \mathbb{R}^+$ satisfies

$$f\big((1-\lambda)t+\lambda t'\big) \ge (1-\lambda)\left(\frac{\sin\big((1-\lambda)\alpha|t-t'|\big)}{(1-\lambda)\sin\big(\alpha|t-t'|\big)}\right)^\theta f(t) + \lambda\left(\frac{\sin\big(\lambda\alpha|t-t'|\big)}{\lambda\sin\big(\alpha|t-t'|\big)}\right)^\theta f(t')$$

for $|t-t'|$ small. Then the above inequality holds true for all $t, t' \in [0,1]$.

6.3 Ricci and Diffusion Equations

The non-negativity of the Ricci curvature is important to get contraction properties of solutions of the heat equation on a manifold. Indeed the following result holds (see [32–34]):

Theorem 6.4. *(M,g) satisfies $\mathrm{Ric} \ge 0$ if and only if, for all μ_t and $\tilde{\mu}_t$ solutions of the heat equation, $W_2(\mu_t,\tilde{\mu}_t)$ is non-increasing in time.*

Another way to state this theorem is to say that $\mathrm{Ric} \ge 0$ if and only if $(e^{t\Delta})_{t\ge 0}$ is a contraction in $P_2(M)$.

Recalling that Ric ≥ 0 is equivalent to the convexity of the entropy functional $H(\mu_t)$ (see Sect. 4.4), the above equivalence if formally explained by the Jordan–Kinderlehrer–Otto theorem [15]: *the heat flow is the gradient flow of H in $P_2(M)$.* A generalization of the above result has been given McCann and Topping [28] (see also [24]):

Theorem 6.5. *A family of Riemannian metric $g(t)$ are super-solutions of the backward Ricci-flow $\frac{\partial g_t}{\partial t} \leq 2\mathrm{Ric}_{g_t}$ if and only if, for all μ_t and $\tilde{\mu}_t$ solutions of the heat equation in (M, g_t) (i.e. $\partial_t \mu_t = \Delta_{g_t}\mu_t$), $W_2(\mu_t, \tilde{\mu}_t)$ is non-increasing in time.*

Let us also notice that Lott [24] and Topping [37] have recently studied properties of the Ricci flow with the help of the optimal transport, and they can for instance recover Perelman's monotonicity formula.

6.4 Discretization

A natural question in probability theory and statistical mechanics is how to define a notion of curvature on discrete spaces. The optimal transport allows to answer to this question: the idea is to "discretize the synthetic formulation":

- Replace length space by δ-length space, etc.
- Allow for errors, either in the heat formulation (Markov kernels, etc.) as done by Ollivier [31] and Joulin [16], or in the optimal transport formulation as done by Bonciocat and Sturm [2].

Example 6.6. Consider the metric space $\{0,1\}^N$ endowed with the Hamming metric (i.e. each edge is of length 1). Then Ric $\gtrsim \frac{1}{N}$ at scale $O(1)$ (see [31, Example 8]).

6.5 Smoothness

Regarding the smoothness of the optimal transport map, two main questions arise:

- Find further examples of manifolds satisfying the MTW conditions. (Recall $\mathbb{S}^n$ and its quotients, $\mathbb{RP}^n$ and its perturbations, perturbations of $\mathbb{S}^2$, products of spheres.)
- Find results yielding regularity in terms of the MTW condition, the shape of the cut locus, assumptions on μ and ν...

As an example, the following theorem was proven by Loeper and the second author [23]:

Theorem 6.7. *Let (M, g) be a compact Riemannian manifold such that there is no focalization at the cut locus (i.e., $d_{t_c(x,v)v}\exp_x$ is invertible for* *all x, v). Assume that (M, g) satisfies* MTW(K) *for some $K > 0$, and let f and g be two probability densities on M such that $f \leq A$ and $g \geq a$ for some $A, a > 0$. Then the optimal map T is $C^{0,\alpha}$, with $\alpha = \frac{1}{4n-1}$.*

Examples of manifold satisfying the assumptions of the above theorem are the projective space $\mathbb{RP}^n$ and its perturbations, and a challenge is to understand what happens when one removes the "no-focalization" assumption.

The following conjecture was formulated in [23]:

Conjecture 6.8. MTW implies CTIL.

This conjecture has been proved by Loeper and the second author [23] under MTW(K) with $K > 0$, and the no-focalization assumption. Using a variant of the MTW condition, Rifford and the first author proved the following result [12]:

Theorem 6.9. *If (M, g) is a C^4-perturbation of $\mathbb{S}^2$, then* CTIL *holds.*

Through these considerations we see that the MTW condition, originally introduced as a way to explore the regularity of optimal transport, turns out to give a new kind of geometric information on the manifold (compare with the Bonnet–Myers theorem, or with classical theorems on the rectifiability or local description of the cut locus).

7 Selected References

This section provides a list of references which seem to us the most significant. To this list should be added the book [40], which is attempts at providing a synthetic overview of all the links between geometry and optimal transport.

7.1 Links Between Optimal Transport and Ricci Curvature

The reference [33] by Otto–Villani may be considered as the founding paper for this topic. A body of technical tools was developed by Cordero-Erausquin–McCann–Schmuckenschläger [9] to make progress on these issues, and solve open problems stated in [33]. At that time the emphasis was rather on applications from calculus of variations and functional inequalities.

Later the interest of these links for geometric applications was realized, and explicitly noted by von Renesse–Sturm [34]. The synthetic theory of Ricci curvature bounds in the general setting of metric-measure spaces was developed independently by Sturm [35, 36] and by Lott–Villani [25].

7.2 *Optimal Transport and Ricci Flow*

The links between these objects were suspected for some time, and hinted for by a preliminary result by McCann–Topping [28]. Finally this link was made precise in contributions of Lott [24] and Topping [37].

7.3 *Discrete Ricci Curvature*

This theory is in construction with preliminary works by Joulin and Ollivier; one may consult in particular [31] by the latter.

7.4 *Smoothness of Optimal Transport, Cut Locus and MTW Tensor*

After the landmark papers by Ma–Trudinger–Wang [27] and Loeper [21], the theory was partly simplified and rewritten by Kim–McCann [20]. Applications to the geometry of the cut locus were first investigated by Loeper–Villani [23], and further developed by Figalli–Rifford [12].

References

1. L. Ambrosio, P. Tilli, Topics on analysis in metric spaces. *Oxford Lecture Series in Mathematics and its Applications*, vol. 25 (Oxford University Press, Oxford, 2004)
2. A.-I. Bonciocat, K.-T. Sturm, Mass transportation and rough curvature bounds for discrete spaces. J. Funct. Anal. 256(9), 2944–2966 (2009)
3. Y. Brenier, Polar factorization and monotone rearrangement of vector-valued functions. Comm. Pure Appl. Math. 44(4), 375–417 (1991)
4. D. Burago, Y. Burago, S. Ivanov, A course in metric geometry. *Graduate Studies in Mathematics*, vol. 33 (American Mathematical Society, Providence, RI, 2001)
5. L.A. Caffarelli, The regularity of mappings with a convex potential. J. Am. Math. Soc. 5(1), 99–104 (1992)
6. L.A. Caffarelli, Boundary regularity of maps with convex potentials. Comm. Pure Appl. Math. 45(9), 1141–1151 (1992)
7. L.A. Caffarelli, Boundary regularity of maps with convex potentials. II. Ann. Math. 2 144(3), 453–496 (1996)
8. D. Cordero-Erausquin, Sur le transport de mesures périodiques. C. R. Acad. Sci. Paris Sèr. I Math. 329(3), 199–202 (1999)
9. D. Cordero-Erasquin, R.J. McCann, M. Schmuckenschlager, A Riemannian interpolation inequality à la Borell, Brascamp and Lieb. Invent. Math. 146(2), 219–257 (2001)
10. P. Delanoë, Classical solvability in dimension two of the second boundary-value problem associated with the Monge–Ampère operator. Ann. Inst. H. Poincaré Anal. Non Linéaire 8(5), 443–457 (1991)

11. A. Fathi, A. Figalli, Optimal transportation on non-compact manifolds. Israel J. Math. 175(1), 1–59 (2010)
12. A. Figalli, L. Rifford, Continuity of optimal transport maps on small deformations of $\mathbb{S}^2$. Comm. Pure Appl. Math. 62(12), 1670–1706 (2009)
13. A. Figalli, C. Villani, An approximation lemma about the cut locus, with applications in optimal transport theory. Methods Appl. Anal. 15(2), 149–154 (2008)
14. W. Gangbo, R.J. McCann, The geometry of optimal transportation. Acta Math. 177(2), 113–161 (1996)
15. R. Jordan, D. Kinderlehrer, F. Otto, The variational formulation of the Fokker-Planck equation. SIAM J. Math. Anal. 29(1), 1–17 (1998)
16. A. Joulin, A new Poisson-type deviation inequality for Markov jump process with positive Wasserstein curvature. Bernoulli 15(2), 532–549 (2009)
17. L.V. Kantorovich, On mass transportation. C. R. (Doklady) Acad. Sci. URSS (N.S.) (Reprinted) 37, 7-8 (1942)
18. L.V. Kantorovich, On a problem of Monge. C. R. (Doklady) Acad. Sci. URSS (N.S.) (Reprinted) 3, 2 (1948)
19. Y.H. Kim, Counterexamples to continuity of optimal transportation on positively curved Riemannian manifolds. Int. Math. Res. Not. IMRN, Art. ID rnn120, p.15 (2008)
20. Y.H. Kim, R.J. McCann, Continuity, curvature, and the general covariance of optimal transportation. J. Eur. Math. Soc. (JEMS) 12, 1009–1040 (2010)
21. G. Loeper, On the regularity of solutions of optimal transportation problems. Acta Math. 202(2), 241–283 (2009)
22. G. Loeper, Regularity of optimal maps on the sphere: The quadratic cost and the reflector antenna. Arch. Ration. Mech. Anal. (to appear)
23. G. Loeper, C. Villani, Regularity of optimal transport in curved geometry: the nonfocal case. Duke Matk. J. 151(3), 431–485 (2010)
24. J. Lott, Optimal transport and Perelman's reduced volume. Calc. Var. Part. Differ. Equat. 36(1), 49–84 (2009)
25. J. Lott, C. Villani, Ricci curvature via optimal transport. Ann. Math. 169, 903–991 (2009)
26. J. Lott, C. Villani, Weak curvature conditions and functional inequalities. J. Funct. Anal. 245(1), 311–333 (2007)
27. X.N. Ma, N.S. Trudinger, X.J. Wang, Regularity of potential functions of the optimal transportation problem. Arch. Ration. Mech. Anal. 177(2), 151–183 (2005)
28. R.J. McCann, P. Topping, Ricci flow, entropy and optimal transportation. Amer. J. Math. 132, 711–730 (2010)
29. R.J. McCann, Polar factorization of maps on Riemannian manifolds. Geom. Funct. Anal. 11(3), 589–608 (2001)
30. S.-I. Ohta, Finsler interpolation inequalities. Calc. Var. Part. Differ. Equat. 36, 211–249 (2009)
31. Y. Ollivier, Ricci curvature of Markov chains on metric spaces. J. Funct. Anal. 256(3), 810–864 (2009)
32. F. Otto, The geometry of dissipative evolution equations: the porous medium equation. Comm. Part. Differ. Equat. 26(1-2), 101–174 (2001)
33. F. Otto, C. Villani, Generalization of an inequality by Talagrand and links with the logarithmic Sobolev inequality. J. Funct. Anal. 173(2), 361–400 (2000)
34. M.-K. von Renesse, K.-T. Sturm, Transport inequalities, gradient estimates, entropy, and Ricci curvature. Comm. Pure Appl. Math. 58(7), 923–940 (2005)
35. K.-T. Sturm, On the geometry of metric measure spaces. I. Acta Math. 196(1), 65–131 (2006)
36. K.-T. Sturm, On the geometry of metric measure spaces. II. Acta Math. 196(1), 133–177 (2006)
37. P. Topping, L-optimal transportation for Ricci flow. J. Reine Angew. Math. 636, 93–122 (2009)

38. J.I.E. Urbas, Regularity of generalized solutions of Monge–Ampére equations. Math. Z. 197(3), 365–393 (1988)
39. J. I. E. Urbas: On the second boundary value problem for equations of Monge–Ampère type. J. Reine Angew. Math. 487, 115–124 (1997)
40. C. Villani, Optimal transport, old and new. Notes for the 2005 Saint-Flour summer school. To appear in *Grundlehren der mathematischen Wissenschaften.* Preliminary version available at www.umpa.ens-lyon.fr/~cvillani
41. C. Villani, Stability of a 4th-order curvature condition arising in optimal transport theory. J. Funct. Anal. 255(9), 2683–2708 (2008)

List of Participants

1. Al-hassem Nayam
   University of Pisa, Italy
   alhassem_n@yahoo.fr
2. Ancona Fabio
   University of Bologna, Italy
   ancona@ciram.unibo.it
3. Barbagallo Annamaria
   University of Catania, Italy
   barbagallo@dmi.unict.it
4. Bianchini Stefano
   S.I.S.S.A., Italy
   bianchin@sissa.it
5. Bigolin Francesco
   University of Trento, Italy
   bigolin@science.unitn.it
6. Bonforte Matteo
   Universidad Autonoma de Madrid, Spain
   matteo.bonforte
7. Brancolini Alessio
   University of Bari, Politecnico, Italy
   a.brancolini@poliba.it
8. Calvez Vincent
   Ecole Normale Supérieure, Paris, France
   vincent.calvez@ens.fr
9. Caravenna Laura
   S.I.S.S.A.-I.S.A.S., Italy
   l.caravenna@sissa.it
10. Carlen Eric
    Rutgers University, USA
    carlen@math.gatech.edu

S. Bianchini et al., *Nonlinear PDE's and Applications*, Lecture Notes in Mathematics 2028, DOI 10.1007/978-3-642-21861-3,

11. Carvalho Maria Conceicao
CMAF, University of Lisbon, Portugal
mcarvalh@cii.fc.ul.pt
12. Cavalletti Fabio
S.I.S.S.A., Italy
cavallet@sissa.it
13. Ciomaga Adina
CMLA, Ecole Normale Supérieure, Cachan, France
ciomaga@cmla.ens-cachan.fr
14. Coclite Giuseppe
University of Bari, Italy
coclitegm@dm.uniba.it
15. Coglitore Federico
Albert-Ludwigs Univesity, Freiburg im Breisgau, Germany
federico.coglitore@math.uni-freiburg.de
16. Crasta Graziano
University of Rome I, Italy
crasta@mat.uniroma1.it
17. Daneri Sara
S.I.S.S.A, Italy
daneri@sissa.it
18. Dolera Emanuele
University of Pavia, Italy
emanuele.dolera@unipv.it
19. Donadello Carlotta
S.I.S.S.A-I.S.A.S, Italy
donadel@sissa.it
20. Esposito Pierpaolo
University of Rome III, Italy
esposito@mat.uniroma3.it
21. Fedeli Livio,
S.I.S.S.A, Italy
livio_fedeli@hotmail.com
22. Figalli Alessio
CNRS University of Nice, France
figalli@unice.fr
23. Fonte Massimo
RICAM - Austrian Academy of Sciences, Austria
massimo.fonte@ricam.oeaw.ac.at
24. Franek Marzena
Westfälische Wilhelms University Münster, Germany
marzena.franek@uni-muenster.de
25. Gallouet Thomas
Ecole Normale Supérieure, Lyon, France
thomas.gallouet@unpu.ens-lyon.fr

26. Gelli Maria Stella
University of Pisa, Italy
gelli@dm.unipi.it
27. Ghiraldin Francesco
Scuola Normale Superiore, Pisa, Italy
f.ghiraldin@sns.it
28. Gigli Nicola
Scuola Normale Superiore, Pisa, Italy
nicolagigli@googlemail.com
29. Gloyer Matteo
S.I.S.S.A, Italy
mgloyer@gmail.com
30. Grillo Gabriele
University of Torino, Politecnico, Italy
gabriele.grillo@polito.it
31. Helmers Michael
University of Oxford, United Kingdom
helmers@maths.ox.ac.uk
32. Iagar Razvan Gabriel
Universidad Autónoma de Madrid, Spain
razvan.iagar@uam.es
33. Kurtzmann Aline
Mathematical Institute, Oxford, United Kingdom
kurtzmann@maths.ox.ac.uk
34. Langer Andreas
RICAM, Austria
andreas.langer@oeaw.ac.at
35. Lazzaroni Giuliano
S.I.S.S.A, Italy
giuliano.lazzaroni@gmail.com
36. Lisini Stefano
University of Piemonte Orientale, Italy
stefano.lisini@unipv.it
37. Loreti Paola
University of Rome, La Sapienza, Italy
loreti@dmmm.uniroma1.it
38. Lussardi Luca
University of Brescia (Italy), Italy
luca.lussardi@ing.unibs.it
39. Maas Jan
TU Delft, Netherlands
j.maas@tudelft.nl
40. Mainini Edoardo
Scuola Normale Superiore, Pisa, Italy
edoardo.mainini@sns.it

41. Malusa Annalisa
University of Rome, La Sapienza, Italy
malusa@mat.uniroma1.it
42. Marcati Pierangelo
University of L'Aquila, Italia
marcati@univaq.it
43. Marcellini Francesca
University of Milano Bicocca, Italy
f.marcellini@campus.unimib.it
44. Marigonda Antonio
University of Pavia, Italy
amarigo@math.unipd.it
45. Marson Andrea
University of Padova, Italy
marson@math.unipd.it
46. Mascolo Elvira
Università of Firenze, Italy
mascolo@math.unifi.it
47. Matthes Daniel
Universita of Pavia, Italy
matthes@mathematik.uni-mainz.de
48. Mielke Alexander
Weierstrass Inst. Applied Analysis and Stochastics, Berlin, Germany
mielke@wias-berlin.de
49. Monti Francesca
University of Milano-Bicocca, Italy
f.monti3@campus.unimib.it
50. Morandotti Marco
S.I.S.S.A-I.S.A.S., Italy
marco.morandotti@sissa.it
51. Muntean Adrian
TU Eindhoven, Netherlands
a.muntean@tue.nl
52. Natile Luca
University of Pavia, Italy
luca.natile@unipv.it
53. Nguyen Khai
University of Padova, Italy
khai@math.unipd.it
54. Otto Felix
University of Bonn, Germany
otto@iam.uni-bonn.de
55. Ouyang Shun-Xiang
Bielefeld University, Germany
ouyangshx@gmail.com

56. Paronetto Fabio
University of Salento, Italy
fabio.paronetto@unile.lit
57. Pass Brendan
University of Toronto, Canada
brendan.pass@utoronto.ca
58. Petrov Adrien
Weierstrass Inst. Applied Analysis and Stochastics, Berlin, Germany
petrov@wias-berlin.de
59. Pezzotti Federica
University of L'Aquila, Italy
federica.pezzotti@univaq.it
60. Pietra Paola
IMATI, CNR, Pavia, Italia
pietra@imati.cnr.it
61. Pisante Adriano
University of Rome, La Sapienza, Italy
pisante@mat.uniroma1.it
62. Pisante Giovanni
Second University of Naples, Italy
giovanni.pisante@unina2.it
63. Priuli Fabio
N.T.N.U, Norway
priuli@math.ntnu.no
64. Pulvirenti Mario
University of Rome, La Sapienza, Italy
pulvirenti@mat.uniroma1.it
65. Recupero Vincenzo
University of Torino, Politecnico, Italy
vincenzo.recupero@polito.it
66. Rossi Riccarda
University of Brescia, Italy
riccarda.rossi@ing.unibs.it
67. Savaré Giuseppe
University of Pavia, Italy
giuseppe.savare@unipv.it
68. Schö nlieb Carola-Bibiane
University of Cambridge, United Kingdom
CBS31@cam.ac.uk
69. Shao Jinghai
Beijing Normal University and University of Bonn, Germany
shaojh@bnu.edu.cn
70. Solombrino Francesco
S.I.S.S.A., Italy
fsolombr@sissa.it

71. Spinolo Laura Valentina
Northwestern University, USA
spinolo@math.northwestern.edu
72. Stojkovic Igor
Leiden University, Netherlands
stojkov@math.leidenuniv.nl
73. Tonon Daniela
S.I.S.S.A., Italy
tonon@sissa.it
74. Tubaro Luciano
University of Trento, Italy
tubaro@science.unitn.it
75. Ulusoy Suleyman
University of Oslo, Norway
suleyman.ulusoy@cma.uio.no
76. van Gennip Yves
Technische Universiteit Eindhoven, Netherlands
y.v.gennip@tue.nl
77. Veneroni Marco
Technische Universiteit Eindhoven, Netherlands
m.veneroni@tue.nl
78. Villani Cedric
Ecole Normale Supérieure, Lyon, France
cvillani@umpa.ens-lyon.fr
79. Volzone Bruno
University of Naples, Italy
bruno.volzone@uniparthenope.it

LECTURE NOTES IN MATHEMATICS

Springer

Edited by J.-M. Morel, B. Teissier; P.K. Maini

Editorial Policy (for Multi-Author Publications: Summer Schools / Intensive Courses)

1. Lecture Notes aim to report new developments in all areas of mathematics and their applications - quickly, informally and at a high level. Mathematical texts analysing new developments in modelling and numerical simulation are welcome. Manuscripts should be reasonably selfcontained and rounded off. Thus they may, and often will, present not only results of the author but also related work by other people. They should provide sufficient motivation, examples and applications. There should also be an introduction making the text comprehensible to a wider audience. This clearly distinguishes Lecture Notes from journal articles or technical reports which normally are very concise. Articles intended for a journal but too long to be accepted by most journals, usually do not have this "lecture notes" character.

2. In general SUMMER SCHOOLS and other similar INTENSIVE COURSES are held to present mathematical topics that are close to the frontiers of recent research to an audience at the beginning or intermediate graduate level, who may want to continue with this area of work, for a thesis or later. This makes demands on the didactic aspects of the presentation. Because the subjects of such schools are advanced, there often exists no textbook, and so ideally, the publication resulting from such a school could be a first approximation to such a textbook. Usually several authors are involved in the writing, so it is not always simple to obtain a unified approach to the presentation.
   For prospective publication in LNM, the resulting manuscript should not be just a collection of course notes, each of which has been developed by an individual author with little or no coordination with the others, and with little or no common concept. The subject matter should dictate the structure of the book, and the authorship of each part or chapter should take secondary importance. Of course the choice of authors is crucial to the quality of the material at the school and in the book, and the intention here is not to belittle their impact, but simply to say that the book should be planned to be written by these authors jointly, and not just assembled as a result of what these authors happen to submit.
   This represents considerable preparatory work (as it is imperative to ensure that the authors know these criteria before they invest work on a manuscript), and also considerable editing work afterwards, to get the book into final shape. Still it is the form that holds the most promise of a successful book that will be used by its intended audience, rather than yet another volume of proceedings for the library shelf.

3. Manuscripts should be submitted either online at www.editorialmanager.com/lnm/ to Springer's mathematics editorial, or to one of the series editors. Volume editors are expected to arrange for the refereeing, to the usual scientific standards, of the individual contributions. If the resulting reports can be forwarded to us (series editors or Springer) this is very helpful. If no reports are forwarded or if other questions remain unclear in respect of homogeneity etc, the series editors may wish to consult external referees for an overall evaluation of the volume. A final decision to publish can be made only on the basis of the complete manuscript; however a preliminary decision can be based on a pre-final or incomplete manuscript. The strict minimum amount of material that will be considered should include a detailed outline describing the planned contents of each chapter.
   Volume editors and authors should be aware that incomplete or insufficiently close to final manuscripts almost always result in longer evaluation times. They should also be aware that parallel submission of their manuscript to another publisher while under consideration for LNM will in general lead to immediate rejection.

4. Manuscripts should in general be submitted in English. Final manuscripts should contain at least 100 pages of mathematical text and should always include
   - a general table of contents;
   - an informative introduction, with adequate motivation and perhaps some historical remarks: it should be accessible to a reader not intimately familiar with the topic treated;
   - a global subject index: as a rule this is genuinely helpful for the reader.

   Lecture Notes volumes are, as a rule, printed digitally from the authors' files. We strongly recommend that all contributions in a volume be written in the same LaTeX version, preferably LaTeX2e. To ensure best results, authors are asked to use the LaTeX2e style files available from Springer's web-server at
   ftp://ftp.springer.de/pub/tex/latex/svmonot1/ (for monographs) and
   ftp://ftp.springer.de/pub/tex/latex/svmultt1/ (for summer schools/tutorials).
   Additional technical instructions, if necessary, are available on request from:
   lnm@springer.com.

5. Careful preparation of the manuscripts will help keep production time short besides ensuring satisfactory appearance of the finished book in print and online. After acceptance of the manuscript authors will be asked to prepare the final LaTeX source files and also the corresponding dvi-, pdf- or zipped ps-file. The LaTeX source files are essential for producing the full-text online version of the book. For the existing online volumes of LNM see:
   http://www.springerlink.com/openurl.asp?genre=journal&issn=0075-8434.
   The actual production of a Lecture Notes volume takes approximately 12 weeks.

6. Volume editors receive a total of 50 free copies of their volume to be shared with the authors, but no royalties. They and the authors are entitled to a discount of 33.3 % on the price of Springer books purchased for their personal use, if ordering directly from Springer.

7. Commitment to publish is made by letter of intent rather than by signing a formal contract. Springer-Verlag secures the copyright for each volume. Authors are free to reuse material contained in their LNM volumes in later publications: a brief written (or e-mail) request for formal permission is sufficient.

Addresses:

Professor J.-M. Morel, CMLA,
École Normale Supérieure de Cachan,
61 Avenue du Président Wilson, 94235 Cachan Cedex, France
E-mail: morel@cmla.ens-cachan.fr

Professor B. Teissier, Institut Mathématique de Jussieu,
UMR 7586 du CNRS, Équipe "Géométrie et Dynamique",
175 rue du Chevaleret,
75013 Paris, France
E-mail: teissier@math.jussieu.fr

For the "Mathematical Biosciences Subseries" of LNM:

Professor P. K. Maini, Center for Mathematical Biology,
Mathematical Institute, 24-29 St Giles,
Oxford OX1 3LP, UK
E-mail : maini@maths.ox.ac.uk

Springer, Mathematics Editorial I,
Tiergartenstr. 17,
69121 Heidelberg, Germany,
Tel.: +49 (6221) 487-8259
Fax: +49 (6221) 4876-8259
E-mail: lnm@springer.com